Tourism Analysis: A Handbook

Tourism Analysis:

A Handbook

2nd edition

Stephen L J Smith

LONGMAN

Longman Group Limited
Longman House, Burnt Mill, Harlow
Essex CM20 2JE, England
and Associated Companies throughout the world

First published 1989
Second edition 1995

British Library Cataloguing in Publication Data
A catalogue entry for this title is available from the British Library.

ISBN 0-582-25160-5

Library of Congress Cataloging-in-Publication data
A catalog entry for this title is available from the Library of Congress.

Set by 7 in Palatino 10/11pt

Printed and bound in Great Britain by Bookcraft (Bath) Ltd

Contents

Preface

This book is about questions. It is about asking and, especially, answering questions in tourism. Questions are, of course, central to the study of any field. Not only do they drive inquiry, the quality of inquiry depends on the quality of questions asked and the methods used to answer them.

The quantity of questions being asked by students, teachers, researchers, managers, planners, policy analysts, and government leaders clearly is increasing. Several new journals are launched and literally scores of new books are published each year. Dozens of conferences are held around the world, with thousands of delegates from virtually every nation on earth. Tens of thousands of post-secondary students enrol in tourism courses.

The increase in the quality of questions and answers is less certain. For example, some questions, as they are currently worded, appear to defy resolution. A common example of this is, 'What is tourism?' The wording of this question implies the belief that something called tourism has an independent, verifiable, and objective existence and that there is a single correct answer to the question if only we are clever enough to figure it out. As Chapter 1 in this text argues, there are many different and equally legitimate approaches to the definition and the study of tourism. The question of 'What is tourism?' might be better reworded as, 'What is the best way to define operationally the phenomena related to temporary travel by individuals away from their home environment?' That may seem to be a wordy question, but it is one that can be meaningfully answered. The answer will be different depending on whether your interest is in counting international tourists, better understanding the motivations of pleasure travellers, or some other objective – but a meaningful, useable answer can be obtained because the question itself is meaningful.

The task of improving the quality of questions and answers is immense and never-ending. The first edition of this book was offered as one small contribution to that large and important task. The approach was to bring together, in one volume, a variety of quantitative methods drawn from numerous sources in the social sciences and to present them in a clear, step-by-step style. This second edition continues to pursue this goal.

The new edition offers an expanded selection of methods with up-to-date background information and examples. In particular, the second edition includes new material on the latest WTO definitions, survey research, the development of indicators for sustainable tourism, and techniques for improved decision-making such as benefit–cost analysis and conversion studies.

I would like to express my gratitude to the many students, faculty, consultants, and government researchers who have read **Tourism Analysis** and especially to those who wrote with comments and suggestions. It was not possible to incorporate all your ideas, but all of you made a difference.

I am particularly grateful for the intellectual support and friendship of the tourism researchers and managers in Tourism Canada and Statistics Canada. They are world leaders in their fields. As a result, they not only provide invaluable assistance in my own research and teaching, they continue to challenge, test, and refine my understanding of tourism. They have given me much and I hope, through my work and this book, I am able to provide a little something in return.

Finally, I would like to express my appreciation for the continuing support and assistance of the editorial and production team at Longman Group. These talented individuals, with whom I am proud to be associated, have placed Longman as the world's leading publisher of scholarly tourism texts.

Stephen L J Smith
University of Waterloo

Acknowledgements

We are grateful to the following for permission to reproduce copyright material:

Crane, Russak and Co. for table 8.1 from table 15, p. 274, figs 8.4 & 8.5 from figs 18, p. 276 & 24, p. 292 (Gunn 1979); Department of Tourism and Small Business, Saskatchewan for fig 8.10 from Map 1, p. 5 (Balmer, Crapo & Associates); the Literary Executor of the late Sir Ronald A. Fisher, F. R. S., Dr Frank Yates, F. R. S. and Longman Group Ltd. London for table A.1 from table IIi of *Statistical Tables for Biological, Agricultural and Medical Research*, 6th edition, 1974 (Fisher & Yates); The Geographical Association for fig. A.2 from fig. 3 (Pinder & Witherick 1975); the Editor, *Journal of Travel Research* for figs 8.8 & 8.9 from figs 1 & 2 (Smith 1983a); Regional Science Association (University of Illinois) for fig. 10.1 from table V by Dunn on p. 111 of *Papers and Proceedings of the Regional Science Association* Vol. 6, 1960; Texas Agricultural Experiment Station (Texas A & M University) for fig. 8.6 from fig. 1, p. 8 (Gunn & Worms 1973); the Society of American Archaeology and the author, D. A. Pinder for fig. A.1 from *American Antiquity* Vol, 44, No. 3, 1979; Taylor & Francis, Inc. (Washington D C, USA) for fig 8.7 from *Leisure Science* Vol. 9, p. 105, 1987 (Fridgen 1987); World Tourism Organization (WTO) for fig. 2.1 from *Recommendations on Tourism Statistics*, 1994.

Whilst every effort has been made to trace the owners of copyright material, in a few cases this has proved impossible and we take this opportunity to offer our apologies to any copyright holders whose rights we may have unwittingly infringed.

CHAPTER 1

An overview of tourism research

Introduction

Tourism is a major force in global trade. It plays a vital role in the social, cultural and economic development of most nations, and has the potential both to preserve heritage and to destroy it. Despite the importance of the industry, reliable, verifiable and objective information can be frustratingly difficult to obtain.

Consider a simple question: how big is the global tourism industry? Is it, as the World Travel and Tourism Council (WTTC) asserts, the world's largest industry? WTTC (1993), in co-operation with the Wharton Economic Forecasting Association (WEFA) Group, estimated that: (1) tourism provided direct and indirect employment for more than 200 million people (one in every nine workers globally); (2) the number of tourism-related jobs would rise to 350 million by 2005; (3) tourism generates 10.3 per cent of global wages (US$1.7 trillion); (4) tourism is responsible for 10.1 per cent of direct and indirect world GDP (US$3.4 trillion); and (5) tourism accounts for 11.0 per cent of all consumer spending.

These estimates have been questioned by a group of Canadian and American researchers on several grounds. For example, WTTC/WEFA's estimates include capital expenditures, such as investment in highway construction, as part of tourism's economic impact. It is unclear how one can reliably estimate what portion of such investments can be attributed to tourism. WTTC/WEFA's estimates of tourism's share of the GDP in most countries are based on matrices derived from a small number of highly developed countries. The Canadian, Australian, and New Zealand estimates are based on input–output matrices for the US economy, while estimates for the United Kingdom, Scandinavia, Japan and France are based on a German matrix. WTTC/WEFA were also constrained by whatever definition their secondary data sources adopted. These definitions were not always consistent, as noted in Table 1.1.

Table 1.1 Varying trip definitions in secondary data sources used by WTTC

Data source	Definition of trip	Page[a]
US Bureau of Labor Statistics Survey of Consumer Expenditures	75 miles or more	7
US Statistical Abstract	100 miles or more	18
US National Restaurant Association	'non-local trips'	18
US Department of Transportation Nationwide Personal Transportation Survey	'social and recreational trips'	22
Photo Marketing Association International Consumer Survey	'trips to selected events'	25

[a] Page in *Measuring the Size of the Global Travel and Tourism Industry*, by the WEFA Group, World Travel and Tourism Council (1993), on which the source is described.

The point here is not the actual magnitude of global tourism or the accuracy of the WTTC/WEFA estimates. Rather, it is that the industry lacks consistent, credible, and coherent data for many important decisions. The need for better data extends from the classroom to the boardroom, from local 'mom-and-pop' businesses to the United Nations. Entrepreneurs, business leaders, policy analysts, planners, elected officials, and voters need better information if they are to manage the forces that affect tourism and, in turn, affect them.

Tourism is, in the minds of many in the industry, merely a matter of marketing – in the sense of advertising and public relations. While marketing is undeniably important, the industry requires much more. Management, planning, development, and policy issues transcend the concerns of marketers. The future success of the industry requires better information and tools to assist with product development, regional planning, impact assessment and control, industry evaluation, and resource allocation as well as marketing. But the development of better information and tools requires, first, that we appreciate the various research perspectives, issues, challenges and strategies facing the industry. This introductory chapter explores these matters.

Let us begin with the following example. A couple about to be married is planning their honeymoon. A tourism researcher, depending on his or her perspective, may be inspired to ask one or more of the following questions. What type of product is the couple looking for? For how long will they go? Will they travel domestically or internationally? How do they collect information about their alternatives? Are they more influenced by advertisements, travel agents, or family and friends? How do they make the final decision? How will they book their trip? How will they travel to the destination? What activities do they wish to engage in? What levels of service are they expecting? How much will they spend? Is it possible to convince them to extend their honeymoon over a longer period of time? If the travel component of the trip is enjoyable, are they likely to return to the same destination in future years? Why are some destinations more popular with honeymooners than others? Is the honeymoon market, generally, worth pursuing by a particular destination?

These are just a few of many questions that can be explored about a single type of travel. The diversity of questions that can be asked about a phenomenon as familiar as a honeymoon hints at the diversity of perspectives possible on tourism research. These perspectives include tourism as a: (1) human experience, (2) social behaviour, (3) geographic phenomenon, (4) resource, (5) business, (6) industry, and (7) intellectual debate.

Perspectives of tourism research

Tourism as a human experience

Tourism is an activity that individuals enjoy. To understand much of the tourism phenomenon, we thus must understand individual behaviour – the psychology of tourists and potential tourists. The development and testing of models that help explain the antecedents and consequences of human behaviour are an important research priority in tourism. Such information can be of particular value in designing new tourism products and marketing initiatives to promote these products. Knowledge of how individuals make decisions about alternative products, what sources of information they require at various stages of their decision-making, how they evaluate that information, and how they structure and interpret their experiences can offer important benefits to businesses as well as to our general understanding of the tourism experience.

When human beings travel, they use objects as part of their experience. Artefacts such as cameras, tour books, and casual clothing are part of the tourist stereotype. The role of these artefacts as mediators or filters between the tourist and the landscape or society being visited is poorly understood.

Travellers are motivated by a wide range of desires. Our hypothetical honeymoon couple seeks a destination that offers romance and escape. These are qualities that will be readily recognized when the couple is enjoying them, but they can be difficult for a resort operator to define operationally. Is 'romance' a secluded bungalow on a tropical beach with swaying palm trees or a luxurious ski lodge with a crackling fireplace in private rooms? Is a hotel's potential for 'romance' best advertised through images of a couple enjoying each other's company, or through images of intimate decor and discrete service? The conditions that create a feeling of 'romance' – or of any other desired experience – are obviously of relevance to tourism operators who benefit from the artful creation of an environment that promotes the desired experience. Studying how environments are created and manipulated by operators is also of interest to scholars who seek a better theoretical understanding of human motivations and perceptions. The perspectives of psychology and social psychology have much to offer to tourism research. Tourism, in turn, offers these social sciences an opportunity to examine a form of behaviour that involves issues not often found in other realms of human activity.

Tourism as a social behaviour

While tourism is an individual experience, it is usually shared with other people. A honeymoon involves a couple travelling together; some honeymooners, such as many Japanese newly-weds, prefer also to tour with a group. Whether travelling alone, with one other person, or with a large group, tourists come into contact with other people and social institutions. Many of the decisions related to a tourism experience are influenced by both the individual's psychology as well as his socialization experiences and self-perceived social role.

When a tourist comes into a new area, he or she meets local residents and other travellers. These meetings range from the briefest and most casual anonymous encounters to intense interpersonal interaction. Friendships may be formed or conflicts may occur. Tourists can unintentionally contribute to the deterioration of the social fabric if they come in large numbers, inject unprecedented wealth into an economy, or display forms of public behaviours radically different from local norms.

Tourists may be the victims of crime or may commit socially unacceptable acts themselves. Knowledge of the social interaction of tourists with each other and with residents and local institutions can contribute to ameliorating potential conflicts. This knowledge may also lead to better understanding of general social behaviour and of the structure and functioning of different societies.

Tourism also implies economic decisions. Individuals and families assess their desire for a vacation against their financial resources. Our honeymoon couple balances the allocation of their finances among various wedding expenditures, the honeymoon, and start-up costs of their new household. A city council weighs the merits of funding a new local facility that will draw tourists against other uses for public funds. A tourism operator considers the potential risks and profits from expanding his business.

As these questions suggest, the study of economics is the study of the allocation of scarce resources among competing uses. The actual allocations involve personal values, business goals, industrial strategies, or politics – but in all cases the allocations have social implications. The average allocation of personal resources for honeymoons tells us something of the social values of honeymoons in a culture. A city council's decision about the level of support for tourism development can indicate how that community views economic development. The entrepreneur's decision about expansion of his business reflects his perceptions of social trends that influence the demand for his product.

Tourism as a geographic phenomenon

Travel from an origin to a destination is an inherent characteristic of tourism. Our honeymoon couple are likely to want to get far away from their usual environment in order to enjoy their time together and to make special memories. Other tourists travel to see new places, experience new cultures, or just to enjoy the journey. Regardless of the actual motivation, tourism is geography in action.

The industry is often organized according to geographical divisions such as counties or provinces. These organizations range from local visitor and convention bureaux to multinational groupings such as the World Tourism Organization's classification of world regions.

Tourism destinations often rely on conveying a sense of place in their advertising. Images of sunny beaches, the attractions of local cultural groups, the dynamism of an urban skyline, and the solitude of a mountain range are all geographic concepts used to promote destinations. Geographic research contributes to tourism knowledge in several other ways. It delineates tourism regions for the effective structuring of an industry association. It provides the basis for forecasting visitor volumes and expenditures. The analysis of the morphology of tourism destinations assists in planning and impact studies. Geography is fundamental to the selection of a development site for any tourism business. It may help explain why our honeymoon couple may choose destinations like Hawaii or the Poconos (USA) that are popular honeymoon destinations, while other locales are normally ignored as honeymoon venues. More generally, the geographic perspective reminds other social scientists studying tourism that space and the traveller's response to space are a distinctive aspect of tourism research.

Tourism as a resource

Communities seek to develop a local tourism industry to diversify their economic base. Some will attempt to attract our honeymoon couple, because honeymooners tend to spend more than many other segments and offer potential for significant repeat business. Regardless of the market segment involved, tourism can be an important source of export income and an efficient creator of new jobs with fewer negative environmental impacts than from many other industries.

Tourism can also be a positive force for the preservation of local architecture, heritage and environmental resources. Income from tourists helps finance the maintenance and renovation of English cathedrals and pays for game wardens to protect wildlife from poachers in Kenya. Many small towns would be virtual ghost towns if it were not for the jobs that tourists support.

A town's success in drawing tourists to enjoy local attractions can, ironically, carry the seed of its destruction. The daily arrival of tens of thousands of visitors in communities whose permanent population may be only a few thousand can overwhelm the qualities that originally drew tourists. The perspective of tourism as a community resource is one that requires a realistic assessment of both the benefits that tourism may produce and the costs it may impose. Planning and management strategies that combine business interests, environmental concerns, and social awareness can often optimize the returns from tourism to communities.

Tourism offers nations more than economic benefits. The Japanese Government has explicitly recognized the potential for international tourism into Japan to improve the understanding and appreciation of Japanese culture by foreigners. In fact, this 'educational' or public relations benefit is seen as outweighing the economic benefits of inbound tourism for Japan.

Tourism as a business

Tourism is a source of income to hundreds of millions of individuals world-wide. Workers and employers alike can benefit from research that: (1) improves the efficiency of business structures and administrative arrangements; (2) improves the strategies for coping with the risk and uncertainty inherent in the industry; (3) offers a firmer basis for profitable marketing; (4) offers improvements in the terms and conditions of employment; and (5) promotes pride, professional development and rewarding careers in the industry.

Consumer expectations and, in many nations, rights are changing rapidly. Travellers, including our hypothetical honeymooners, are becoming more sophisticated and demanding in their tastes. Travellers with disabilities are pressing for greater accessibility and services to permit them to travel. Legislation to promote competition, such as airline deregulation or the extension of anti-trust legislation to computer reservation systems, is being enacted in more and more countries. Consumer protection legislation, such as requiring tour companies to carry some form of insurance to protect customers in the event of bankruptcy, is also becoming more common.

There is a growing sense of professionalism in many tourism businesses. Travel agents in several countries now have certification programmes. Careers in certain hospitality businesses have long enjoyed a degree of respect. The quality of service in Swiss hotels, for example, reflects the tradition of professionalism in that nation's hotels. Canada is implementing national competency-based standards for dozens of tourism jobs. Research is needed on human resources to determine: (1) the demand for various types of employees, (2) the knowledge, skills and attitudes required to perform different tourism jobs, and (3) the most effective delivery mechanisms for skill training and professional education.

Tourism enterprises and entrepreneurs may pursue any of several different business goals. Bull (1991: 55–8) identifies the following as the main objectives of tourism enterprises. The classical objective is profit maximization, i.e. maximizing the difference between long-run revenues and total costs. Sales revenue maximization is a related but more simplistic goal, in which gross revenues are maximized although net profits are not necessarily maximized. A goal even more simplistic than revenue maximization is output maximization. Here market share or some measure of the volume of goods and services is paramount to maximizing either gross revenues or net profits.

As enterprises grow and ownership becomes more distant from day-to-day management, managers may begin to allow their personal agendas to influence business operations. In such circumstances, empire-building or prestige-enhancement may become the pre-eminent goal. More traditional business objectives, such as profit-maximization, are then reduced to a subordinate role. In contrast to the goal of career aggrandizement, some owner-operators of small firms may seek to optimize the total quality of their life. They seek neither maximum profits, sales, output, market share, nor prestige – but rather that mix of business and leisure that produces the greatest overall happiness in their life. This phenomenon is sometimes known as 'satisficing' (Simon 1959). It reflects a behaviourial approach to business decisions rather than a financial approach. Meeting people, combining a hobby with a job, or the potential to enjoy part-time or seasonal employment while having sufficient leisure and income to pursue other interests, are typical goals for 'satisficing' entrepreneurs. Although these various objectives are recognized by analysts, research is needed to determine how and why different business leaders select different objectives. Research may also help provide practical guidelines for business leaders to meet their preferred objective.

Tourism businesses are especially vulnerable to exogenous forces. Political instability or terrorism will harm businesses in one country by diverting tourists to other destinations and, in turn, benefit businesses in those nations that the tourists turn to as substitutes. Disease, natural disasters, unfavourable changes in currency exchange, new taxes or fees, or alterations in border-crossing formalities, can quickly and dramatically alter the relative appeal of competing destinations. Very often tourism operators rely only on personal experience to help them cope with these problems. Their coping mechanisms, however, can be improved if they have access to professional development courses, professional literature, or consultants. The effectiveness of these information sources depends on the degree to which they are based on empirically tested knowledge. Both practical tourism research and the extension of basic scholarly research into the problems of individual businesses are needed.

Tourism as an industry

Tourism is not just hundreds of thousands of businesses, but a global industry with major policy implications. One important characteristic of the tourism industry is that it is labour-intensive. A given level of revenue or capital investment creates many more jobs in tourism than the same level of revenue or investment would in agriculture, automobile manufacturing or petrochemicals.

Tourism is also an important source of interregional and international cash flows. Governments at all levels encourage tourism development because it generates new wealth through export sales, which creates new jobs. Governments are also concerned about the social and environmental impacts of tourism. Policies designed to maximize benefits and to minimize problems work best if they are based on empirical research.

In many nations, tourism is strongly linked to advertising. National advertising budgets frequently run into tens of millions of US dollars. Expenditures of this magnitude from general tax revenues can be sensitive political decisions. Competition by advertising agencies for tourism contracts is intense; the scrutiny of policies and practices governing the awarding of those contracts can be equally intense.

Tourism research can play an important role in governmental advertising in several ways. Although creative advertising is still often seen as an 'art', promotional campaigns will be more successful if they are based on market research. Research can guide the choice of advertising media, messages and audience. Consumer tastes for tourism products vary significantly among nations, socio-economic classes within one nation, and over time. Researchers can identify these variations and track trends in tastes as well as the nature and relative sizes of different market segments. They also provide information about whether the allocation of a portion of scarce public budgets for tourism advertising was a prudent decision.

Many facilities used by tourists are owned and operated by public agencies: national, provincial and state parks; game preserves; historic sites; ferries; museums and galleries; and sports stadia. Management of these facilities may sometimes place tourism far down on the list of objectives. In national parks, for example, tourism may be seen as secondary to environmental preservation. In local sports facilities, a tourism agency may find itself ranked behind local sports groups for access to space. Residents may find themselves fighting crowds of tourists at local beaches the residents consider their own.

Research is needed to develop tools for guiding the efficient allocation of resources among conflicting claims, and to set politically acceptable management guidelines of public facilities. A basic principle in economics is that resources should be allocated in accordance with their value in alternative uses. Anglo-American economies operate on the assumption that prices, as they develop under the condition of pure competition, are the best indicator of the value of resources in alternative uses. Pure competition, however, rarely exists. Other market-driven economies, such as those of Germany, Japan and Korea are based on the assumption that 'pure competition' can be 'excessive competition', and that unfettered market forces can produce unacceptable social and business consequences. In all national economies, therefore, industrial policies shaping resource allocations to tourism are based on a mix of market and political forces. Further, the use of publicly owned resources for tourism, such as national parks, are usually made without any reference to market prices. Government planners and managers, especially, need research tools to help them estimate appropriate values of resources in alternative uses.

Tourism as an intellectual debate

The perspectives described above are equally legitimate and important. Each represents a different set of questions and issues of relevance to different individuals or groups involved in tourism. Any discussion of which perspective is 'right' or 'best' would be misconceived. However, there is still substantial intellectual debate among analysts interested in tourism. This debate reflects fundamental differences in assumptions and value judgements made about tourism. Jafari (1992) summarizes these debates as a series of 'platforms' from which authors, commentators and policy analysts speak and write. These platforms emerged in the order in which they are presented below; however, more recent platforms have not replaced earlier platforms. Rather, they represent a growing dialectic among different schools of thought on the nature of tourism.

The earliest platform is that of 'advocacy', the concern for maximizing the economic benefits of tourism. Supporters of this platform also emphasize the potential of tourism to preserve natural and cultural heritage and to promote international understanding. This perspective is shared both by businesses that profit from tourism as well as governments that are seeking tax revenues, economic diversification, foreign exchange and job creation.

In time, certain groups who did not share in the benefits of tourism began to describe tourism as a blight rather than as a blessing. Jafari labels this perspective as the 'cautionary platform'. The cautionary platform emphasizes negative social and environmental impacts of tourism and downplays its economic benefits.

The advocacy and cautionary platforms represent polar views on the nature of tourism, yet they share a concern with assessing the impacts of tourism. The third platform to emerge attempted to find a compromise between the extremes of the two earlier platforms by: (1) acknowledging that tourism may create problems in some circumstance, but (2) asserting that appropriate forms of tourism development are possible. 'Appropriate' tourism development balances the demands of tourists, the needs and values of host communities, environmental constraints, and the goals of business. This platform has given rise to terms such as alternative tourism, ecotourism, sustainable tourism, green tourism and community-based tourism.

The most recent platform attempts to understand the perspectives, assumptions and values of the two extreme platforms (advocacy and cautionary) as well as the compromise platform (adaptancy). This platform is the 'knowledge-based' or 'scientific' platform. This perspective asserts that a more comprehensive understanding of tourism is both possible and necessary. While the advocacy and cautionary platforms focus on impacts and adaptancy focuses on development, the scientific platform focuses on the systematic study of tourism. This study is necessarily interdisciplinary, and recognizes the legitimacy of the perspectives identified previously. Jafari traces the roots of the scientific platform to a paper by Wahab (1972) in which he argued for a systems view of tourism encompassing personal, group, business and governmental perspectives. The growth of objective studies of tourism in scholarly journals is one expression of the emergence of the knowledge-base platform. This platform also led to the creation of the International Academy for the Study of Tourism, an international organization devoted to the scientific study of tourism.

Contemporary issues in tourism research

The critical issues in tourism vary with time and locale. In 1993, the International Travel and Tourism Research Association produced a research agenda for the North American tourism industry (Smith *et al.* 1993). This agenda, based on surveys of industry leaders as well as a review of industrial, governmental and scholarly publications, identified the ten most important issues facing the North American industry in the early 1990s. Each issue was then recast as a series of specific research questions.

Although the issues are specific to Canada and the United States, they may illustrate the diversity of issues that tourism analysts are wrestling with around the world. The order of issues is random; it does not represent any sort of priority ranking. Each issue is seen as being of equal importance with all other issues from a broad industrial perspective. Individual sectors, businesses, agencies or researchers, however, may impose their own priorities on the list.

1 Social structures, communities, businesses, families and individuals are experiencing profound change. Social and family values are being redefined, average family size is decreasing, communities are becoming more pluralistic, the population as a whole is changing, and lifestyles are becoming more diverse and often more complicated. Many individuals are showing greater interest the role a healthy lifestyle – including travel – can play in enhancing their overall well-being.

   Questions

   a. What specific changes are affecting different market segments?
   b. What product needs are developing because of these shifts?
   c. How are these shifts affecting seasonality?
   d. What are the marketing implications of these changes?
   e. To what degree are the recent increases of interest in exercise and healthier food long-term shifts that should be reflected in tourism products?
   f. To what degree do concerns over potential exposure to disease influence travel behaviour?

2 Consumer expectations, experiences and legal rights are expanding, posing significant challenges to tourism businesses. The concept of 'universal access', in particular, is of growing importance.

Questions

a. *Are firms aware of these changes? If not, how can they be kept abreast of change?*
b. *What changes in operations will be required for firms to conform to new consumer expectations and rights?*
c. *What are the perceptions of employees and employers in different sectors of disabled travellers?*
d. *What are the current levels of accessibility in different facilities?*
e. *What is needed to improve accessibility – in terms of modified services, modified products, and training programmes?*
f. *Who will pay for modifications and training?*

3 Tourism businesses have a vested interest in protecting the environment that draws tourists, but some firms lack the long-term vision to adopt environmentally appropriate management strategies. However, more and more businesses are being forced to adopt such strategies as they cope with rising costs of energy and waste disposal, as well as government regulations. Many tourists and consumer groups express concern for environmental protection, but there are doubts about the willingness of consumers to pay for a healthier environment through higher prices.

Questions

a. *To what extent are tourists willing to pay more for environmentally friendly products?*
b. *How can higher prices generated by environmental protection action be best explained to tourists?*
c. *Does environmental protection necessarily raise prices?*
d. *What are the best methods for promoting environmentally appropriate management strategies?*
e. *What current practices are not environmentally sound? What alternatives can be developed?*
f. *What are the costs of alternative technologies?*
g. *How can one monitor the impacts of different tourism management techniques in time to avoid irreparable damage?*
h. *Are government regulations the best mechanism for protecting the environment?*

4 The tourism infrastructure in many areas is believed to be congested and/or deteriorating to the point that it is constraining tourism growth.

Questions

a. *What standards should be developed or used to assess the state of tourism infrastructure?*

b. *What is the level of congestion or deterioration of the infrastructure in specific locations?*
c. *How should improvements be funded?*
d. *What is the impact of these problems on tourist satisfaction, economic growth and business profitability?*

5 Social perceptions of tourism are quite mixed. Residents of some communities, especially rural and aboriginal communities, view tourism as a potential force for economic revitalization. Residents of other communities see tourists as insensitive hordes invading their privacy, creating traffic congestion, and generally inflicting social damage on the places they visit.

Questions
a. *Are the perceptions of residents who complain about tourists accurate?*
b. *Would tourist codes of behaviour be accepted by tourists and would they have any impact? What role could education or interpretive programmes play in improving relationships between hosts and guests?*
c. *How can tourism areas better manage visitor impacts and volumes?*
d. *Under what conditions does tourism offer net economic growth for rural and aboriginal communities?*
e. *What products offer the greatest benefits in different communities?*
f. *How can information and case studies on 'best practices' in community tourism be effectively distributed to businesses and community leaders?*

6 Crimes against tourists have gained international news coverage. Regrettably, crime against visitors is neither new nor limited to one or two destinations. Law enforcement agencies as well as destinations are struggling to find ways to cope with this problem. Tourists are also potential targets of politically motivated terrorist attacks.

Questions
a. *How extensive are crimes against tourists? Are tourists significantly more at risk than locals?*
b. *If tourists face a higher risk, what factors contribute to the greater risk?*
c. *What can be done to reduce their risk?*

7 Many economies are moving towards greater reliance on 'knowledge workers'. The competitiveness of businesses is increasingly determined by the knowledge and skill of the labour force. However, available statistics suggest that very few tourism businesses are investing in education for employees. The results of insufficient training include lower customer satisfaction and lower employee productivity. Many tourism businesses complain that

they are unable to find adequately skilled employees and that they suffer from high levels of employee turnover and low employee morale. Tourism jobs have a reputation for being low-skilled and low-paying, with no career possibilities.

Questions

a. *How can productivity, competitiveness and customer satisfaction best be measured in various tourism businesses?*

b. *Is there a demonstrable link between education, productivity and customer satisfaction with company performance or profitability?*

c. *What specific skills do employees lack?*

d. *What training programmes would be supported by industry, particularly in terms of sponsorship of development and delivery costs?*

e. *What is the impact of tourism education advisory councils on improving labour quality and supply?*

f. *How can tourism educational programmes attract the best qualified students in competition with law, business and other professional fields and disciplines?*

g. *To what degree can the conditions of employment in tourism, especially front-line positions, be changed to provide for greater employee retention and loyalty?*

h. *How do compensation and working conditions in tourism compare with those in other industries?*

i. *What do labour force entrants look for in a job? What messages need to be delivered in career-awareness programmes to change the industry's poor image?*

j. *What are the costs to employers per 'employee-turnover'? How does this vary by occupation, size of business and location of business?*

k. *What degree of turnover should be considered inevitable?*

l. *Is a high degree of turnover among part-time employees actually more cost-efficient for some businesses than an equivalent number of permanent staff with no turnover?*

8 Tourism bears a disproportionate load of taxes – hotel room taxes, gasoline taxes, alcohol excise taxes, airport surcharge and so on. Value-added taxes can negatively affect a country's competitiveness in international tourism.

Questions

a. *Are these perceptions accurate? How does tourism fare in comparison to other sectors in terms of tax burden?*

b. *If the industry does carry a disproportionate load, how did this situation arise?*

c. *To what degree do taxes hamper economic growth and job creation in tourism?*

d. *What are the awareness and acceptance levels of taxes among tourists?*

9 The industry is highly fragmented; its sectors tend not to co-operate with each other. One result is that the industry is unable

to speak with a united voice to government. Another result is that many industry associations are often ignorant of each other's initiatives and activities. Most associations have only a low penetration into their potential membership base. In addition to these problems, associations are expected to perform potentially conflicting goals such as providing better quality business services, acting as lobbyists for their members, and providing policy recommendations to government.

Questions

a. *What changes are needed to improve the effectiveness of industry associations?*
b. *Whose responsibility is it to unite the industry?*
c. *Are businesses willing to pay for improved association effectiveness and communication through higher membership dues?*
d. *Are there too many associations in tourism?*

10 Virtually all aspects of tourism firms are being dramatically changed by new technology. Despite the proliferation of new technology, the industry appears reluctant to adopt new tools and methods. Further, the industry generally does not take an active role in developing or adapting new technology.

Questions

a. *Are these perceptions correct? If so, why are tourism firms slow in picking up new technology?*
b. *Is the industry too fragmented to support research and development costs for new technology?*
c. *What new technologies being used now by larger firms could be scaled down for use by small firms?*
d. *What new technologies are being adopted by various sub-sectors within the industry, and what effects do they have on productivity?*
e. *Are there barriers to the development or application of new technology in the industry? If so, how can these be overcome?*

Challenges of tourism research

One may feel excited about the prospect of doing original, meaningful research in tourism just from considering the wide range of perspectives and pressing issues confronting analysts, but researchers face a number of challenges in trying to understand tourism. These include: (1) the lack of credible measurements for describing the size and impact of tourism; (2) great diversity in the industry, with some analysts questioning whether tourism is a single industry or group of related industries (see, e.g. Leiper 1993); (3) spatial and regional complexities; and (4) a high degree of fragmentation.

This section discusses each of the four challenges in greater detail. Although they are presented separately, they are related. The lack of credible measures about the size of the industry is due in part to its diversity. Its diversity and spatial complexity contribute to its lack of organization. The fragmentation of the industry makes the development of credible data more difficult. Much of the appeal of destinations is in their regional characteristics and diversity, but overcoming the diversity and fragmentation in the industry must not be accomplished by forcing a boring homogeneity on places and attractions.

Lack of credible measures

Tourism often lacks credibility in the eyes of decision-makers because the field is ambiguously defined and because the data used to estimate the size of the industry are suspect. When one attempts to measure the magnitude of tourism in a national economy, for example, some percentage of revenue from automobile fuel sales can be reasonably attributed to tourists. But exactly what percentage is reasonable? Should capital investments in airports be considered part of tourism's contribution to the economy? What about capital investment in roads, sewer systems or telecommunications – all of which serve tourism (as WTTC/WEFA argues)?

The potential for double-counting contributes to the lack of credibility in some tourism data. A hotel with a restaurant and bar may report restaurant and bar sales together in one document and separately in another document. If a researcher is not careful about the purpose, coverage and conventions used in secondary data sources, food and beverage sales (in this example) could be counted twice.

A data-collection problem more frustrating for the industry than double-counting is the omission of data. This problem stems from the use of Standard Industrial Classification codes (SICs) to label establishments. A discussion of these codes and their application to tourism is described in Chapter 2. It is sufficient for now to note that one difficulty with SICs is that any establishment can be classified into only one category. The decision of which category is based on the business's primary source of income. If the hotel mentioned in the last paragraph derived most of its revenues from the letting of its rooms on a nightly basis, it would be classified in the SIC category of 7011, 'Hotels, motels and tourist courts' (this is the US code; other nations have other codes). This classification would apply whether the hotel offered only rooms, or a restaurant, convention services, a fitness club and a gift shop along with the rooms.

Data can be missed when an establishment conducts a significant volume of tourism-related business but earns most of its revenues from other sources. The West Edmonton Mall in Edmonton, Alberta, is the world's largest shopping mall. It is classified as a retail establishment because of its large volume of retail sales. With over 4 million visitors a year who come from origins well over 160 km (100 miles) away, many of whom stay at the attached hotel and use the mall's indoor wave pool, beach, theme park and golf course, it is also a major tourist attraction in Alberta. Estimation of the size of the tourism industry in Alberta, if based on receipts associated with those firms who have tourism-related SIC codes only, would miss a significant portion of the total receipts in the province.

Simply counting the number of visitors to a destination is not as simple as you might initially think. Different governments have different definitions of 'tourist'. Some are more restrictive than others. Totals obtained by adding reported figures together from different sources that use different definitions will produce misleading results. The potential confusion associated with varying definitions may be seen in estimates of Mexico's volume of international arrivals. Mexico once defined an international tourist as someone who had obtained a tourist card. This definition excluded many business and convention travellers as well as overnight visitors to many border areas. With the implementation of the North American Free Trade Agreement in 1994, Mexico switched definitions to match the more inclusive definition used by its new trading partners, Canada and the United States, and which conforms to World Tourism Organization (WTO) guidelines. WTO defines an 'international tourist' as any individual who crosses a border and stays at least one night but not more than one year. The result is that Mexico's estimates of tourist arrivals increased by over 50 per cent – due solely to a change in definitions.

Tourism analysts occasionally use the concept of a multiplier as part of their estimation of the economic impact of tourism. Multipliers are difficult to calculate precisely under the best circumstances. They require substantial amounts of very detailed data. The methods used are also difficult and require a high level of statistical and/or macro-economic expertise. Some of the dangers in using multipliers are described in Brian Archer's now-dated but still excellent book chapter, 'The uses and abuses of multipliers' (Archer 1976). Regrettably, the abuses of multipliers often seem to be as frequent as legitimate uses – thus contributing further to the industry's lack of credibility.

Diverse industry

Consider your last vacation. You may have purchased a guide book from a local bookseller and read about a new resort that appealed to you. You booked both the resort and airline reservations through your travel agent. While waiting for your departure date, you bought some new luggage and perhaps some resort clothing. When the departure date arrived, you called a taxi for a ride to the airport. You then rented a car at the other end. While on the trip, you ate at local restaurants, danced in local clubs, bought local crafts and visited some attractions. You purchased gasoline for your rental car and picked up a few sundries at a drugstore. You may have even visited a clinic for treatment of your sunburn. Part way through your trip you called home to tell everyone how wonderful the vacation was.

Between the time you began planning your vacation and the time you arrived back home to share your photographs and travel tales, you made tourism-related purchases from a bookstore, a travel agency, three different commercial transportation firms, an accommodation establishment, several food-service operations, different retail shops, a medical clinic and a telecommunications firm. On your next holiday you may use an entirely different set of tourism commodities.

In addition to these direct providers of tourism commodities, one can identify other types of organizations and firms that help to support the tourism industry. Gee *et al.* (1984) suggest classifying tourism organizations into three levels: (1) direct providers; (2) support services; and (3) tourism development agencies. Direct providers refer to the most obvious type of tourism firms: hotels, airlines, rental car agencies and so on. Support services are less obvious but still essential: tour organizers, travel publishers, contract laundry services, contract food services and the like. Tourism development organizations include visitor and convention bureaux, municipal planning offices, government ministries, and universities offering advanced education in tourism.

One can note still other firms that contribute to the industry that do not fall within the Gee–Choy–Makens typology. These include business services such as accounting, legal and insurance firms. Manufacturers, normally not included in any definition of tourism, also play an important role. They provide the supplies and equipment for direct providers and tourists: luggage, cameras and film, aircraft, motor coaches, automobiles, crafts, linens and more. Agriculture and fisheries also have close ties with tourism in many regions. The point here is not to argue for a wildly inclusive definition of tourism, but to make clear the diversity of firms and organizations associated with tourism and the difficulty of trying precisely to delimit the economic impact of the industry.

Spatial and regional complexities

Tourism implies travel from one place to another. The movement of people in space means that tourism is fundamentally a spatial phenomenon. Beyond this is the fact that tourism varies dramatically in form and function from place to place. The tourism attractions in Kelowna, British Columbia are quite different from those in Kauai, Hawaii, or Kyoto, Japan. Yet all three places are popular tourism destinations. An understanding of the role and impacts of tourism in each locale requires an appreciation of the differences in the geographic situation of each. These differences include local customs, climate, level of economic development and tourism infrastructure, types of attractions, and the area's previous experience with tourism.

Tourism also means different things at different geographic scales. The questions asked when planning, developing and evaluating tourism in a local community are different from those asked for a province or state. They are different again when the focus is on a nation. Analytic and planning questions, data precision, definitions and even the purposes of research and planning vary according to the scale of the problem and region being studied.

Unorganized industry

The diversity of tourism enterprises as well as their widespread geographic distribution and the many scales at which tourism operates discourages co-ordinated planning, marketing and research. Most tourism industry sectors do not attempt to co-ordinate either marketing or product development (there are notable exceptions among some individual businesses, but they are just that – exceptions). Co-operative data-collection and data-sharing agreements are especially rare. This is not due to the lack of any compelling reason for co-operation. On the contrary, significant benefits can be derived when different firms and organizations work together. The major reasons for the lack of organization include the fact that the great majority of tourism enterprises in most countries are small businesses. Many small business operators lack experience in forming strategic partnerships with other operators. As a result, they often fail to perceive that benefits from carefully selected and managed partnerships can outweigh any associated risk.

Partnerships in the form of tourism industry associations do exist, to be sure. However, the associations themselves are often in competition with each other. The situation in the province of Ontario, Canada, illustrates the fracturing of the industry. (Ontario is selected because it is a typical, not an extreme, case.) At the provincial level alone, more than 50 separate sector associations represent the interests of attractions, hotels, resorts, museums and galleries, restaurants, outfitters, tour companies and so on. On top of these are dozens of national associations, some of which serve the same constituencies. Add to these the separate marketing programmes of Tourism Canada; the Ontario Ministry of Culture, Tourism, and Recreation; 12 different Ontario Tourism Association Partners (multi-county regions), and hundreds of local visitor and convention bureaux and comparable organizations, and you can begin to appreciate the confusion surrounding any attempts to understand the structure of the tourism industry.

Summary

The tourism industry is facing many challenges and opportunities. Among the issues that analysts must wrestle with are: (1) the lack of credible data, (2) a high degree of diversity among businesses in the industry, (3) complexities created by the geographical nature of the field, and (4) the lack of coherent industry organization.

Among the opportunities facing tourism researchers is the potential to make substantial contributions to helping the industry manage complex and critical issues, some of which were identified earlier in this chapter. Tourism researchers also have the opportunity to apply concepts and tools from a wide range of social sciences to improve the scientific understanding of tourism from a number of perspectives.

There is also the opportunity to improve research for tourism planning and policy analysis. Some of the more important areas in which research is needed are: (1) defining and describing tourism, (2) assessing and forecasting tourism market trends, (3) determining the locational forces that affect business success and the geographic structure of the industry, (4) identification of optimal resource allocations for tourism development, and (5) assessment of the impacts of tourism marketing and development. These themes form the basis for this book.

Chapter 2 examines definitions in the industry, with particular attention given to the latest World Tourism Organization guidelines. The development of a Standard Industrial Classification of Tourism Activities (SICTA) is also described.

Chapter 3 considers how data may be collected in tourism, particularly through the use of surveys. The chapter provides advice on the use of secondary data sets and outlines how the quality of these data sets may be assessed.

Chapters 4, 5 and 6 cover various aspects of the demand-side of the industry. Chapter 4 describes several motivational models and scaling techniques used to study tourists, while Chapter 5 examines methods for grouping tourists into market segments. Chapter 6 presents some tools for forecasting demand and market trends, including the application of basic economic concepts to the analysis of tourism demand.

Geographic aspects of tourism are covered in Chapters 7, 8 and 9 These aspects range from site selection (Chapter 7) through descriptive spatial statistics (Chapter 8) to regionalization (Chapter 9).

Some of the more important tools for guiding resource allocations and assessing impacts of tourism are covered in the last two chapters. Chapter 10 explores methods for determining the values of public resources as well as intangible business assets such as good will. Chapter 11 addresses impact and evaluation methods, including monitoring the sustainability of tourism development, benefit–cost analysis, and conversion studies for determining advertising effectiveness.

CHAPTER 2

Defining and describing tourism

Introduction

Samuel Pegge reported the use of 'tour-ist' as a new word for travellers *c.* 1800; England's *Sporting Magazine* introduced the word 'tourism' in 1811. Despite the fact that both words have now been part of the English language for almost two centuries, there is still no universally accepted operational definition for either. The lack of consistent and accepted definitions is a continuing source of frustration for tourism planners and analysts. Inconsistent definitions reduce the credibility of the field in the eyes of its critics and is, in part, behind the reluctance of some analysts and policy-makers even to accept the notion that tourism is an industry.

From an analytical perspective, definitional inconsistency makes comparisons of tourist flows and other related phenomena among jurisdictions difficult. The development of cumulative data sources and of data banks to support tourism analysis depends on the development of a consensus about working definitions of 'tourism', 'tourist' and similar terms. Further, the limited resources for developing statistical databases in most nations combined with a growing need for data means that co-operative data-sharing agreements between governments and between governments and the tourism industry are increasingly important. Fortunately, there is progress.

Tourism analysts have a variety of perspectives and units of analysis for measuring tourism. Two of the more common perspectives are those of the person and the trip. Analysts may focus on tourists, same-day visitors and other types of individuals or they may focus on international trips, domestic trips, business trips, pleasure trips and other trip types. Although the distinction between the person and the trip is useful, a more fundamental distinction is that between demand side and the supply side of tourism: the consumer (whether measured in terms of the person or the trip) and the suppliers of travel services (information, transportation, accommodations, food and beverage, attractions and so on).

Despite the validity of the distinction between supply and demand, there is also a need to retain some sense of the totality of tourism as an industry and human experience that combines both demand and supply. This chapter presents not only the most current demand-side and supply-side definitions developed by the World Tourism Organization (the international standard for tourism data collection) but also describes an innovative strategy for combining both supply and demand perspectives in a statistical framework.

A brief history of tourism definitions

The World Tourism Organization (WTO) is the lead agency responsible for the development of standardized tourism definitions. Its guidelines have an extensive history. The first recommendations for standardization of international tourism definitions were made in 1937 by the Committee of Statistical Experts of the short-lived League of Nations (OECD Tourism Committee 1973). They defined an 'international tourist' as anyone visiting a country, other than that which is his usual residence, for more than 24 hours. The Committee of Statistical Experts excluded by direct reference individuals arriving to take up work or residence, students attending schools, commuters who cross borders on their way to work, and travellers who do not stop *en route* through a country regardless of the length of time physically present in that country.

The International Union of Official Travel Organizations (IUOTO) resurrected and modified the Committee's definition in 1950 by including students on study tours as tourists and by specifying a new type of traveller called an 'international excursionist': an individual travelling for pleasure who visits another country for less than 24 hours. Also, IUOTO defined another category, 'transit travellers', as those individuals who pass through a country without stopping regardless of the time they spend in the country, or as those individuals who travel through a country in less than 24 hours and make only brief, non-tourism stops.

In the early 1950s, the United Nations formulated a **Convention Concerning Customs Facilities for Tourism** that expanded the earlier definition of a tourist by specifying a maximum stay of six months in addition to the existing criteria of the IUOTO definition. A decade later, in 1963, the United Nations Conference on International Travel and Tourism drew a distinction between 'tourists', who stayed for more than 24 hours, and 'visitors', who stayed for less than 24 hours. The distinction is identical to that made by IUOTO in 1950 between tourists and excursionists. Terminology referring to this distinction was examined in 1967 by the Expert Statistical Group working under the United Nations Statistical Commission. They suggested that the distinction be made between 'tourists', who stayed overnight, and 'day visitors' or 'excursionists', who did not. This latter category, according to the Group, also included those individuals previously classified as transit travellers. The United Nations Statistical Commission convened an international conference in 1976 involving representatives of the World Tourism Organization, the United Nations Conference on Trade and Development, the Conference of European Statisticians, the East Caribbean Common Market, and the Caribbean Community, which ratified and refined the 1963 definitions.

The most recent international conference on travel and tourism statistics was convened by WTO in Ottawa in 1991. The Ottawa Conference organizers opened the conference by noting that: (1) the requirements for tourism statistics are exceptionally diverse among national administrations, industry associations, local communities, academia, and individual businesses, (2) some governments have already established statistical systems to provide tourism data while others have barely begun, and (3) there is still a lack of agreement on basic definitions associated with tourism (World Tourism Organization 1991).

The conference thus focused on developing definitions and classifications that would: (1) be of world-wide practical application in both developed and developing nations, (2) emphasize simplicity and clarity, (3) be limited to statistical purposes, and (4) be consistent with current international standards and classifications in areas such as demography, transportation, and national accounts to the maximum practical extent. These recommendations were then submitted to the United Nations Statistical Commission, which accepted them in March 1993 at its 27th Session. The following sections are based on the recommendations of the Ottawa Conference and the 1993 UN statistical conventions (World Tourism Organization 1994).

Basic tourism concepts

Tourism is the set of activities of a person travelling to a place outside his or her usual environment for less than a year and whose main purpose of travel is other than the exercise of an activity remunerated from *within* the place visited. The phrase, 'usual environment', excludes trips within the person's community of residence and routine commuting trips. The phrase, 'exercise of an activity remunerated from *within* the place visited', excludes migration for temporary work paid for by an economic agent resident in the place visited. This exclusion does not apply to business-related travel such as sales calls, the installation of equipment, or conventions where the traveller's employer is located elsewhere than the place visited.

Tourism is divided into the following categories:

1 **Domestic tourism**: residents of a country visiting destinations in their own country. (It should be noted that 'domestic' in this context is used with a marketing perspective; in macro-economic and national accounts perspectives 'domestic' refers to expenditures by both residents and non-residents within the reference country, e.g. as in gross domestic product.)
2 **Inbound tourism**: visits to a country by non-residents.

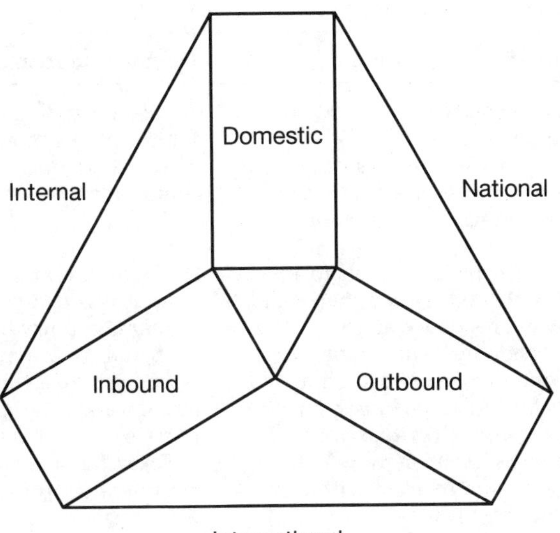

Figure 2.1 Forms of tourism
Note: The term 'domestic' used in a tourism context differs from its use in a national accounting context. In tourism, it has a marketing connotation of residents travelling within their own country. In national accounting, it refers to activities and expenditures of both residents and non-residents travelling within the reference country, i.e., both 'domestic' and inbound tourism.

3 **Outbound tourism**: residents of a country visiting destinations in other countries.
4 **Internal tourism**: the combination of domestic tourism and inbound tourism.
5 **National tourism**: the combination of internal and outbound tourism.
6 **International tourism**: the combination of inbound tourism and outbound tourism.

Figure 2.1 provides a graphic representation of the relationships among these terms.

Demand-side concepts

Demand-side concepts include classifications of individuals and of trips. One of the important accomplishments of the WTO's Ottawa Conference was consensus on demand-side definitions for domestic tourism; prior to the Ottawa Conference, the WTO's definitions were limited to international travel only. A further accomplishment was agreement on the concept of 'visitor' as the basic concept for the whole system of tourism statistics. As noted in Table 2.1, visitors may be classified into international visitors and domestic visitors. Each of these may, in turn, be divided into tourists (overnight visitors) and same-day visitors.

Table 2.1 Travellers' classifications

Terms	International tourism	Domestic tourism
Resident	A person who has lived in a country for at least 12 consecutive months prior to arrival in another country for a period not exceeding one year.	A person residing in a country for at least six months prior to arrival at another place in the same country.
Visitor	A person who travels to a country other than that of his usual residence and that is outside his usual environment for a period not exceeding one year, and whose main purpose of visit is other than the exercise of an activity remunerated from within the country visited.	A person residing in a country, who travels to a place within the same country but outside his usual environment for a period not exceeding six months and whose main purpose of visit is other than the exercise of an activity remunerated from within the place visited.
Tourist	A visitor who travels to a country other than that in which he has his usual residence for at least one night but not for more than one year, and whose main purpose of visit is other than the exercise of an activity remunerated from within the country visited.	A visitor residing in a country who travels to a place within the same country but outside his usual environment for at least one night but not more than six months and whose main purpose of visit is other than the exercise of an activity remunerated from within the place visited.
Same-day visitor	A visitor who travels to a country other than that in which he has his usual residence for less than 24 hours and without spending the night in the country visited, and whose main purpose of visit is other than the exercise of an activity remunerated from within the country visited.	A visitor residing in a country who travels to a place within the same country but outside his usual environment for less than 24 hours and without spending the night in the place visited and whose main purpose of visit is other than the exercise of an activity remunerated from within the place visited.

Source: World Tourism Organization (1991).

Note: The following classes of individuals are excluded from tourism statistics: border workers; temporary immigrants; permanent immigrants; nomads; transit passengers; refugees; military personnel; consular representatives; diplomats; and dependents and household servants accompanying military personnel, consular representatives, and diplomats.

Special attention was given to the question of measuring same-day travel. As the WTO notes in its resolutions from the Ottawa Conference (World Tourism Organization 1991) the question of measuring same-day visits is of increasing importance given the changing behaviour of consumers and improvements in technology that makes travel over greater distances easier. The removal of trade barriers among many nations, such as among the members of the European Union or the North American Free Trade Agreement, also encourages same-day visits between neighbouring countries. The concept of same-day travel as a form of tourism suggests that any operational definition of the phenomenon should be similar to that for other forms of tourism, viz., a non-routine trip away from the usual environment. The WTO recommends that measures be kept separately for same-day business trips and other types of same-day trips, and that international same-day travel and domestic same-day travel also be tabulated separately. Same-day trips are part of tourism, although same-day travellers are not tourists, *per se*, because tourists, by definition, are overnight visitors.

There are three types of same-day travel: (1) round-trips starting from the place of usual residence (or other places where one is not a visitor), (2) round-trips from a place where one is an overnight visitor, and (3) stopovers as part of transit travel. Only the first of these three is considered a same-day trip for the purpose of tourism statistics. The other categories usually are captured in other measures (for example, a day outing from a hotel to a folk village would be considered part of the larger trip that included the overnight stay at the hotel).

'Usual environment'

An important concept in contemporary tourism definitions is that of 'usual environment'. This is the first criterion that distinguishes tourism from other travel. While 'usual environment' is intuitively understood by most people – travel to work, school, a day care centre, a favourite local restaurant, a local movie theatre – it is difficult to develop an objective operational definition. The WTO suggests that two dimensions be considered when defining usual environment (den Hoedt 1994: 19). The first is frequency. Places visited frequently and on a routine basis, even if they are relatively distant from one's residence, would normally be considered as part of the usual environment. In practice, a place visited at least once a week would be considered to be visited frequently.

The second dimension is distance. Places close to one's residence are considered within one's usual environment even if the specific sites are not often visited. The issue of 'closeness' is a difficult one, and can be subject to substantial local variations. Places on the Canadian prairies that are 200 km apart are often thought of as being 'close' because of the mobility of the population and the low population densities. In contrast, in eastern China, where many individuals have much less mobility and many regions are densely populated, a distance of 50 km can represent a significant distance. As a compromise, the WTO (den Hoedt 1994: 19) recommends that a threshold of 160 km be used for defining the 'usual environment' for domestic same-day travel. Thus, non-routine trips longer than 160 km away from the origin should normally be included in tourism statistics for domestic travel.

Length of stay

The second criterion for defining tourism is length of stay. A person who stays in a place for more than 12 months is not considered a visitor but a resident. People who have spent less than 12 months in a place but consider it to be their residence also are excluded as visitors to that place. Thus a person who has spent less than 12 months in a place but intends to return within 12 months to live there is not considered to be a tourist there. This criterion would exclude someone who has recently taken up residence in a new country but returns to his former home for a short visit as being a visitor in the country to which he has just moved.

A concept related to length of stay is 'overnight'. Staying overnight distinguishes tourists from same-day travellers. While there is usually no confusion about whether one spent the night in a place or not, uncertainty can arise in special situations. For example, has an air traveller who has a lay-over during a long-haul flight, checks into a hotel at 6:00 pm to refresh and nap and then leaves at midnight spent the night? What about someone who visits a friend outside his usual environment and returns after midnight without sleeping? The WTO (den Hoedt 1994: 34) proposes two operational definitions:

- Destination-based: someone spends a night in a place if the date of his arrival and departure are different. In practice, a guest at a hotel registering after midnight is usually registered as arriving the previous day, i.e. before midnight.
- Origin-based: someone spends a night on a qualifying trip if the dates of departure and return are different and he sleeps during his absence.

These definitions, applied literally, are still vulnerable to peculiar situations. Consider a person who travels across a time zone boundary, from west to east, at 10:55 pm in the western zone on 30 June. After crossing into the eastern zone, where it is now 11:55 pm, he spends ten minutes and then leaves the eastern zone at 12:05 am, 1 July. He returns to the western zone, where it is now 11:05 pm, 30 June. Using the destination-based definition, he has spent the night; using the origin-based definition, he has not.

The point of presenting this scenario is not to argue against these operational definitions but to acknowledge that *any* operational definition for defining 'tourists' will have limitations or weaknesses that can lead to 'nonsensical' conclusions in odd situations. Operational definitions must be judged on their ability to deal with the great majority of cases, not highly unusual situations.

Cruise ship passengers who arrive in a country onboard ship and return to the ship to sleep present a special case that must be addressed. They are considered same-day visitors to the country or port visited, even if the ship stays in port for several days. The same is true of yachts and crew members of merchant ships.

Remuneration

Visitors are also characterized by the fact that remuneration – wages, salaries, payments-in-kind (but excluding travel allowances or small participation fees) – from within the place visited is not a main purpose of travel or that they are not travelling with someone seeking such remuneration. If a traveller is seeking remuneration, he is considered a migrant and not a visitor. This distinction is true for both international and domestic tourism. Thus, seasonal workers at resorts or boarding schools, performing artists, consultants, lecturers, individuals working as au pairs, and professional athletes receiving remuneration from within the place visited are excluded from tourism statistics – even though they will not become residents of the place from which they are remunerated.

Classifications of travellers

Four major classifications of individuals are relevant in tourism analysis: residents, visitors, tourists and excursionists. The distinctions among these are given in Table 2.1. Tourism analysts often find it desirable to have information about the socio-economic characteristics of the various categories of travellers as well. Table 2.2 lists the socio-economic variables that the WTO recommends data be collected on, and Table 2.3 lists some additional socio-economic and trip characteristics that may be desirable for specific situations. The list of supplemental measures in Table 2.3 should be considered provisional only. Note that all, with the exception of household income, apply to the individual traveller rather than a group or household.

Classifications of trips

Trips may be classified according to the purpose of travel. The WTO recognizes six main purposes: pleasure, visiting friends or relatives, business, health, religion and other. Each of these may consist of a number of more specific motivations, as shown in Table 2.4.

Length of stay

Length of stay is a particularly important variable in measuring tourism. The temporal dimension of a trip is a distinguishing characteristic between same-day visitors and tourists. The levels of aggregation for recording lengths of stay, as suggested by the WTO, are shown in Tables 2.5 and 2.6.

Table 2.2 Socio-economic variables for describing travellers

Characteristics	Divisions	Subdivisions (optional)
Sex	Male	
	Female	
Age	0–14 years	
	15–24	
	25–44	25–29
		30–34
		35–39
		40–44
	45–64	45–49
		50–54
		55–59
		60–64
	65 or older	65–69
		70–74
		75–79
		80–84
		etc.
Education	Primary education not completed	
	Completed primary education	
	Completed secondary education	
	Completed vocational or technical training	
	Completed college, university, or graduate studies	
	Other	
Economic activity status	Economically active	Employed
		Unemployed but looking for work
	Not economically active	Students
		Homemakers
		Income recipients (e.g. retirees)
		Others
Occupation[a]	Legislators, senior officials, managers	
	Professionals	
	Technicians and associate professionals	
	Clerks	
	Service workers; shop and market sales workers	
	Skilled agricultural and fishery workers	
	Craft and related trades workers	
	Plant and machine operators and assemblers	
	Elementary occupations	
	Armed forces	
Income	No standard classification is suggested, but income figures should ideally be reported. Guidelines: (i) use annual gross income (i.e. before taxes), total of **all** income should be reported, for subsistence farming workers, record cash equivalent relevant to the reporting country; (ii) report household income, even if traveller is travelling alone; (iii) collect income in currency of country of residence and then convert to local currency using current exchange rate in effect at time of survey; (iv) present results in quintiles or deciles, categories may be rounded to substantial but mutually exclusive categories	

[a] Detailed descriptions of these categories can be found in the International Labour Organization publication on ISCO-88.

Table 2.3 Supplemental traveller characteristics

Characteristics	Divisions	Subdivisions
1 Marital status	a. Married	
	b. Separated/divorced	
	c. Never married	
	d. Widowed	
2 Household size and composition	a. One-person household	
	b. Multi-person household with no children under age (legal age of majority in nation)	
	c. Multi-person household with child(ren) under age	i. Youngest child is under 6
		ii. Youngest child is 6 or older
3 Trip organization	a. No advanced booking	
	b. Advanced booking of accommodation only	i. Directly with property
		ii. Through intermediary such as tour operator or travel agent
	c. Advanced booking of transport only	i. Directly with provider
		ii. Through intermediary
	d. Inclusive booking	
4 Activities	a. Participant sports	E.g. golf, tennis, skiing, boating, swimming, hiking, hunting, fishing, picnicking
	b. Spectator sports	E.g. soccer, football, baseball, hockey, cricket, polo, Olympics
	c. Cultural activities	E.g. theatre, museums, zoos, botanical gardens, amusement parks, circuses, historic sites
	d. Meetings and conventions	E.g. congresses, conferences, trade shows, classes.
	e. Religious activities	E.g. pilgrimages, attending religious events
	f. Sightseeing	E.g. self-guided tours, guided tours
	g. Health-related activities	E.g. spas, hospitals, clinics, exercise classes, gymnasia
	h. Shopping	
	i. Other	E.g. restaurants, night clubs and bars, casinos

Table 2.4 Purpose of trip – classifications

Major group	Examples
1 Leisure, recreation and holidays	Sightseeing, shopping, attending sporting and cultural events, recreation and cultural activities, non-professional sports, trekking and mountaineers, use of beaches, cruises, gambling, rest and recreation for armed forces, summer camps, honeymooning
2 Visiting friends and relatives	Visits to relatives or friends, home leave, attending funerals, care of invalids, weddings
3 Business and professional	a. Installing equipment, inspection, purchases, sales for foreign enterprises b. Attending meetings, conferences or congresses, trade fairs, exhibitions c. Giving lectures or concerts d. Programming tourist travel, contracting of accommodation and transport, working as guides or other tourism professionals e. Participation in professional sports activities f. Government missions including diplomatic, military or international organization personnel except when stationed on duty in the country visited g. Paid stay, education, research such as university sabbatical leaves h. Language, professional or other specialized courses in connection with and supported by visitor's business or profession
4 Health treatment	Spas, fitness, thalassotherapy, health resorts, other treatments
5 Religion, pilgrimages	Attending religious events, going on a pilgrimage
6 Other	Aircraft and ship crews on public carriers, transit activities, other or unknown activities

Source: World Tourism Organization (1994).

Expenditures

Governments promote tourism development in their jurisdictions in large part because of the potential for revenue-generation. Information on tourist expenditures is therefore of great interest to governments at all levels. It is often desirable to record expenditures by same-day visitors separately from those of tourists. Expenditures for international travel fares should also be recorded separately from other expenditures. The WTO suggests the classifications in Table 2.7 for expenditures.

Table 2.5 Classification of same-day visitors by duration of trip (in hours)

Major groups	Minor groups
1 Less than 3	1.1 Less than 2
	1.2 2–3
2 3–5	2.1 3
	2.2 4
	2.3 5
3 6–8	3.1 6
	3.2 7
	3.3 8
4 9–11	
5 12+	

Source: World Tourism Organization (1994).

Table 2.6 Classification of tourists by duration of trip (in number of nights)

Major groups	Minor groups
1 1–3	1.1 1
	1.2 2–3
2 4–7	
3 8–28	3.1 8–14
	3.2 15–21
	3.3 22–28
4 29–91	4.1 29–42
	4.2 43–56
	4.3 57–70
	4.4 71–91
5 92–365	5.1 92–182
	5.3 183–365

Source: World Tourism Organization (1994).

Transport

The mode of travel is also of substantial interest to both industry and government. Travel often constitutes the single greatest expenditure item in a traveller's budget. Modes of travel are first classified as being air- , water- or land-based. Each of these is then further divided, as shown in Table 2.8.

Accommodation

Being a tourist, by definition, implies the need for accommodation. WTO suggests dividing accommodation first into 'collective' establishments (those serving the public) and private. Both these major groups are then disaggregated into 'minor groups' and 'unit groups'

Table 2.7 Expenditure classifications

- 1 Package travel, package holidays, and package tours
- 2 Accommodation
  - 2.1 Collective tourism accommodation
    - 2.1.1 Hotels and similar accommodation
      - 2.1.1.1 Hotels
      - 2.1.1.2 Similar accommodation
    - 2.1.2 Specialized accommodation
      - 2.1.2.1 Health facilities
      - 2.1.2.2 Work and holiday camps
      - 2.1.2.3 Public means of transport
      - 2.1.2.4 Conference centres
    - 2.1.3 Other collective accommodation
      - 2.1.3.1 Holiday dwellings
      - 2.1.3.2 Tourist campsites
      - 2.1.3.3 Other collective accommodation
  - 2.2 Private tourism accommodation
    - 2.2.1 Private tourism accommodation
      - 2.2.1.1 Owned dwellings
      - 2.2.1.2 Rented rooms in family homes
      - 2.2.1.3 Dwellings rented from private individuals or professional agencies
      - 2.2.1.4 Accommodation provided without charge by friends or relatives
      - 2.2.1.5 Other private accommodation
- 3 Food and drink
  - 3.1 Prepared food consumed on premises
  - 3.2 Beverages consumed on premises
  - 3.3 Food and beverages for preparation and/or consumption elsewhere
- 4 Transport
  - 4.1 Air
    - 4.1.1 Scheduled flights
    - 4.1.2 Non-scheduled flights
    - 4.1.3 Other air services
  - 4.2 Waterway
    - 4.2.1 Passenger lines and ferries
    - 4.2.2 Cruise
    - 4.2.3 Other waterway transport
  - 4.3 Land
    - 4.3.1 Railway transport
    - 4.3.2 Motorcoach or bus and other public road transport
    - 4.3.3 Private vehicle transport
      - 4.3.3.1 Gasoline and oil
      - 4.3.3.2 Repair services
      - 4.3.3.3 Parking fees, tolls, fines
    - 4.3.4 Vehicle rental
    - 4.3.5 Other means of land transport
  - 4.4 Other transport items

Table 2.7 *continued*

5 Recreation, culture and sporting activities
5.1 Recreation and sporting activities
5.2 Cultural activities
5.3 Entertainment
6 Shopping
6.1 Souvenirs
6.2 Duty-free goods
6.3 Clothing and footwear
6.4 Luggage
6.5 Tobacco products
6.6 Personal care products
6.7 Other goods
7 Other
7.1 Financial services
7.2 Travel items, charges not elsewhere classified
7.3 Health or medical services
7.4 Education or training services
7.5 Other services not elsewhere classified

Source: World Tourism Organization (1994).

Table 2.8 Classification of means of transport

Major groups	Minor groups
1 Air	1.1 Scheduled flights
	1.2 Non-scheduled flights
	1.3 Other services
2 Waterway	2.1 Passenger lines and ferries
	2.2 Cruise
	2.3 Other
3 Land	3.1 Railway
	3.2 Motor coach or bus and other public road transport
	3.3 Private vehicles
	3.4 Vehicle rental
	3.5 Other

Source: World Tourism Orgnization, (1994).

representing increasingly precise classifications. Table 2.9 presents the guidelines for classifying accommodation. The unit groups, in particular, may be subject to modification by individual nations to reflect local supply. For example, Chinese analysts may wish to divide 'hotels and similar establishments' into 'starred hotels' (serving international visitors) and 'other hotels' (serving Chinese nationals). A Japanese analyst may want to distinguish *ryokans*, a traditional Japanese

Table 2.9 Classification of tourism accommodation

Major groups	Minor groups	Unit groups
1 Collective tourism establishments	1.1 Hotels and similar establishments	1.1.1 Hotels
		1.1.2 Similar establishments
	1.2 Specialized establishments	1.2.1 Health establishments
		1.2.2 Work and holiday camps
		1.2.3 Public means of transport
		1.2.4 Conference centres
	1.3 Other collective establishments	1.3.1 Holiday dwellings
		1.3.2 Tourist campsites
		1.3.3 Other collective establishments
2 Private tourism accommodation	2.1 Private tourism accommodations	2.1.1 Owned buildings
		2.1.2 Rented rooms in family homes
		2.1.3 Dwellings rented from private individuals
		2.1.4 Accommodation provided without charge by relatives or friends
		2.1.5 Other private accommodation

Source: World Tourism Organization (1994).

Note: This classification may need to vary from country to country to reflect national variations in accommodation types, particularly at the unit level. The system is intended to be exhaustive, so that a specific type of accommodation may be assigned to a minor group or at least a major group and so have a generic title and conform to a standard classification and definition. Countries may adapt this classification to the structure of their supply of tourism accommodation without detracting from its international comparability.

inn, as a distinct tourist accommodation category. Canadian tourism researchers may wish to have explicit categories for 'self-catered cabins' and 'bed-and-breakfasts'.

Supply-side concepts

The supply side of tourism comprises the businesses and agencies that provide services to tourism. These include both profit-oriented firms as well as non-profit organizations such as destination marketing organizations and educational institutions. The importance of the supply-side perspective, however, goes far beyond just the classification of tourism businesses. Tourism is traditionally conceived in terms of demand-side or consumer characteristics (e.g. the duration and geographic extent of a trip taken by a traveller), which places the industry in a difficult political and statistical position. By tradition, industries are defined in terms of their products, not their consumers. For example, the agricultural industry consists of those enterprises and organizations involved in the production, processing and distribution of food and fibre. If agriculture were to use a demand-side definition as tourism traditionally does, the agricultural industry would be defined in terms of someone who eats.

The tourism industry may be defined as those enterprises and organizations involved in facilitating travel and activity away from one's usual environment. One challenge in this approach to defining tourism is, of course, the fact that many enterprises which produce commodities for tourists also serve non-tourists. For example, more restaurant meals are consumed by local residents than by tourists; taxis may be used by local business people and residents more frequently than by out-of-town visitors. Local attractions may draw local residents as well as tourists.

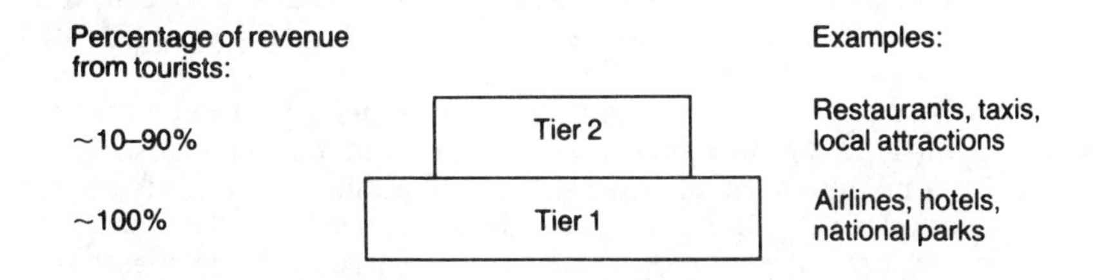

Figure 2.2 A supply-side definition of the tourism industry

A conceptual solution to this problem was proposed by the (Canadian) National Task Force on Tourism Data (1985). Building on the definition presented in the paragraph above, the Task Force went one step further. They proposed dividing tourism businesses into two tiers (see Figure 2.2). Tier 1 firms are those that would not exist in the absence of tourism. Examples include hotels, airlines, cruise ships and travel agents. Tier 2 firms are those that would continue to exist in the absence of tourism, but in a diminished form. These businesses include taxis, restaurants, rental car agencies, gift shops, and attractions and events.

The inclusion of particular businesses in either tier 1 or tier 2 depends on the scale of analysis as well as regional characteristics of the tourism industry. Some restaurants located in popular tourism destinations may derive virtually all of their revenue from tourists. In such cases, these businesses would be classified as belonging in tier 1. If the level of analysis were national, however, restaurants would be classified in tier 2. Drugstores and video rental establishments in cottage communities may properly be classified as tier 2 businesses at a local scale, whereas they would not be considered tourism businesses at all nationally.

This definition depends on a sense of reasonableness and the willingness to overlook minor inconsistencies. The case of hotels is a good example of this need. Hotels are tier 1 firms because they exist to serve the travelling public, but many will derive some of their revenues from local residents who might eat or drink at the hotel, or who might check-in for a 'get-away' weekend. Thus many hotels will not derive 100 per cent of their revenue from tourists.

A numerical example may be seen in Table 2.10. The data shown here are drawn from a variety of Canadian business and consumer surveys. The table consists of several columns. The first is a list of tourism commodities. These roughly parallel SIC divisions, although the focus in this table is on the commodity, not the industry sector. For example, the data in the table reflect supply of and expenditures on parking regardless of which industry sector operates the pay parking lot. The second column reports 'total domestic supply', which is the total value of the output of Canadian suppliers of this commodity. The next column, 'total tourism demand', represents total expenditure on these commodities by Canadian residents as well as exports (i.e. purchases made by international visitors to Canada). The last column, 'tourism's share of total domestic supply', reflects the percentage of expenditure on a given commodity made by tourists. This is the ratio between 'total tourism demand' and 'total domestic supply'. Thus, for example, 92.1 per cent of all receipts for the sale of passenger air services was derived from tourists. The balance, 7.9 per cent, represents sales to non-tourists: same-day visitors and to travellers not included in tourism statistics.

Finally, one must set an operational definition for classifying the various commodity groups as either tier 1 or tier 2. In this particular example, we use a limit of 80 per cent to classify a category as tier 1. This threshold is somewhat arbitrary; the threshold should be reasonably high, but need not be 100 per cent, because – as noted previously – even 'pure' tourism businesses may derive a percentage of revenues from non-tourists. As noted, air transportation is purchased not only by tourists but also by same-day visitors, members of the armed forces, immigrants, and other non-tourist travellers.

The line of reasoning behind tiers 1 and 2 has been extended and formalized in a new development many industrialized nations are exploring, called the **Standard International Classification of Tourism Activities (SICTA)**.

Standard International Classification of Tourism Activities

The SICTA is a formal attempt to link tourism with the International Standard Industrial Classification system (ISIC). ISIC, like all national Standard Industrial Classifications (SICs), is a taxonomic system that classifies all elements of economic activity in a comprehensive and consistent framework. SICs serve several functions. They help government agencies aggregate data from diverse industries in a consistent format; provide a structure for national business surveys and censuses; order the development and application of industrial policies and regulations across industries; and guide industry associations, consultants, academics, and other analysts in their study of industries and industrial policies.

Table 2.10 An illustration of the two-tier tourism industry model

Commodities	Total domestic supply	Total tourism demand	Tourism ratio	Tier
Passenger air	6 566	6 044	92.1	1
Passenger rail	239	200	83.8	1
Passenger water	308	288	93.5	1
Interurban, charter and tour bus	456	402	88.1	1
Urban transit	1 138	34	3.0	2
Taxis	1,381	398	28.8	2
Vehicle rental	728	604	83.0	1
Vehicle repairs and parts	8 893	1 692	19.0	2
Vehicle fuel	11 103	3 592	32.4	2
Parking	561	39	7.0	2
TOTAL TRANSPORTATION	31 372	13 294	42.4	2
Hotels	2 959	2 704	91.4	1
Motels	617	589	95.4	1
Camping	255	213	83.6	2
Private cottages	n.a.	166	n.a.	n.a.
Outfitters	153	135	88.2	1
Other accommodation	147	69	46.6	2
TOTAL ACCOMMODATION	4 131	3 875	89.8	1
Meals from:				
accommodation	1 792	928	51.8	2
restaurants and bars	14 365	3 726	25.9	2
other tourism businesses	519	144	27.7	2
Alcoholic beverages from:				
accommodation	1 841	299	16.2	2
restaurants and bars	3 341	491	14.7	2
other tourism businesses	348	96	27.7	2
TOTAL FOOD AND BEVERAGE SERVICES	22 206	5 685	25.6	2
Recreation and entertainment	7 235	2 005	27.7	2
Travel agency services	469	459	97.8	1
Convention fees	n.a.	92	n.a.	n.a.
TOTAL OTHER TOURISM COMMODITIES	n.a.	2 556	n.a.	n.a.
Groceries	38 542	1 082	2.8	2
Beer, wine and liquor from stores	7 595	124	1.6	2
Other commodities	n.a.	2 304	n.a.	n.a.
Pre-trip expenses	n.a.	1 420	n.a.	n.a.
TOTAL OTHER COMMODITIES	n.a.	4 930	n.a.	n.a.

Source: Statistics Canada (1994).

In practice, an SIC is a hierarchical system. It begins with a small number of very general categories of economic activity that are then broken down into two to four levels, depending on the level of precision required for a particular industry. Any firm is classified in one and only one category, based on where its primary activity (usually in terms of gross revenue or value-added) is. Hotels, for example, often derive some revenue from the sale of tobacco products, clothing and other retail items through convenience stores and gift shops. However, their primary source of revenue is the letting of rooms – and this is the basis for classifying hotels as part of the accommodation sector in an SIC. The order and number of divisions and hierarchical levels is not theoretically based; rather it reflects idiosyncratic decisions or needs specific to the economy being classified. Thus SICs are not always comparable between nations. SICs are periodically updated to reflect the evolution in the economic structure of the nation.

The ISIC is a generic SIC developed by the United Nations to provide a model for national SICs and to encourage consistency for international comparisons. It is based on the 'enterprise', the smallest legal entity that encloses and controls all functions associated with its business. The enterprise is the owner of a commodity-producing unit. It may involve multiple locations and multiple products. For example, a hotel chain with hundreds of properties as well as separate food services and catering, management services and perhaps entertainment divisions could be a single enterprise. The fundamental characteristic of the enterprise is unity of ownership.

Enterprises may be divided into 'kind-of-activity-units': divisions of an enterprise that produce predominantly the same commodity. Continuing the example of a hotel, the property management division properties would be one kind-of-activity-unit a separate catering division would be another. Each unit is autonomous with respect to its own activities, although it may be owned by an enterprise with other units.

The kind-of-activity unit is subdivided into 'local units' that are distinguished by their geographical location. The set of hotel properties operated by the hotel corporation in one city or region would be the local unit. The local unit may be either the actual location of the unit or the area served by the unit. Local units, in turn, are disaggregated into 'establishments', which are distinct units of activity in terms of both kind of activity and geographic location. They are the smallest observable statistical unit for the purpose of supply-side statistics. In our example, the establishment would be an individual hotel property.

Pisarski (1992), at the request of the WTO, began a tourism-based modification of the ISIC in 1990. This modification has been named the Standard International Classification of Tourism Activities (SICTA). The SICTA is essentially a subset of the ISIC consisting of those divisions that draw a substantial portion of their revenue from tourists. It is ideally based on the establishment, as defined above, rather than the enterprise because the establishment provides more precise data. The enterprise level is less useful because many enterprises that produce tourism commodities are also involved in the production of non-tourism commodities. Pisarski (1992: 17) notes that the two intermediate unit categories are rarely used, which is regrettable because they could often be of value to tourism analysis. As noted previously, the distinction between a tier 1 and tier 2 establishment is a geographic phenomenon. In some locales, restaurants and gas stations may derive 90 per cent or more of their revenue from tourists, while elsewhere the same type of establishments may draw 5 per cent or less.

The SICTA combines the supply-side and demand-side of tourism. The basic structure of the SICTA is drawn from a supply-side concept: establishments. However, these establishments are selected on the basis of a demand-side concept: the nature of their customers. Data reported for each establishment should, ideally, be disaggregated to permit estimation of the contributions from tourists and non-tourists. At a minimum, the SICTA can conceptually indicate whether any individual entry in the hierarchy (at each level) draws its revenue partially or primarily from tourists. It should be noted that at higher levels in the SICTA hierarchy, there will be relatively few classifications that are predominantly tourist-related; however, as one works down into more disaggregated levels, the number of predominantly tourism categories will increase.

Pisarski's SICTA consists of four levels: the main division, the group, the main class, and the sub-class. (This parallels many national SICs, which are divided into a 'division', or major activity (such as agriculture or manufacturing), a 'major group', an 'industry group', and finally an 'industry'). The system also notes whether each category draws most of its revenues from tourists (denoted by a 'T') or only partially from tourists (denoted by a 'P'). The name of each category is also identified. The complete SICTA is too long to reproduce here, but a typical portion is depicted in Table 2.11.

Summary

The definition of tourism and related words continues to challenge the industry. The problem is not simply one of trying to find a definition, but of sorting through a myriad of competing definitions to find those definitions for tourism-related phenomena that can be accepted across political jurisdictions. The need for accurate and comparative data grows annually, and that need is helping to drive the global tourism industry towards greater consistency in definitions and data collection procedures. This chapter has described some of the more important steps in the evolution of standardized definitions.

We have also acknowledged explicitly that operational definitions are often incapable of dealing with peculiar or extreme cases. While this may be seen as a weakness, it is also a fact of analytic life. SICs, national accounts, international conventions on definitions must be designed to handle normal situations and to reflect or monitor typical cases. A degree of reasonableness and a sense of tolerance is needed to accept

Table 2.11 A portion of the Standard International Classification of Tourism Activities

Div	Group	Class main-Sub	P/T	Name
I				TRANSPORT, STORAGE AND COMMUNICATIONS
60				LAND TRANSPORT
	601	6010	P	Railways
		6010-1	T	Inter-urban rail passenger services
		6010-2	T	Special retail services
	602			OTHER LAND TRANSPORT
		6021	P	Other scheduled passenger land service
		6021-1	T	Scheduled inter-urban buses
		6021-2	T	Long-distance tour buses
		6021-3	P	Scheduled local and urban transit services
		6021-4	P	Specialized scheduled vehicles, e.g. airport buses
		6022	P	Other non-scheduled passenger land transport
		6022-1	P	Taxis
		6022-2	P	Chauffeured vehicles
		6022-3	T	Local tour vehicles
		6022-4	P	Charter buses, excursions
		6022-5	P	Man- or animal-drawn vehicles
61				WATER TRANSPORT
	611	6110	P	Sea and coastal transport
		6110-1	T	Cruise ships
		6110-2	T	Ship rental with crew
	612	6120		Inland water transport
		6120-1	T	Inland water passenger transport with accommodation
		6120-2	T	Inland water local tours
		6120-3	P	Inland water taxis, ferries

Source: World Tourism Organization (1994).

the fact that macro-scale tools, such as those described in this chapter, will not be able to apply to unusual situations that one might dream up as a logical challenge.

Definitions not only guide data collection but also shape how analysts and policy-makers conceptualize tourism. Thus, defining tourism as an industry is more than just defining the phenomena – tourists, visitors, excursionists, etc. – that are studied. A more comprehensive definitional model is needed – one that defines tourism in the supply-side terms comparable to those of other industries and yet can accommodate the traditional demand-side focus in much tourism analysis. The two-tier model of tourism and the Standard Industrial Classification of Tourism Activities as described in this chapter may provide such a model. Further information on tourism definitions and some associated debates may be found in the sources listed under the Further reading section of this chapter.

Further reading

Cohen E 1974 Who is a tourist? *Sociological Review* **22**: 527–53.

Hunt J D and **Layne D** 1991 Evolution of travel and tourism terminology and definitions. *Journal of Travel Research* **29**(4): 7–11.

Leiper N 1979 The framework of tourism: towards a definition of tourism, tourist, and the tourist industry. *Annals of Tourism Research* **6**: 390–407.

Leiper N 1983 An etymology of tourism. *Annals of Tourism Research* **10**: 277–80.

Leiper N 1990 Partial industrialization of tourism systems. *Annals of Tourism Research* **17**: 600–5.

Leiper N 1992 Industrial entropy in tourism systems. *Annals of Tourism Research* **19**: 221–6.

Smith S L J 1991 The supply-side definition of tourism: Reply to Leiper. *Annals of Tourism Research* **18**: 312–15.

Smith S L J 1992 Return to the supply-side. *Annals of Tourism Research* **19**: 226–9.

CHAPTER 3

Collecting data on tourism

Introduction

Data for tourism analyses are obtained from many different sources, but are collected in only three basic ways: observation, administrative record-keeping, and surveys (either original or those done by someone else). This chapter focuses on survey data because surveys are, arguably, the most important source of information for tourism analysis, planning and decision-making. Before examining issues associated with surveys, however, we will consider briefly some of the characteristics of the other two sources of information: observation and administrative records.

Observation refers to the structured collection of information through systematic observation and measurement of empirical phenomena, such as numbers of visitors pausing to look at a museum exhibit, lengths of queues and/or time hotel guests spend waiting at the front desk to register, or an inventory of the contents and weight of garbage generated by a hotel or restaurant. Observation has the potential to provide valuable data that are grounded in real behaviour rather than opinion or official records. However, it requires a careful protocol to ensure that the information collected is not biased as a result of arbitrary or careless selection of observation periods, the observer's own prejudice of the meaning of the phenomenon being observed, and the potential influence of an obvious observer on tourists' behaviour.

Administrative data are obtained in the course of routine business operations but have potential applications beyond the original administrative applications. For example, vehicle counts at border crossings are often maintained by immigration and customs officials as part of their routine activities, but this information also provides an estimate of international visitors. Credit card records, containing information about the amount and location of an expenditure as well as the specific commodity purchased, are another potentially rich source of data related to travellers. Administrative data can also refer to resource or event inventories compiled as a normal part of running a business or agency. The registration records from accommodation as well as the names and addresses of individuals who wrote to a destination marketing organization for information are also types of administrative data.

Administrative data are often overlooked as sources of information about tourism because they were not collected with analysis in mind and because they are usually proprietary. However, they can provide valuable information unavailable from any other source. Much data could be made available if basic procedures were implemented to ensure the confidentiality of the records, such as removal of any code or information identifying actual individuals or enterprises.

As noted, the single most important source of information for tourism analysts is probably survey research. Surveys may be conducted specifically by or for the analyst, or by another party for an independent purpose and then made available to an analyst for further analysis. This chapter examines a number of the fundamental issues associated with survey research, including questionnaire design, sample selection, coding questionnaire data, and principles for the use of secondary data sets (data collected by someone else for their own purposes).

It should be emphasized that issues of questionnaire design and sample selection can be quite complex. This chapter gives only a superficial overview of some issues as well as some general practical advice. However, the reader interested in developing a major survey is encouraged to consult some of the books listed at the conclusion of the chapter and/or to discuss his needs with a professional statistician or survey researcher.

Principles of questionnaire design

Format

One of the first questions that must be considered when planning a survey is what format the survey will take. 'Format' here refers to the method of delivering the survey. There are three basic formats used in tourism: personal interviews, telephone surveys and mail-back questionnaires. These can be combined into a multi-stage interview. You might briefly survey people waiting in line to board a ferry in a personal interview and then provide them with a mail-back questionnaire to complete at a later date. Or you might solicit participants in a survey project on travel expenditures through random telephone calls, followed by a brief personal interview, and then provide the respondents with a diary to fill out to record trip expenditures for subsequent mail-back. Although there are many permutations possible, most surveys in tourism will be based on personal, telephone or mail-back questionnaires.

The choice of format is based on several considerations, including the purpose of your study, the nature of the questions you wish to ask, the

Table 3.1 Characteristics of survey formats

Issue	Personal interviews	Telephone surveys	Mail-back questionnaires
Budget	High cost	Medium cost	Relatively lower cost
Sample size	Small	Medium	Large
Need for trained interviewers	Very high	High	No
Need to probe responses	Good	Limited	Not possible
Rating of long lists of items	Acceptable	Difficult	Good
Open-ended questions	Easily handled	Easily handled	Difficult
Need to control who answers questionnaire	Possible	Possible	Not possible
Difficulty in delivery of questionnaire	High	High	Low
Typical response rates	High	Medium	Low
Time required for conducting survey	Medium	Short	Extended

characteristics of your sample, and the time and money you have available to do the survey. Table 3.1 summarizes some of the issues that will help you select the questionnaire format most appropriate for your survey project.

As Table 3.1 suggests, there is often no clear answer to the question about which medium is the best. Further, issues other than those listed can also influence your final decision. For example, in the US, telemarketers often represent themselves as doing a 'telephone survey' in order to get past the respondent's understandable lack of interest in hearing a telephone sales pitch. As a result of such unethical practices, many residents now refuse to respond to any telephone survey – legitimate or not. The proliferation of answering machines, and their use to screen unwanted calls, has created a further impediment to completing telephone surveys. In the late 1980s, telephone surveyors found that, typically, they needed to place three calls for every completed survey. By the mid-1990s, the ratio had increased to ten calls for every completed survey. The difficulty of conducting telephone surveys in such conditions is not likely to improve in the foreseeable future.

Confidentiality and anonymity

Confidentiality of respondents is expected in virtually all tourism surveys. First of all, there is usually no need to know the identity of individual respondents. The interest in most surveys is not in the patterns of individual responses but of aggregate patterns that may be generalized to the larger population. But at a more basic level, most societies place value on the privacy of individual citizens. And this translates in the context of survey research into expectations that responses of private citizens to most surveys should be kept private.

Confidentiality is not synonymous with anonymity. Confidentiality implies that the identities of respondents will not be made public or shared with individuals outside the immediate research team. Anonymity, in contrast, means that no one, including the researcher, knows anything about the identity of respondents. This is not always practical or even desirable. First of all, it is difficult to select respondents for many types of surveys without knowing their identity and address or telephone number for the initial contact. Follow-ups, either to increase response rate or to assess the extent of non-response bias, require knowing the identities of both respondents and non-respondents to a survey. In practice, respondents may be identified by a code that can be linked to a confidential list of names. The list is maintained by the project manager and should normally not be released to other researchers or to firms interested in using the list for direct marketing.

Sample selection

You usually cannot survey every member of the group in which you are interested. Whether members of an organization, local businesses, or residents of a community, analysts often lack the time and money to interview everyone. So, a sample of people is chosen for study. A critical objective in sample selection is to ensure that it represents the population (sometimes called the universe) from which the sample is drawn. The concept of the population is not as simple as it might seem. First, there is a difference between the survey population and the target population. The survey population is the population from which the sample is actually drawn, while the target population is the population you are studying. In principle, they should be the same but, in practice, may differ. For example, social surveys in Canada usually exclude residents of the Yukon and Northwest Territories because population densities are so low. Members of the armed forces, members of the First Nations (aboriginal populations) living on reserves, inmates in correctional institutions, and other institutionalized individuals are also routinely excluded.

The sample population is operationalized as a sampling frame. The sampling frame is the statistical universe from which the sample population is drawn. The sampling frame can be a significant source of bias. A survey of visitors concerning their perceptions of the importance of tourism in a local community might be based on a random sample drawn from the telephone book, especially if the survey is to be conducted as a phone survey. However, not every household in a community has a telephone and not everyone who has a telephone will be listed in the directory. As a result, the use of telephone listings as the sampling frame will bias your results towards emphasizing the views of those individuals who have telephones and who have published telephone numbers. The exact nature of this bias usually is not known.

The sampling frame may also involve your 'qualifying' potential respondents before administering the full questionnaire to them. This refers to refining the pool of prospective respondents to just those who possess the desired attributes for your sample. Common criteria for qualifying potential respondents in tourism surveys are to select only those individuals who, for example:

- are 18 years of age or older;
- have taken a pleasure trip of a minimum duration and/or minimum distance within the last year;
- purchase and consume alcoholic beverages;
- used commercial accommodation on their last trip; or
- are members of a frequent flier programme.

Sometimes you will need to select an individual within a randomly selected household or travel party for a telephone or personal interview only after you have actually contacted the group. One practical way of doing this is (if you are interviewing adults only) is to ask to interview the person, over the age of 18 (or whatever age is appropriate), who will have the next birthday. This technique should normally provide you with a representative mix of females and males and provides an easily understood mechanism for objectively and randomly selecting an interviewee from a household or travel party.

Once you have a clear understanding of your target and survey populations and your sampling frame, the next step is to begin planning how to draw a sample from your sampling frame. There are several ways in which a sample can be drawn, but the three most important are: (1) simple random sampling, (2) stratified sampling, and (3) cluster sampling.

Simple random sampling implies that everyone in the sampling frame has a known and equal opportunity to be selected. There are two traditional ways of selecting a simple random sample. The first step in each is to obtain a list of every individual in your sampling frame and then number each one. Then determine the number of individuals you wish to sample (see the section on sample size below).

For the first method: divide the number of people on your list by the number you need and round, if necessary, to the nearest integer, *n*. Using a table of random numbers, select a random starting point on the list and then select every *n*th individual. For example, if you need 350 people from a population of 1000, you would select every third individual ($1000 \div 350 \cong 3$). If you selected 052 from the table of random numbers, you would select the fifty-second person on your list and then select every third individual after that. If you reach the bottom of your list before completing your sample, return to the top of the list, continuing every *n*th selection.

This simple procedure produces a reasonably representative sample. However, examine the original list to determine whether there are any cyclical patterns that might produce unexpected biases. For example, some municipal enumeration lists order the entries by address. If there are an average of 10 houses on each block, and you are selecting every tenth entry, you might inadvertently pick up a series of residences located on corners of blocks. Residents who live in corner houses tend to differ systematically from those who live in the middle of a block.

For the second method, again consult a table of random numbers. Selecting an arbitrary starting point in the table, pull out one random number for each individual in your sample. Use only the number of digits necessary to cover your list. In other words, if you are pulling a sample from a list of 1000 people, use only four-digit random numbers between 0001 and 1000. Select a series of random numbers long enough to yield your sample. Thus, if you want a sample of 50, select 50 random numbers. Match each random number with the number of each respondent in your list. If a number comes up twice (as is possible in tables of random numbers), simply skip the second occurrence and draw another number.

This approach is more likely to yield a true random sample than the first method but is more cumbersome to implement (if done manually) than taking every *n*th person. If your list exists in a machine-readable database, computer programs such as SPSS are available to draw a random sample.

A simple random sample has some disadvantages. One of the more important is that a random sample might, through chance, miss important sub-groups. This is especially likely if the proportion of respondents drawn from your population for the sample is very small. Further, respondents in some groups such as seniors or recent immigrants are sometimes less likely to agree to participate in a survey. A stratified sample is a systematic procedure to help ensure that your final sample is representative.

The first step in a stratified sampling design is to identify the range of relevant sub-populations or strata, the number of individuals in each, and the proportion of the total population each stratum represents. The set of strata (sub-populations) should be exhaustive and mutually exclusive. In other words, the total population should be divided into strata and there should be no overlap between strata.

Determine the number of respondents needed for your sample, as described below. Multiply that number by the proportion of the total population represented by each stratum. This provides the number of respondents needed from each stratum in your total sample. Using one of the two simple random selection procedures described earlier, draw a sample from each. Because individuals in some sub-populations are less likely to answer surveys, the number identified for each stratum becomes a quota for that stratum. It may be necessary, therefore, to actually over-sample some sub-populations in order to ensure that the final sample is representative of the population.

The following example will illustrate this process. Assume you want to sample certified travel agencies. You believe there are differences among the views of small, medium and large agencies, so you want to ensure they are all adequately represented. You obtain a list of certified travel agencies from an industry association and divide it into the following categories:

Stratum	Employees	Number	Proportion
Large agencies	11+ employees	48	7.0%
Medium agencies	6–10 employees	467	68.0%
Small agencies	1–5 employees	172	25.0%

Assume you want to draw a sample of 50 agencies. You would select 4 individuals ($50 \times 7.0\% = 3.5$, rounded to 4) from large agencies; 34 medium-sized agencies, and 12 small agencies.

The third type of sampling is cluster sampling (sometimes referred to as multi-stage sampling). To understand the logic of cluster sampling, consider the following situation. You want to conduct a survey of the travel patterns of adult Canadians. You also have determined that a sample of 5000 people is appropriate. You could, in principle, select the 5000 using a simple random sampling approach. Alternatively, you could use a stratified sampling approach, with the strata defined in terms of age–sex cohorts to ensure adequate demographic representativeness. However, either approach requires having some sort of a list of all (approximately 20 million) adult Canadians from which individuals can be randomly identified. This is unrealistic. Instead, you can – for the sake of illustration – randomly select 50 postal code areas across the country and then draw households randomly from within each postal code. The number of household respondents from each postal code would be proportional to the population resident in each code area. Finally, you select from within each household the member, 18 years of age or older, who will celebrate the next birthday.

Cluster sampling thus is a way to simplify the collection of data from a dispersed population and to reduce field costs. Clusters do not necessarily need to be administrative units such as postal code areas. A region can be divided into a set of sampling units by imposing a grid over a map. A series of grid cells could then be chosen randomly, with sample sizes from each cell proportional to the population in them.

Multiple cluster levels also can be defined. For example, Canadian postal codes are subdivided into 'forward sortation areas' (FSAs) that are subdivided into postal walks. You might first select a number of postal code areas, then several FSAs within each postal code area and, finally, a postal walk within each FSA. Every household on the postal walk might then be studied. The principle of giving every household in Canada an equal chance of being selected is achieved through random selection of postal codes and FSAs, but the actual administration of the survey is conducted as a census (the surveying of every potential respondent) within the final clusters.

Table 3.2 Examples of sample sources

Target sample	Potential sample frame
General population	• telephone directories • voter enumeration lists • door-to-door cluster sample
Specific businesses, e.g. tour operators	• industry directories • membership directories from associations • general telephone directories
Specific market segments, e.g. seniors	• special interest clubs • subscription lists for special interest magazines and newsletters
Visitors to specific attractions or facilities	• on-site sampling • reservation or registration records • enquiry records (mailing addresses obtained for information follow-up)
Visitors to destinations	• exit surveys at airports, train stations, ferry ports, etc. • random sampling at information centres • enquiry records (addresses obtained for information follow-up)

Where to find the sample

Your sampling frame operationally defines where you find your sample. The sampling frame, in turn, will be shaped by your research objectives. If you are interested in the opinions of tour operators, you will seek a representative list of operators. If you want to measure the expenditure of visitors to an art gallery, you will seek to tap a representative sample of gallery visitors. Table 3.2 identifies some typical sampling frames for different target samples.

Sample size

Another fundamental question is how many people to interview. Indeed, this is one of the first questions many managers ask of their analysts. An accurate answer to that question can be difficult to obtain, especially if nothing is known about the characteristics of the sample population. None the less, a clear and defensible answer is expected and needed. Let us begin with a couple of observations.

First, if the sample population is reasonably large, the precision of survey results is a function of sample size, not population size. To put this into statistical terms: the standard error of a sample finding can be weighted by a ratio, the finite population correction, to reflect the effects of population size:

$$\frac{(N-n)}{N}$$

When the population (N) is large with respect to sample size (n), the value of the correction approaches 1.00. As a rule of thumb, if n is less than 5 per cent of N, population size is irrelevant in determining sample size or the precision of survey results.

Second, the sample size depends on how the data are to be analysed and, especially, disaggregated. If you wish to divide your respondents into male and female, you will need a larger sample size than if you were not to disaggregate respondents by sex. In effect, each sub-group for analysis must be treated as a separate sample.

Next, you will rarely obtain a 100 per cent response rate. If you know that you want a sample of 500 respondents and you suspect you will achieve about a 30 per cent response rate, start with 1600 or 1700 potential respondents. Finally, in order to determine sample size, you need to have some estimate of the variance in the sample population on variables of interest and on the confidence levels you require. Assume you are interested in the characteristics of who visited a national park. If 20 per cent of your respondents visited a national park, and your sample was 350 respondents, your confidence interval and confidence level will be based on 20 per cent of 350 (or 70 people) – not 350 people.

We have just used two terms: confidence interval and confidence level. These should be defined. Sampling provides only an approximation of the correct answer that could be obtained if you had unlimited time and money by conducting a census of the entire population. The quality of this approximation is assessed on two criteria: (1) how likely is the estimate to be right and (2) how precise is the estimate. We might observe that 50 per cent of national park visitors, ±30 per cent, engaged in photography during their visit. That estimate is probably accurate but the margin of error is wide. On the other hand, we might estimate that 50 per cent of national park visitors, ±2 per cent, engaged in photography. The margin of error is narrower, but we are less confident that the real value falls within such that range.

Confidence intervals are usually expressed as a certain percentage range around a mean value: ±5.0 per cent, 7.5 per cent, and so on. Confidence levels are expressed in probabilistic terms: a particular finding (the mean, plus or minus a certain percentage) is accurate at the 95 per cent level; or, to express the same finding differently, it is accurate 19 times out of 20.

Generally, higher confidence levels (e.g. 99 per cent rather than 95 per cent) are preferable. However, as you increase the confidence level with a fixed sample size, you necessarily broaden the confidence interval. You can achieve a greater probability of being right in any given sample only by broadening your conception of what the right answer is. The choice of the values of confidence intervals and confidence levels is arbitrary. A confidence level of 95 per cent ('19 times out of 20') is standard in much tourism research. There is generally no consensus of confidence intervals; however, anything more than perhaps plus or minus 10 per cent is often too wide to be of practical value.

The interested reader is encouraged to consult texts such as Moser and Kalton (1972) and Fleiss (1973) for a detailed discussion of sample size calculation. Tables 3.3 and 3.4 can be used to provide an estimate of the needed sample sizes for various combinations of confidence intervals and levels.

Table 3.3 Approximate confidence intervals for selected percentages (95% confidence level). For example, in a sample of 100 respondents, if you observe a result of 10%, the probability is that the true answer is within ±6% (4% to 16%) 95 times out of 100

Sample size	10% or 90%	20% or 80%	30% or 70%	40% or 60%	50%
100	6%	8%	9%	10%	10%
150	5%	6%	7%	8%	8%
200	4%	6%	6%	7%	7%
250	4%	5%	6%	6%	6%
300	3%	5%	5%	6%	6%
350	3%	4%	5%	5%	5%
400	3%	4%	4%	5%	5%
450	3%	4%	4%	5%	5%
500	3%	4%	4%	4%	4%
600	2%	3%	4%	4%	4%
700	2%	3%	3%	4%	4%
800	2%	3%	3%	3%	3%
900	2%	3%	3%	3%	3%
1000	2%	2%	3%	3%	3%
1500	2%	2%	2%	2%	3%
2000	1%	2%	2%	2%	2%

Questionnaire design

Format

The design of questionnaires is at least as important as the selection of a sample in determining the quality of your data. There are a number of matters that must be considered in designing a questionnaire. We have already considered the issue of format: personal interview, telephone interview or mail-back survey. Personal interviews allow for lengthy interviews (perhaps up to an hour, if the respondent is motivated) involving open-ended, complex questions, and the use of visual aids such as maps or photographs. Telephone surveys, while permitting open-ended questions and the possibility of clarifying questions and probing answers, tend to have more severe time restrictions. In fact, many people become restless after 10 or 15 minutes in a telephone survey.

Mail-back questionnaires are less constrained by time although a complex questionnaire with many pages is more likely to be thrown into the trash than a short, attractive questionnaire form. Mail-back

Table 3.4 Approximate confidence intervals for two survey percentage differences (95% confidence level). For example, in two samples of 2000 each, if you observe a difference of two or more percentage points around 10% or 90%, the probability is 95% that the difference is real and not due to chance

Sample sizes	10% or 90%	20% or 80%	30% or 70%	40% or 60%	50%
2000 and:					
2000	2%	2%	3%	3%	3%
1000	2%	3%	3%	4%	4%
500	3%	4%	4%	5%	5%
100	6%	8%	9%	10%	10%
1500 and:					
1500	2%	3%	3%	4%	4%
750	3%	4%	4%	4%	4%
500	3%	4%	5%	5%	5%
100	6%	8%	9%	10%	10%
1000 and:					
1000	3%	4%	4%	4%	4%
750	3%	4%	4%	5%	5%
500	3%	4%	5%	5%	5%
100	6%	8%	9%	10%	10%
750 and:					
750	3%	4%	5%	5%	5%
500	3%	5%	5%	6%	6%
100	6%	8%	10%	10%	10%
500 and:					
500	4%	5%	6%	6%	6%
100	6%	9%	10%	11%	11%
250 and:					
250	5%	7%	8%	9%	9%
100	7%	9%	11%	11%	12%
100 and:					
100	8%	11%	13%	14%	14%

questionnaires do not permit clarification or probing and thus are best suited for simple, closed-ended questions. Response rates tend to be lowest with a mail-back format; 30 per cent rates for general population surveys are common, although rates can be as low as 10 per cent, depending on questionnaire content and design.

Phrasing and content of questions

Regardless of questionnaire format, certain types of questions can cause problems. The following identifies some of the more common problems in all survey design.

'Loaded' words and 'leading questions' cause unintended biases in responses (they can also be used to create intended biases, but let us assume you really do want to be fair). Words such as 'vandalism', 'pollution', 'prostitution' and 'exploitation' bias a survey by providing cues as to how the respondent is expected to answer. Questions such as, 'Are you in favour of ensuring that the rights of disabled travellers are respected by air carriers?', are 'leading' in that they encourage a politically correct answer.

You can also 'load' questions by carefully structuring response categories of closed-ended questions. Consider the following:

- *How would you describe the changes in the number of weekend get-aways you have taken over the last three years?*

  ____ *Decreased*
  ____ *Stayed the same*
  ____ *Increased slightly*
  ____ *Increased moderately*
  ____ *Increased substantially*

The response categories are 'loaded' in that there are three categories for increase, but only one for decrease.

Not only can you load individual questions, you can load an entire questionnaire. This is done by expressing every question in such a way as to insinuate a bias into the respondent's answers. Consider the following brief example of questions related to residents' perceptions of tourism impacts on a local community:

- *Tourists are responsible for increased traffic congestion.*

  ____ *strongly agree*
  ____ *agree*
  ____ *undecided*
  ____ *disagree*
  ____ *strongly disagree*

- *Crime has increased as a result of tourism development.*

  ____ *strongly agree*
  ____ *agree*
  ____ *undecided*
  ____ *disagree*
  ____ *strongly disagree*

- *Tourists are insensitive to local residents' rights to privacy.*

  ____ *strongly agree*
  ____ *agree*
  ____ *undecided*
  ____ *disagree*
  ____ *strongly disagree*

If you were to continue this line of questioning, you would quickly convey the impression that tourism is to be viewed as the source of a significant number of social problems. Ideally, the contents of a questionnaire should be seen as balanced. This can be done by alternating the wording of questions so that half are worded 'positively' and half 'negatively'. Varying the wording of questions also helps avoid 'response set bias'.

'Response set bias' refers to the tendency of some respondents to checking off the same category for every question. Some respondents, for example, will tend to automatically check off 'Agree'. This tendency becomes particularly pronounced when every question is worded from the same perspective (either for or against tourism). Varying the perspective of questions can discourage this bias. Further, if you vary the perspective among questions (some positive and some negative), you can readily identify if a respondent tends to provide the same response for every question, including those that logically should prompt opposite responses. If someone exhibits this pattern, remove the respondent from your data set.

Use imprecise and vague words with caution. There are times when such words are probably unavoidable. Consider questions of the form:

Different people use different criteria when selecting a destination for a vacation. How important is each of the following to you when selecting where to go on a vacation?

Criteria	*Very Important*	*Somewhat Important*	*Not Important*
Beaches	___	___	___

Measures of importance such as 'very' or 'somewhat' are necessarily imprecise. Even if you assign ordinal scale values to them, the terms lack precision – which is usually acceptable for both respondents and analysts. However, if you were to ask:

How often do you take a vacation?

___ *Often*
___ *Sometimes*
___ *Rarely*
___ *Never*

The imprecision here is annoying, misleading and unnecessary. Your question would be clearer and would yield more useful results if you were to phrase the question something like:

How many times did you take a vacation (a pleasure trip lasting three nights or more) last year?

___ *0 (did not take a vacation)*
___ *1*

____ *2 or 3 times*
____ *4 or more times*

Apply common sense to what you ask. Some questions, however valuable valid answers might be, are unrealistic to ask under certain conditions. The author once participated in an economic impact study for the Kitchener–Waterloo Oktoberfest, the largest Bavarian festival outside of Munich, Germany. Part of the research design called for interviews with visitors on-site. This meant attempting to interview festhall patrons on both past and anticipated expenditure while they were drinking beer with a near-deafening roar of thousands of other drinkers, dancers and polka bands in the background. Recall of expenditure is difficult under the best of circumstances; data on anticipated expenditure is even more questionable; and this particular environment as well as the mental acuity of the respondents biased the data in an unpredictable way.

Anticipate your respondents' emotions and sensitivities, and use tact with sensitive questions or avoid asking the questions at all. Questions related to illegal acts, such as not declaring all goods being imported when returning home from an international trip, the use of drugs or the hiring of prostitutes are sensitive issues to some people. Respondents may lie, refuse to answer, or become offended and complain to your employer or other authority about your questions. Other presumably innocuous but personal questions can also trigger problems. Questions related to alcohol consumption, smoking, exercise, attitudes towards minorities, attitudes towards sexual activity, income, even the respondent's address can elicit lies or refusal to answer. If you must ask such questions, use one or more of the following strategies:

1. Provide broad categories for responses. If asking about income, for example, divide the income range into five or six broad categories. If asking about address, limit your enquiry to just a postal code or a portion of the code.
2. Allow the respondent to provide answers to sensitive questions by writing rather than orally, especially if the interview is conducted in a public place.
3. Explain why you need the information, as an introduction to the questions. Avoid any hint of judgement about the respondent's answers. Begin with some 'soft' general questions related to the topic, such as:

   - *As you may know, there is some concern about prostitution in the downtown core of our city. Do you consider prostitution to be a problem? Check the response that best describes your view.*

     ____ *Not a problem*
     ____ *Slight problem*
     ____ *Moderate problem*
     ____ *Serious problem*
     ____ *No opinion/Don't know*

- *Do you think the level of prostitution in this area has changed over the last year?*

  ____ *Decreased*
  ____ *No change*
  ____ *Increased*
  ____ *No opinion/Don't know*

- *Do you know anyone who has ever hired the services of a prostitute in this community?*

  ____ *Yes*
  ____ *No*

- *Have you ever hired the services of a prostitute in this community?*

  ____ *Yes*
  ____ *No*

4 Place sensitive questions, including demographic questions, at the end of the questionnaire.

Question format

Questions are either open-ended or closed-ended. Closed-ended questions provide fixed options for the respondent; open-ended questions allow the respondent to express his answers in whatever way he chooses. Closed-ended questions provide for fast response and simple coding and they are appropriate when the range of possible answers is known, relatively limited, and factual. These questions are useful for demographic information, the number of previous trips to a destination, how reservations were made and so on. Open-ended questions are used when the answers are complex, when the range of answers is not known, or you wish to explore answers in depth.

Although closed-ended questions simplify coding, they can still present problems, especially if the questionnaire is self-administered. The range of problems is probably unlimited, but a few examples will help to illustrate potential trouble-spots. Tourism Canada and the US Travel and Tourism Administration conducted a series of long-haul pleasure travel market surveys in the major international markets for North American tourism. The questionnaires were conducted as personal interviews, but with cards for the respondent to indicate his answers to certain questions. One of these concerned activities engaged in on the last vacation. Although the questionnaires were conducted in the respondent's language, certain activities still presented problems in terms of their cultural meaning. For example, in a survey of Japanese tourists who had been on an international visit, fewer than 15 per cent checked off 'photography' from the list of activities. The taking of pictures is a stereotype of Japanese travellers, and such a low percentage was seen as unusual. Upon a debriefing of some respondents, it was learned that 'photography' was interpreted as meaning serious involvement in photography as an art form; the casual taking of snapshots was not seen as 'photography'.

A survey conducted in Hong Kong found that over 60 per cent of respondents reported liking outdoor sports – a finding that again surprised the research team. Upon further discussion, they learned that the phrase 'outdoor sports' had a different connotation for the respondents than for the researchers. Researchers were interpreting 'outdoor sports' as activities such as golf, tennis, skiing, fishing and hunting. The respondents generally took the term to mean horse and dog racing.

Even without the challenges of linguistic and cultural differences, self-administered closed-ended questions can yield unexpected difficulties. For example, you might ask,

> *'Which of the following reasons was the most important in influencing your choice of this hotel? Please check one only.*

Be prepared for a certain percentage of your respondents to check two or more. If this happens, one strategy is to select one answer randomly and to not code the others. Or you may allow the checking off of multiple answers and might even ask respondents to rank their answers by relative importance. Experience suggests that few respondents provide rankings – unless the question design forces it.

Survey designers often include 'Other' as a final category in a list of responses to a closed-ended question. While this category may capture unanticipated but important answers, usual answers tend to be: (1) too vague to be useful, (2) identical to one of the listed categories, or (3) unrelated to the question actually asked. As a rule, 'other' is a 'throw-away' category; the results not only increase the work of coding but, in the final analysis, are frequently useless.

The collection of information is only the first step; once the questionnaires have been completed, you have to record the results, usually in a computer file for analysis. Design your survey to facilitate coding. This includes, of course, the use of simple, closed-ended questions as often as possible. But it also implies giving some thought to the way you will analyse your data. Responses from surveys may be coded either as dichotomous (yes/no) or as multiple values. Questions such as one's sex or whether or not one went on a vacation in the last year can be coded as a simple dichotomous variable. However, if you ask about the person's occupation, you have a more complex set of possible answers. Consider the following partial list of occupations:

____ *clerical/secretarial*
____ *manager*
____ *professional*
____ *skilled labourer*
____ *owner*
____ *student*
____ *home-maker*
____ *retired*

You can consider each category as a separate question (i.e. is the respondent a student or not) and code each category as yes/no. However, in the case of occupation, presumably the respondent will check off only one category. If so, numbers can be assigned to each category (e.g. 'student' might be coded as a '6' in the above list); so if your respondent is a student, you would simply enter '6' in the appropriate field for respondent's occupation.

Simplicity in coding is important for both your subsequent analysis and for any one else who might wish to work with your data set later. Consider the following example, drawn from an exit survey of Japanese visitors to Canada. This particular question was aimed at identifying activities the visitors engaged in during the trip just completed as well as attempting to discover what activities they might enjoy on a return trip.

Please indicate which of the following activities you: (1) engaged in during this current trip and (2) might like to try on a future trip.

	This trip	*Future trip*	*Not interested*
White water rafting	______	______	______
Bicycling	______	______	______
Hunting	______	______	______

The full list contained approximately 50 activities. The consultant responsible for coding the questionnaire used the following scheme:

If the respondent checked 'this trip' only:	1
If the respondent checked 'future trip only':	2
If the respondent check both 'this trip' and 'future trip':	12
If the respondent checked 'not interested':	3
If the respondent checked 'this trip' and 'not interested':	13
If the respondent checked all three:	123

(Note: the last response is illogical but, in fact, several respondents apparently adopted a 'response set' and checked all three categories.

The problem with this strategy (beyond the fact that illogical answers were accepted uncritically and coded) is that if you wanted to identify those activities respondents actually engaged in, you would have to look for values: '1', '12', '13' and (maybe) '123'. It would be simpler to treat each activity and each category ('This trip', 'Future trip' and 'Not interested') as separate dichotomous variables.

Enhancing response rates

Response rates determine the reliability of your results. You will usually want as large a sample as you can afford to obtain in order to reduce the error associated with your sample estimates. You will also want to reduce non-response bias – distortions in your conclusions created by having too few people in certain categories respond to your questionnaire. The following suggestions outline a few basic techniques that can be employed to increase your response rate and thus reduce non-response bias.

1 Test your instrument. Administer it to colleagues who can provide expert advice on its content and structure and to individuals representative of your target population to identify areas of confusion or other problems.
2 If the information you are seeking is valuable, especially commercially, offer compensation to respondents. Budgets will constrain how much you can offer but some compensation should be considered, especially if the survey is sponsored by a profit-making firm. Information has a price, and if you are able to use information obtained from respondents to make money, you should be willing to share some of the proceeds with those who helped make your profit possible.

   This does not mean that you have to write a cheque for a share of your profits or operating budget to every respondent, although some surveyors do provide cash payment for particularly onerous survey instruments such as time-budget diaries. At a minimum, you might consider providing a pen (that could be used to complete the survey, as well), a souvenir pin, a voucher for a free drink or meal, or a gift certificate redeemable at your store.
3 Explain the reason for the survey. Respondents are usually more willing to answer your questions if they understand their significance. This can be done orally in a personal interview or telephone survey, or in a cover letter for a mail-back.
4 Respect your respondents' time and right to privacy. You are asking a favour, a gift of their time and opinions. Make sure they know that you know this. Be respectful, tactful and direct. Acknowledge their participation is voluntary. Let them know how much time you are asking of them and what, if any, compensation you will be able to provide them.
5 Keep the survey as short as possible. Ensure the design is attractive, clear and logical.
6 For mail-back surveys, follow a multi-stage strategy. Contact your pool of respondents with a card or letter noting that they have been selected to participate in a survey, provide them with the information suggested in steps 1, 2 and 3 above, and note that they will receive the questionnaire in about a week. A week later, send the questionnaire to them along with a cover letter reminding them that this is the survey you had described in your card.

   If they do not return the questionnaire by the deadline, send a reminder to them. If this does not result in a return, follow-up about a week later with another questionnaire and a new cover letter acknowledging they may have misplaced the earlier questionnaire, that the survey is important, and that you value their answers. Experiments with this multi-stage approach have produced response rates as high as 80 per cent – in contrast to single mailings, which often yield no more than 30 per cent and sometimes less.

7. If possible, print the questionnaire on business or agency letterhead to add credibility. If your sample is aimed at business leaders, public officials or another select group, it is helpful to have a letter of support from a recognized and respected individual endorsing the survey.
8. Include an addressed, postage-paid return envelope with the questionnaire. Some researchers suggest that the use of postage stamps rather than business-reply envelopes yields better results. However, the use of stamps increases your costs significantly; business-reply envelopes result in your paying postage only on those surveys actually returned.
9. Softer colours, especially blue, appear to yield a higher response rate. Bright and hard colours such as yellow and orange yield lower response rates.
10. The questionnaire should be printed on good quality, heavy stock (within the constraints of your budget, of course).
11. Conclude the questionnaire with an expression of thanks and a reminder about where to send the questionnaire.

Use of secondary data sources

You do not always need to conduct your own survey to obtain information. Many tourism agencies conduct surveys for their own purposes and, because the surveys have been conducted at public expense, are willing to make the data available for further analysis. Databases that have been developed by someone else for their own purposes but are available for others to use are known as secondary data sources. (The term 'secondary analysis' is sometimes used to refer to analysis of data from secondary sources. This terminology is incorrect; only the data are secondary. Your analysis, presumably, is original.)

Secondary data sources can be rich sources of information. Typically, only a small percentage of the total information is ever harvested from the data set by the original analysts. Analysts can often glean substantial information from further analysis of the original data set. And because the initial objectives for conducting the survey have presumably been met, secondary data sources may provide the researcher with more time to focus on analysis because less (or no) time need be spent on data collection, coding, inputting and editing.

There are disadvantages with the use of secondary data sources, of course. The most obvious is that you are constrained by the contents of the original survey and any peculiarities of the original sample. You must also deal with any problems in quality control associated with data inputting and editing. Although you do not usually have the freedom to go back and re-input the original data, you can still exercise some degree of control over the quality of the secondary data set. This is done through a technical audit of the data set. A technical audit is, in essence, a detailed examination of the original data to identify any potential problems such as mis-codes or documentation errors.

Although the specific activities of a technical audit will be tailored to match the characteristics of the data set, there are some general tasks that form the basis for most audits. These include the following.

1 Read all supporting documentation, particularly code books. Also examine copies of research reports based on the data set to identify sample tables for possible replication. The documentation should describe precisely in which columns the data for each question are coded. The documentation should also provide information on:

   a. sample and target populations
   b. the sample frame and method of drawing the sample (e.g. stratified random sample)
   c. sample size and response rate
   d. location and dates of survey
   e. the format of the survey (e.g. telephone survey using random digit dialling)
   f. methods used to reduce and assess the degree of non-response bias, if any
   g. a copy of the questionnaire(s)
   h. the sponsor of the survey and the organization actually conducting the survey
   i. the agency responsible for data coding and inputting (if different to the survey firm), plus the name and contact details (e.g. telephone and fax numbers) of an individual familiar with data coding

2 Tabulate frequencies for all values of all variables. Review these to identify any anomalies, including the presence of values that fall outside the ranges specified in the code book. For example, a variable reporting the sex of the respondent might be coded '1' for females and '2' for males. Check, then, to see if there are any '0's, '3's, or other values that fall outside the legitimate range. Such values should be re-coded as 'missing' for any subsequent analysis.
3 Compare the sample size reported in the documentation with that determined from your frequencies tabulation.
4 Identify the codes used for missing values and check to see if such codes were inadvertently used for actual values as well. If '9' was used for missing data, did a data coder also use '9' to represent some real value?
5 Attempt to replicate the reported findings in selected tables from the original research reports.
6 Identify any additional analyses that might test the internal consistency of the data set, such as cross-tabulations or 'select-if' arguments. This exercise can be particularly useful in identifying logical errors in questionnaire design or in respondents' answers. For example, in the exit survey of Japanese visitors described

above, some respondents indicated they engaged in a certain activity on the current trip, would like to engage in it on a future trip, and were 'not interested' in the activity. This pattern of response is illogical and suggests carelessness or a lack of seriousness in the respondents' answers. The solution in this case would be, at a minimum, to recode the individual's answer to this question as 'missing'. Depending on other patterns in the individual's responses, you might consider eliminating the individual completely from analysis.

Another question asked how many people were in the respondent's travel party. A proportion answered '0' – implying they interpreted the question to be referring to 'the number in the travel party in addition to yourself'. In this case, one can assume with some degree of accuracy that a score of '0' meant the individual travelled alone. But what about those who answered '1' or '2', or some other number? The percentage of those who reported '1' but meant one more in addition to themselves cannot be known. The result is that any conclusions from this question have to be accepted with a great deal of caution.

The results of the technical audit lead to several possible actions. At a minimum, they identify problems in the data set that might be addressed through re-coding or by declaring certain responses to be 'missing'. The results of the audit might also be shared with the original surveyors in order to identify possible answers to anomalous results. In any event, the time spent on a technical audit before beginning your analysis will help avoid embarrassing or frustrating problems later.

Summary

Tourism analysis begins with the collection of tourism data. This chapter has examined a variety of issues associated with the collection of data through surveys. We began by considering the relative merits of different formats of surveys: personal interviews, telephone surveys and mail-back surveys. Questions related to sample and target populations, the selection of a sample and the source of a sample were also explored. We considered different ways of drawing a random sample, ranging from a simple random selection appropriate, for example, to drawing a representative sample from a membership list in an industry association to a complex, multi-stage design for drawing a random sample of residents of a nation.

We also looked at some guidelines associated with how many people you might need to survey. As with questions related to survey format and sampling frame, one is confronted by the need to make trade-offs when deciding on sample size. In particular, you must balance the need for accuracy (the probability of drawing a correct conclusion) with the need for precision. Greater precision requires a larger sample size, given a specified standard of accuracy.

Some guidelines concerning questionnaire design were then offered, including potential problems associated with 'loaded' questions, ambiguous words and phrasings, and low response rates. Finally, we looked at the use of secondary data sources – survey databases collected by someone else for their own purposes. Specific suggestions for conducting a technical audit of a secondary data set were described.

The topic of data collection in tourism is a large one. This chapter has only scratched the surface. However, it has, at the very least, introduced some of the major issues that you will need to address as you begin the long process of turning data into information.

Further reading

Babbie E R 1990 *Social surveys*, 2nd edn. Wadsworth Publishing, Belmont, California.

Bateson N 1984 *Data construction in social surveys.* Allen and Unwin, London.

Bradburn N M and **Sudman S** 1979 *Improving interview method and questionnaire design.* Jossey-Bass, San Francisco, California.

Chaudhuri A and **Mukerjee R** 1988 *Randomized response: theory and techniques.* Marcel Dekker, New York.

De Vaus D A 1990 *Surveys in social research*, 2nd edn. Unwin Hyman, London.

Dillman D A 1978 *Mail and telephone surveys: the total design method.* Chichester, New York.

Fowler F J 1990 *Standardized survey interviewing.* Sage Publications, Newbury Park, California.

Lee E S, Forthofer R N and **Loriman R J** 1989 *Analyzing complex survey data.* Sage Publications, Newbury Park, California.

Oppenheimer A N 1992 *Questionnaire design, interviewing, and attitude measurement.* Pinter, London.

Singer E and **Presser S** (eds) 1989 *Survey research methods: a reader.* University of Chicago Press, Chicago, Illinois.

Sudman S 1974 *Response effects in surveys: a review and synthesis.* Aldine Publishing, Chicago, Illinois.

Sudman S 1976 *Applied sampling.* Academic Press, New York.

CHAPTER 4

Understanding the tourist

Introduction

Tourism marketing problems often require answers to questions about what tourists are thinking and how they make decisions. The owner–operator of a small resort may want to know, 'Are my customers satisfied with my housekeeping?' An airline marketing director may wonder, 'Are we stressing the right qualities in these advertisements to compete successfully?' A restaurateur might ask, 'Are my menu selections and prices attractive?' Such questions require specialized instruments to probe the minds of tourists and potential tourists to better understand their attitudes, tastes, motivations and decision-making criteria. These topics are so complex that entire books are devoted to methods of their measurement and analysis. The purpose of this chapter is to provide an overview of some of the major concepts of attitude measurement and related topics and to examine two procedures often used to obtain a better understanding of tourists' attitudes and actions.

Before beginning our overview, a few definitions may be helpful. An **attitude**, as used in this chapter, is the predisposition of an individual to act or to otherwise respond to an object or stimulus. It is not the response, but rather the tendency towards a consistent response. Attitudes thus persist over time, changing slowly, if at all. They tend to produce repetitive behaviour, although circumstances often cause an individual to act contrary to his normal predispositions.

Attitudes also imply some form of preference for or evaluation of a stimulus. For example, some people prefer low-priced restaurants. Their attitude might be one of 'fiscal restraint' in matters of personal finance. However, there are likely to be circumstances that temporarily override the bias towards 'fiscal restraint' – such as the desire to celebrate a special occasion with friends at an expensive restaurant.

Stimuli that are more directly tied to overt behaviour, such as the decision to go to a specific restaurant, may be termed **motivations**.

These are related to attitudes, but are closer to, and expressed more directly in, actual behaviour. Motivations are goal-directed and operate for a shorter period of time than attitudes. The motivation for going to an expensive restaurant to celebrate a promotion ends once the evening is over. The attitudes about money, career, celebrations and friends endure.

Tastes are more closely tied to sensory stimulation than attitudes. I have a taste for sashimi and for flying first class. However, the former is hard to find in my home town and the second is beyond my budget. Thus, these represent my preferences, but are not necessarily observable in my behaviour. The ability to satisfy your tastes depends on the availability of the products you desire and your ability to afford them. The question of taste also arises when you are asked to evaluate the quality of a product or service, such as a hotel room. The issue is not one of your attitudes or motivations affecting your decision to come, but rather how well the hotel met your tastes, standards and expectations for service.

Researchers, managers and planners are interested in all three concepts: attitudes, motivations and tastes. The first two, though, are of greater importance for marketing and planning. Regardless of the particular quality you want to measure – attitude, motivation or taste – the task is to design an instrument that will allow you to infer conclusions objectively and accurately about the mental processes or states of tourists and potential tourists.

Some may argue that we can never objectively measure anything about others' thoughts or feelings. This is, in fact, true. The point of attitude (or motivation or taste) measurement is not literally one of measuring attitudes; rather it is a matter of counting the number of people who respond in a certain way to certain types of questions. The questions about attitudes are usually designed to generate structured answers; that is, they use scales for respondents to indicate answers. At this point, therefore, we begin our overview by examining different types of scales relevant to tourism: nominal, ordinal, interval and ratio scales.

Types of scales

The most primitive type of scale, the **nominal scale**, is the use of numbers as labels. Identification codes, such as 001, given to each respondent in a survey are an example of this type of scale. Labels can also be given to groups or classes of respondents or objects. All male respondents could be labelled '1', while females could be labelled as '2'. In such cases, the only arithmetic sign of manipulation that can be meaningfully applied is the equality sign, '='. Everyone who has a value of 1 is the same; everyone who is a 2 is the same. No other operations produce meaningful results.

Ordinal scales possess nominal scale qualities as well as the property of indicating order or rank. For example, a listing of preferred destinations, from 1 to 10, or most preferred to least preferred would be an ordinal scale. The relative ranks between any two destinations, say 1 and 5, do not indicate anything about the absolute difference between the attractiveness of the two destinations. The ranks indicate nothing more than the relative attractiveness. You can perform monotonic transformations on ordinal scale data (e.g. multiply each one by 10). A monotonic transformation changes the magnitude of the numbers but does not change their relative positions. Monotonic transformations are often done for cosmetic reasons, such as converting several different ordinal scales into the same range of values. Certain monotonic transformations have analytical value. One of these is conjoint measurement, a tool we will examine later in this chapter.

Interval scales retain the properties of nominal and ordinal scales as well as indicating the distance between objects ranked on the scale. You do not know anything about the absolute magnitude of the objects. A flight that leaves at 12 noon leaves six hours later than a flight that leaves at 6 am, but you cannot say the noon flight is 'twice as late' as the early morning flight. Ratios of interval scales are meaningless. Interval scales remain invariant under most arithmetic transformations. For example, you could transform the flight times into minutes by multiplying each hour by 60. Certain types of attitude scales are expressly designed to have interval scale properties, such as Thurstone's differential and case V scales.

The most powerful scale is the **ratio scale**. This scale measures some phenomena with respect to an absolute zero. The number of tourists, the price of tourism commodities, and the distance between an origin and a destination are all ratio scales. Virtually all types of analysis are possible with ratio scales.

You need to be careful about applying appropriate methods when working with different scales. Obviously, it would be nonsense to add 1 and 2 together to obtain 3, if 1 refers to Canada, 2 refers to the USA, and 3 refers to Mexico. However, it may be less obvious whether one can take the average of ordinal scale data from an opinion survey to obtain a mean ranking of opinion. Nor is it obvious whether it is legitimate to perform a simple regression analysis using Pearson's product-moment correlation coefficient to compare ordinal attitude scores with income (a ratio scale variable). Traditional statisticians would suggest the answer is 'no' to these two situations. One can argue, however, that the potential to gain additional insights from limited analysis on carefully selected 'questionable' scale data may outweigh potential problems. These issues have not been adequately explored in the tourism field, so they tend to persist as nagging concerns. At least one statistician (Nunnally, 1978) suggests that some analyses treating ordinal data as interval data may be acceptable and can lead to accurate conclusions. An opposing view is that of Wilson (1971) who argues there is too much potential for abuse and for falling into logical traps if ordinal data are treated as interval data. It is clear, at least, that the debate about which analyses can be applied to ordinal data is unresolved in tourism. The subject is also reviewed by Sonquist and Dunkelberg (1977: 255–6).

Scaling techniques

Psychologists have developed many different types of scales for attitude measurement, motivational analysis, personality assessment and other behavioural traits. Most of these have, at best, only a remote application to tourism analysis and marketing. Scales in tourism are usually based on some version of five well-established techniques: ranking, Thurstone's differential, Thurstone's case V, Likert, and the semantic differential scale.

Ranking

The simplest of all scaling techniques for attitude measurement is rank ordering. A respondent is given a list of items, such as hotel attributes, and asked to rank them in order of preference. The ranking can be done directly if there are relatively few items, or it can be done in stages if there are a large number of items. A typical procedure for ranking many items is to first sort them into low, medium and high ranks. Each item in these general ranks is then ranked within its category. The highest-ranked item in one category is compared to the lowest-ranked item in the category above it to ensure consistency. When the final ordering for each category is determined, the categories are placed in proper order to obtain overall rankings.

Individual respondents may also be asked to repeat the ranking process two or more times to check for reliability. Although the resulting ranks constitute only an ordinal scale, many researchers calculate arithmetic averages of the ranks of multiple respondents to obtain an overall ranking for each item by their study population.

Thurstone differential scales

The basic Thurstone differential scale (Thurstone 1927) consists of a series of statements that respondents are asked to review. They then indicate which statements they feel express their attitudes. Each statement carries a ranked score, not known to the respondents. The respondents' overall scores are the mean or median score of the statements that they indicate express their own view.

Consider the following statements regarding vacations:

1 Vacations are a proper reward to anyone who has worked hard throughout the year. (4.0)
2 Vacations are absolutely essential for everyone's mental health. (5.0)
3 Vacations are a luxury that most people cannot afford. (1.0)
4 Vacations should be a human right guaranteed by the government. (7.0)
5 Vacation expenses should be subsidized through government or employer-sponsored programmes. (7.0)
6 Vacations are a nice reward, but only after years of hard work and careful saving. (3.0)

These statements express a variety of possible attitudes towards vacations. The score in parentheses (which would not be shown to the respondent) reflects the degree of positive feeling towards vacations. The stronger the positive evaluation, the higher the number. If a respondent agreed with statements 1 and 2, his score would be the average of 4.0 and 5.0, or 4.5.

This type of procedure is quite simple for the respondent. The challenge is in the development of the scale and the associated values. The analyst begins by formulating a large number of statements expressing a range of attitudes towards a particular subject, ranging from very positive to very negative. These statements are then presented to a panel of judges who are asked to group the statements into 7, 9 or 11 categories reflecting their perception of the positiveness or negativeness of each statement. The distribution of each statement is reviewed, and those that received a wide range of rankings are discarded. The items that remain are then combined into a survey instrument for administration to a sample population. The ranking associated with each statement is based on the judges' consensus about the degree of favourableness of the statement.

As the name indicates, this is a differential scale. That means each respondent indicates simply whether or not the statement expresses his attitudes. Two important assumptions behind this type of scale are that attitudes and the statements intended to describe those attitudes are normally distributed and have interval scale properties. The significance of these assumptions will be more clearly seen in the Thurstone case V scale, described next. Details about developing the Thurstone differential scale are described in Edwards (1957); a critique of the scales is in Nunnally (1978). His primary criticism of the scale is the labour involved in developing the scale. Moser and Kalton (1974) disagree with Nunnally's perceptions of the difficulty of developing the scale and suggest the work is not overly demanding. Further, the effort of scale development is offset by the simplicity of the scale from the respondent's perspective.

In comparison to other scales presented here, Thurstone differential scales are relatively rare in tourism research. There is, though, significant potential for their application. One example of their use is Lopez's (1980) analysis of the effects of different tour leaders' styles on the satisfaction of tour members with their trip. A key element in assessing leadership style was the measurement of the degree of the leader's authoritarianism. Lopez used an 80-item Thurstone scale developed by Ezekial (1970) consisting of 40 statements tending towards an authoritarian position and 40 statements tending towards non-authoritarianism. The leaders were asked to fill out the instrument confidentially, checking off all the statements with which they agreed. The scores of the leaders were then correlated with the expressed levels of satisfaction provided by tour members who had travelled under the leader.

Thurstone scaling lends itself well to measuring certain attitudinal characteristics of travellers or potential travellers, such as attitudes towards certain vacation styles. Scales could be developed to assess individuals' desires for excitement, for relaxation, for security or for planning. A Thurstone instrument could also be derived to examine the relative influence or dominance levels of spouses or partners with respect to vacation decision-making. Such an instrument might list a series of specific tasks or decisions such as destination choice, budgeting, mode of travel and duration of the vacation. The spouses would then check off those tasks for which they believed they were responsible. A comparison of the total number of checked statements as well as the identity of those statements could yield useful insights into how couples prepare for vacations.

Thurstone's case V

Thurstone's early work on measuring attitudes led to his development of the law of comparative judgement. This law attempts to explain the inconsistency surrounding the selection of which of two items is preferred by an individual. The following example may help illustrate what the law states as well as how it might be applied to a tourism problem.

Consider three resorts, A, B and C. If we ask a group of individuals to evaluate the attractiveness of each resort on a scale of 1 to 10, we will get a range of rankings for each resort. Thurstone's law of comparative judgement is based on the hypothesis that if we were to ask a large sample of people to rank those resorts, their rankings would form a normal distribution around some mean for each resort.

Assume that resort C has the lowest mean, A has the next highest, and B has the highest, as shown in Figure 4.1. Some individuals, however, gave resort A a higher ranking than B, as can be seen from the overlap of curves for A and B.

If these same individuals were asked to compare pairwise the three resorts, we would find a high proportion selecting B over A. Some, however, prefer A over B. This pattern of responses can also be represented in the form of a normal distribution curve, as shown in

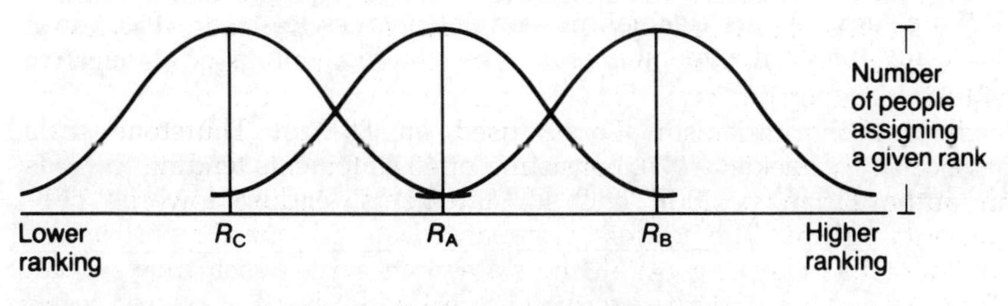

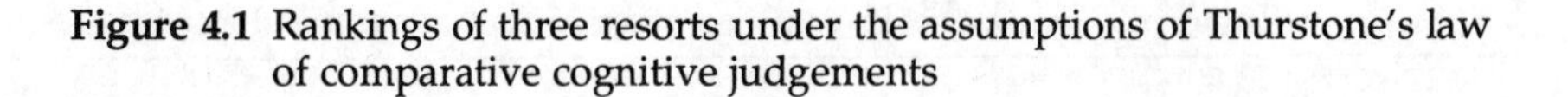

Figure 4.1 Rankings of three resorts under the assumptions of Thurstone's law of comparative cognitive judgements

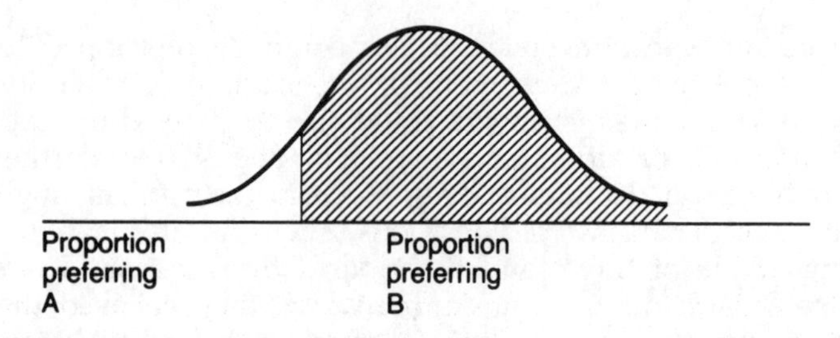

Figure 4.2 Distribution of differential responses for two resorts as implied in Figure 4.1

Figure 4.2. The curve in this figure resembles that in tables of standard normal variables. And, in fact, the standard normal variable, Z, is an integral part of the calculation of Thurstone's case V scale.

Case V is a simple version of the application of the law of comparative judgements. It requires only that: (1) the percentage of respondents choosing each item in a series of pairs be known; (2) that the pattern of responses around their means be identical for each item (as shown in Figure 4.1); and (3) the correlations between the choice processes of the same set of judgements be zero (Green and Tull 1975). Under these conditions you can calculate the discriminal or perceived difference between pairs of items, such as resorts A and B, by the equation:

$$R_A - R_B = Z_{AB}\sqrt{2} \quad [4.1]$$

where: Z_{AB} = the standard normal variable for the percentage of respondents selecting the most preferred item;
$R_A - R_B$ = difference in ranking of resort A over resort B.

Hypothetical data related to the choice of the three resorts by our group of hypothetical respondents is presented in Table 4.1. Note that 76 per cent of the respondents prefer resort B to A, while only 40 per cent preferred C to A. These and the balance of percentages in Table 4.1 were converted to Z values by consulting Table A.1 in the Appendix. The results are shown in Table 4.2. Only percentages equal to or greater than 0.50 are used in this calculation. For percentages less than 0.50, the Z value for the complementary package is noted, but given a negative sign.

Table 4.1 Percentage of respondents preferring resort shown at top of table to resort shown at left

	A	B	C
A	–	0.76	0.40
B	0.24	–	0.05
C	0.60	0.95	–

Table 4.2 *Z*-values for Table 4.1

	A	**B**	**C**
A	–	0.70	–0.25
B	–0.70	–	–1.64
C	–0.45	2.35	–
ΣZ_X	–0.45	2.35	–1.90

Note: These are derived from Table A.1 in the Appendix in the following way. The largest proportion in any pair of resorts is identified (this will be equal to or greater than 0.50). This number is then subtracted from 1.00. For example, 76% of respondents preferred resort B over resort A, so 1.00 – 0.76 = 0.24. The value closest to 0.24 is then located in the body of Table A.1. This particular value is '24 196' in the row labelled '0.7' and in the column labelled '0.00'. The Z-value for 76% is thus about 0.070. The Z value of 24%, the percentage of respondents favouring resort A over B, is the negative value of Z for 76%, or –0.70.

$$R_B - R_A = -Z_{AB}\sqrt{2} \qquad [4.2]$$

Thus Z for 76 per cent, the proportion of people choosing B over A, is 0.70; the associated value for 24 per cent, those who choose A over B, is –0.70.

The scale value for each resort can now be calculated using the equation:

$$R'_X = \left(\frac{\sqrt{2}}{N}\right)\Sigma Z_X \qquad [4.3]$$

where: R'_X = scale value for the item being evaluated
ΣZ_X= column total for Z values for the item being evaluated (see Table 4.2)
N = number of items.

Applying equation [4.3] to our data produces the results shown in Table 4.3. These values are an interval scale of the relative attractiveness of each resort as derived from the percentages of a sample population indicating their preferences for each pair of resorts – an ordinal judgement. Scores calculated with this methodology can then be used in subsequent analyses requiring interval level data.

Table 4.3 Scale values for three resorts

Resort	**Value**
A	–0.212
B	1.108
C	–0.896

Likert scales

A Likert scale (Likert 1932) provides a measure of the degree to which a respondent agrees or disagrees with each of a series of statements. This contrasts with the Thurstone case V scale in which a respondent indicates simple agreement or disagreement. The degree of agreement in a Likert scale is often measured with a five-point scale, ranging from strongly agree to strongly disagree, with a neutral position in the middle. When using a Likert scale, you should seek to have a selection of statements covering a wide range of attitudinal positions so that respondents have a reasonable chance to differentiate themselves from other respondents. There is little value in having a series of statements that are so bland (or extreme) that everyone agrees (or disagrees) with them.

Once a series of statements has been developed, you give them to a sample of respondents representative of the larger group you are studying. These individuals act as judges who indicate the degree of their agreement with each statement (unlike the development phase of a Thurstone scale in which the judges assess only the degree of favourableness of each statement and not their own attitude). A total score for each judge is obtained by summing his responses over all questions (you must make sure that the scoring spectrum is consistent; that is, all negative attitudes should receive low values while all positive attitudes receive high values, or vice versa). The preliminary results are then looked at in detail. Identify, for example, any statements that fail to discriminate between the highest-scoring and lowest-scoring groups of judges. Statements that correlate poorly with the total score of respondents or that fail to contribute to the distinction between high and low scores should be eliminated. The standard method for assessing internal consistency of attitudinal scales is Cronbach's α (Cronbach 1951). Alpha provides a measure of the internal consistency among the items in an attitudinal instrument, with values ranging from 0.00 (no consistency) to 1.00 (perfect consistency). Nunnally (1978) suggests a threshold of 0.80 for accepting an instrument as internally consistent. The statistic is available in most comprehensive statistical software packages such as the SPSS.

Factor analysis (described in Chapter 5, under the section on factor–cluster segmentation) is also used on occasion. Analysts use factor analysis to identify those items (statements) that appear to have confused or ambiguous content (demonstrated by loading highly on more than one factor) or that are poorly correlated with any other items (demonstrated by low loadings or low communalities). A list of 20 to 30 statements is desired in the end.

Once the instrument has been developed, it may be administered to the study population. Scoring is done as with the judges. Means of individual statements as well as overall scores can be determined and studied. Such analysis, it should be noted, is based on the assumption that Likert scales have interval properties. Strictly speaking, this assumption may be unwarranted. The interval between agree and strongly agree is not necessarily the same as the interval between neutral and disagree, for example. As Moser and Kalton (1974) note, however, Likert scales – even when treated as interval scales – are fairly reliable, in the sense of yielding reproducible results. They are also relatively simple to construct and are perhaps the most frequently used scale in tourism research. Detailed guidelines on the development of Likert scales may be found in Likert (1967), Churchill (1979), Tull and Hawkins (1980: 351–2), Dunn-Rankin (1983: 91–2) and Maddox (1985).

Lankford (1994) used Likert scaling in his study of attitudes towards tourism development in the Columbia River Gorge region of Washington and Oregon. Lankford hypothesized that the attitudes of business owners, government staff, elected officials, and residents in 13 communities towards tourism development would differ from each other in consistent ways. A critical step in his research was the development of a reliable attitudinal scale. To accomplish this, he followed a six-step process:

1 Develop a preliminary list of potential scale items using other studies and a panel of experts and local authorities. Attach a five-point Likert scale to each.
2 Test these items using a small random sample of local respondents in two communities.
3 Purify the scale by examining Cronbach's α and item-to-total correlations for the two communities. Items with low item-to-total correlations (defined as 0.50) or that raised α if deleted were eliminated from the instrument. The remaining list of items was factor analysed to identify items that were statistically ambiguous or that correlated poorly with any other statement. The purification process was repeated several times until the scales were deemed reliable. Fifty items remained from the original list.
4 Conduct regional survey in 13 communities. Seventeen socio-economic questions were added to the list of 50 attitudinal questions.
5 Randomly split the survey into two halves. The purification phase (step 3) was repeated with both halves. Only those items meeting the α and factor criteria in both halves were retained. Twenty-seven items remained.
6 Interpret the results.

Lankford found that the attitudinal items consisted of two general themes: 'concern for local tourism development' and 'personal and community benefits'. Statements typical of the 'concern' factor include: 'Columbia Gorge communities should not try to attract more visitors', 'My community should become more of a tourist destination', and 'Tourists are valuable'. Statements typical of the 'benefits' factor included: 'The town has better roads due to tourism', 'I have more money to spend as a result of tourism', and 'Shopping opportunities are better in my community as a result of tourism'.

The results confirmed Lankford's hypothesis of differences among the groups surveyed. Generally, residents were more sceptical about tourism's benefits, perceived more problems, and wanted a smaller level of tourism development than did government officials and business leaders.

Semantic differential

Developed by Osgood, Suci and Tannenbaum (1957), the semantic differential scale examines respondents' attitudes towards one specific topic in terms of bipolar adjective pairs. The pairs are usually separated by five to seven points. Typical statements would include:

Strong	____ :	____ :	____ :	____ :	____	Weak
Bad	____ :	____ :	____ :	____ :	____	Good
Cold	____ :	____ :	____ :	____ :	____	Hot
Friendly	____ :	____ :	____ :	____ :	____	Unfriendly

Respondents are presented with these statements and asked to indicate their feelings about the topic under study by placing a check mark at the appropriate spot on each scale.

Osgood, Suci and Tannenbaum found that attitudes towards many topics consist of three factors: potency (e.g. strong–weak), evaluation (e.g. bad–good) and activity (e.g. cold–hot). Analysts often use these three factors as guides for the development of specific pairs for their research. They also include other pairs relevant to their particular problem. Thus a study of attitudes towards air travel might include 'inexpensive–expensive', 'reliable–unreliable' and 'safe–dangerous'.

If a single dimension, such as evaluation, is used to select the adjectival pairs, a summation system can be used to calculate an individual's score as in Likert scaling. If several independent dimensions are used, separate scores could be calculated for each. You may also find it useful to examine the mean rankings of the sample population on each adjectival pair. So, for example, the ranking of a group of business travellers' perceptions of the reliability of a particular airline could be determined and compared to their perceptions of other airlines.

Milman, Reichel and Pizam (1990) used the semantic differential in their study of the impact of travel in Egypt on Israeli tourists. They hypothesized that visiting Egypt changes the negative attitudes many Israelis hold towards Egypt and that visiting Egypt would reduce the perceived differences between Israelis and Egyptians. The authors administered a survey to a sample of Israeli vacationers visiting Egypt for the first time – both prior to their visit and then at the end of the tour. A control group of Israelis who did not visit Egypt was also surveyed. The survey examined attitudes towards Egyptian people, Egyptian political beliefs and Egyptian institutions. Attitudes towards Egyptian people were measured using a semantic differential instrument, with a seven-point scale. Some of the pairs included in the instrument were: 'coldhearted–warmhearted','war loving–peace loving', 'cruel–kind', 'boastful–modest', 'stupid–intelligent' and 'not at all like me–like myself'.

The authors found that there were only a few differences between those who visited Egypt and those who did not. The differences that did exist suggested that Israelis who were willing to visit Egypt had slightly more positive views of Egyptians. However, there were no differences between the pre-trip and post-trip scores reflecting the perceptions of the Egyptian people by the travellers. The authors concluded that tourism, by itself, was ineffective in changing people's biases against other nations.

The subject of attitude theory and measurement is a very large field. It is impossible to do any more than just touch on some of the major concepts and basic techniques relevant to tourism in this chapter. If you are interested in pursuing the subject of attitude measurement in greater detail, one dated but still useful reference is Guilford's (1954) classic text, *Psychometric Methods*. Green and Tull (1975), Churchill (1979), Tull and Hawkins (1980) and Dunn-Rankin (1983) also provide a useful overview of scaling procedures in marketing research. At this point, we must move on from the subject of measuring attitudes and examine some techniques that can be used to study how people make decisions. Again, these methods do not represent a complete listing of all the techniques that are available. They are, however, important ones for tourism research.

Expectancy-value models

Description

Expectancy-value models represent the single largest group of consumer decision-making analytic tools used in tourism. Most are based on the Fishbein model (described below). Engel, Blackwell and Kollat (1978) have noted, in fact, that Fishbein models accounted for more research in consumer behaviour in the 1970s than any other subject. They continue to be important in the 1990s, as well.

Two names are normally associated with expectancy-value models: Rosenberg (1956) and, of course, Fishbein (1963, 1966, 1967). Although neither author's models were originally intended to be applied to marketing research, their formulations have proven to be important sources of market research hypotheses.

Rosenberg's model describes a hypothetical relationship between the attractiveness of some object or action, A_j, and two variables: (1) the importance of specified characteristics or 'values' associated with the action or object; and (2) the perceived 'instrumentality' of those characteristics. Rosenberg's conceptualization of instrumentality is difficult to describe briefly, but it can be thought of as the degree to which an object or action provides the benefits a consumer associates with a particular choice or option. The mathematical form of the model may be expressed as:

$$A_j = \sum_{i=1}^{N} (V_i)(I_{ij}) \qquad [4.4]$$

where: A_j = attractiveness of some object or action, j;

V_i = importance of the *i*th value or characteristic;
I_{ij} = instrumentality of alternative *j* with respect to *i*;
N = total number of characteristics.

In the context of tourism, A_j might refer to different cruise packages; V_i would represent the importance of a potential cruise passenger attached to the various characteristics of the cruise package, such as ports of call and total price; I_{ij} would be an assessment of the degree to which various packages offered desired ports of call and met desired price levels.

Rosenberg suggested using a 21-point Likert scale to measure V_i, with values ranging from 'gives me maximum satisfaction' to 'gives me maximum dissatisfaction'. He also proposed using an 11-point scale to measure I_{ij}, with the ranks ranging from 'values are completely attained' to 'values are completely blocked or unavailable'. Respondents would be asked to assess the importance of a range of characteristics and their associated instrumentalities for each of a series of alternatives. Each individual's assessment of each alternative is a function of the sum of his scores over all characteristics. The relative attractiveness of the alternatives is determined by the relative scores of each alternative.

Fishbein's model is quite similar to that of Rosenberg. Although some authors, such as Sheth (1972), argue that there are important differences between the two, these differences are subtle and of questionable relevance to tourism analysts. For example, there are methodological differences in how the two authors suggest developing their scales. Further, Fishbein's model was developed with reference to beliefs about the qualities of an object and the associated attitudes people held about the object, whereas Rosenberg's model applies to beliefs about whether an object or action will lead to the attainment of certain goals. For most tourism applications, these are distinctions without a difference.

In any event, Fishbein's model combines two variables, 'belief' and 'effect', to predict the attitude or opinion a person holds about a given choice:

$$A_j = \sum_{i=1}^{N} (E_i)\,(B_{ij}) \qquad [4.5]$$

where: B_{ij} = belief about whether a particular object or action, *j*, possesses a given quality, *i*;
E_i = evaluation of the desirability of the *i*th quality;
and other variables are as described previously.

The Fishbein model in its original form is rarely used in tourism research because it was designed for attitudinal research and not analysis of consumer intentions. A modification of the Fishbein model – actually a hybrid of Fishbein and Rosenberg – is normally used:

$$A_j = \sum_{i=1}^{N} (V_i)\,(B_{ij}) \qquad [4.6]$$

where: A_j = intention to select (or probability of selecting) any particular alternative product, j;
V_i = importance of characteristic, i;
B_{ij} = degree to which alternative j provides characteristic, i.

Despite the popularity of the expectancy-value model in the tourism literature over the last two decades, it does have several weaknesses. Many of these are related to the fundamentally weak connection between the intention to buy, which these models are designed to predict, and actual buying behaviour, which is of greatest interest to marketers. Another limitation is that the context of the decision to buy is normally excluded from consideration. Experience and research confirm that the context of the decision greatly influences behaviour. The problem of context might be handled by explicitly stating the situation in which the respondent is asked to assess alternatives. For example, respondents might be presented with a carefully constructed description of the full context of a particular vacation decision including the time of the year, the social or familial situation of the vacation (e.g. an annual family vacation, a honeymoon, or a combined business–pleasure trip), available budget and available time.

Another problem, related to the issue of context, is the potential to influence family and friends on the decision to make a purchase. An individual consumer has his own preferences and opinions about the importance of various product characteristics and the degree to which various products possess these, but pressure from significant others may lead to a decision contrary to the individual's own preferences.

Expectancy-value models also fail to account for variations in the availability of supply of a product or of the financial resources available to the consumer at the time of the decision. The intention to make a purchase may be strong, but the lack of funds or a shortage in supply can circumvent the intention.

Other weaknesses in these models are related to certain assumptions that must be made in their development. First, these models are normally constructed as a linear combination of variables. The weighted characteristics of each choice are summed to obtain an overall score for each alternative. This implies that there are no significant interactions among the characteristics – such as between expectations of service and the level of price. This assumption is not likely to be valid in all cases.

Expectancy-value models use untransformed data. However, research on some variables affecting travel decisions has shown that certain types of transformations may provide for more precise forecasts. For example, geographers have repeatedly confirmed that travellers tend not to respond to differences in distance among competing destinations in terms of arithmetic differences but rather logarithmic differences. This issue is reminiscent of Fechner's Law (Fechner 1889) which states that psychological responses to physical stimuli vary in direct proportion with the logarithmic magnitude of the change in intensity of the stimulus. This pattern suggests that consumers' responses to product characteristics might involve some sort of transformation of differences beyond simple arithmetic comparisons.

Finally, the development of the weights of each characteristic and the assessment of the degree to which each product possesses those characteristics require the use of ratio scales. You must make a special effort to ensure your data reasonably approximate a ratio scale.

Procedure

1 Identify a number of destination characteristics that are likely to have a significant influence on the perceived attractiveness of the destinations (or other tourism products) being considered. This may be done through a review of other studies, by obtaining expert opinion, or by conducting a series of focus group interviews with potential respondents. There is no absolute minimum or maximum number of characteristics needed for analysis, but more than 20 or 25 often exhausts the patience of respondents. With fewer than four or five, you are likely to have overlooked some important qualities.
2 Establish a sampling design and a set of respondents appropriate to your problem.
3 After drawing your sample, collect the data required for the model as described in equation [4.6]. First, ask each respondent to indicate the relative importance of each attribute. Because this measure of importance will be used as a weighting factor, it must be a ratio scale. A simple way to achieve this is to have the respondent assign a total of 100 points to all attributes so that the number of points represents the proportional significance of each attribute.
4 Next, have each respondent indicate the degree to which he believes each commodity possesses each attribute. This is usually done on a Likert scale, with seven points ranging from '0 – does not possess at all' to '6 – possesses to the maximum degree possible'. This scale, too, should have ratio properties. Respondents might be advised of this by noting that, for example, a rating of 4 implies that the alternative has twice as much of some characteristic as an alternative with a rating of 2.
5 Multiply the weight assigned by the first respondent to the first attribute by that respondent's ranking of the attribute. Repeat for all other attributes. Sum these products to get a commodity weighting for the first respondent.
6 Repeat step 5 with all other respondents. Add and average all their scores for the first commodity to obtain an overall measure of the perceived attractiveness of that commodity.
7 Repeat steps 5 and 6 for all other commodities. The results define the attractiveness of each commodity as perceived by all respondents as a group.

Once you have obtained the relative attractiveness of each commodity, you may be able to estimate the expected market share of each. To do this, sum all A_js, the relative attractiveness of all commodities, to obtain an aggregate measure of the 'total attractiveness' of all commodities. Divide each alternative's A_j by this total. The quotient indicates the relative attractiveness of each alternative in contrast to *all* alternatives. Applying the logic of the Luce choice theorem (Luce 1959), one can hypothesize that the proportional attractiveness of an alternative also represents the expected market share of that alternative:

$$S_j = \frac{A_j}{\Sigma A_j} \qquad [4.7]$$

where: S_j = expected market share of commodity j;
A_j = proportional attraction of commodity j.

Here S_j may be interpreted as either the market share of a commodity alternative or as the probability that the average respondent will select j from among all alternatives.

A word of warning about the usefulness of equation [4.7] in calculating expected market shares is appropriate. The validity of market share estimates depends on how well you have defined the attraction of each commodity. If you do not include some significant component of the utility of a commodity, your estimated market shares are likely to be spurious. This problem is especially apt to happen with estimating market shares of competing destinations. Robelek (1994) applied a Fishbein-type model to estimate the expected market shares of four Ontario cities – Toronto, Hamilton, London and Kitchener–Waterloo – in the meeting and convention market. He defined the attractiveness of each community in terms of a series of attributes specific to meetings and conventions: availability of convention and accommodation services under one roof; adequate accommodation; price of accommodation, meals and meeting facilities; shopping opportunities; cultural attractions; airport access and so on. He estimated the relative shares of the market among these four cities as:

Toronto:	28 per cent
Hamilton:	26 per cent
London:	24 per cent
Kitchener–Waterloo:	23 per cent

Although data on the actual numbers of meetings, delegates, and meeting-related receipts in each community is unavailable, the estimated market shares are unrealistic. Toronto, with a population of nearly 3 million people, is larger than the three other communities combined, with more hotels and meeting facilities than all three. Toronto probably commands at least eight to ten times the number of meetings of any one of the other three cities.

The problem in this analysis of market shares was that no estimate was made of community size or the capacity to host meetings. Robelek focused exclusively on attributes that the literature suggests are important to meeting planners. The scores for each city may thus reflect some measure of their 'abstract' attractiveness to meeting planners, but they missed the overwhelming impact of size and capacity on a community's ability to drawn meetings and conventions.

Examples

Two examples are presented – one hypothetical and one real. The hypothetical example will help clarify the calculations involved in the procedure, while the real example will illustrate how this model is actually used in tourism analysis.

The hypothetical example presents three destinations (A, B and C) and three attributes (X, Y and Z). Table 4.4 is a summary of the responses of a set of hypothetical respondents. The numbers in the matrix represent the score for each attribute–destination pair. The vector of three numbers beneath the matrix contains the weights (out of 100 points) assigned to each attribute.

Table 4.4 Use of the scaling method to model individual destination choices: hypothetical data

Destination	**Attributes**		
	X	Y	Z
A	5	4	4
B	1	3	3
C	4	0	4
Weights	50	20	30

	Relative attractiveness
$A_A = 5(50) + 4(20) + 4(30) = 450$	1
$A_B = 1(50) + 3(20) + 3(30) = 200$	3
$A_C = 4(50) + 0(20) + 4(30) = 320$	2

$S_A = A_A/(A_A + A_B + A_C) = 450/(450 + 200 + 320) = 0.46$

$S_B = 200/(450 + 200 + 320) = 0.21$

$S_C = 320/(450 + 200 + 320) = 0.33$

The individual attractiveness of each destination is calculated with equation [4.6] from the data in Table 4.4. In this case, destination A is the most attractive, followed by destination C and then B. This predicted order could then be compared, if desired, to the order actually expressed by the respondents, either through direct questioning or by examining their visitation rates at each destination. Expected market shares were estimated using equation [4.7]. Destination A is predicted to receive 46 per cent of all visits, while C should receive 33 per cent, and B, 21 per cent. As noted previously, these percentages may also be interpreted as indicating the probability that an average respondent will choose any particular destination: 46 per cent chance of selecting A, 33 per cent of selecting C, and 21 per cent for B.

Hu and Ritchie (1993) were interested in the role of vacation context (whether a trip was taken for pleasure or education) on the assessment of the attractiveness of destinations. They developed a model of destination attractiveness that incorporated 16 attributes: climate, availability and quality of accommodations, scenery, food, shopping and so on. Five different nations were specified as potential destinations: Hawaii, Australia, Greece, France and China.

They conducted a survey of Canadians, asking them to first specify the relative importance of each attribute on a five-point scale. Half the sample was asked to respond to the attributes in the context of an educational trip, half for a recreational trip. Table 4.5 lists the scores of some of the attributes for each group.

Table 4.5 Importance rankings of selected attributes for recreational and educational trips

Attribute	Recreation (rating)[a]	Education (rating)[a]
Climate	4.11	3.32
Accommodation	4.01	3.57
Scenery	4.13	3.83
Food	3.85	3.65
Shopping	2.58	2.75

[a]1 = 'almost no importance'; 5 = 'very important'.

The respondents were also asked to evaluate how well each destination could provide each attribute within the context of the specific trip type. Tables 4.6 and 4.7 present selected responses for both contexts.

Finally, an overall attractiveness score was calculated as the product of a nation's perceived ability to provide an attribute by the attribute's importance over all 16 attributes. Table 4.8 presents the relative attractiveness of each destination for each trip context.

Table 4.6 Destination rankings for selected attributes for recreational trips

Attribute	Destination and rating[a]				
	Hawaii	Australia	Greece	France	China
Climate	4.51	4.35	4.13	3.68	3.53
Accommodation	4.36	4.19	3.63	3.87	3.25
Scenery	4.45	4.53	4.52	4.31	4.44
Food	3.95	3.88	3.62	4.17	3.27
Shopping	3.82	3.83	3.87	4.25	3.89

[a]1 = 'very low ability to provide attribute'; 5 = 'very high ability'.

Table 4.7 Destination rankings for selected attributes for educational trips

Attribute	Destination and rating[a]				
	Hawaii	Australia	Greece	France	China
Climate	4.45	4.32	4.15	3.70	3.60
Accommodation	4.33	4.12	3.69	3.96	3.34
Scenery	4.59	4.50	4.54	4.24	4.45
Food	3.86	3.83	3.92	4.25	3.83
Shopping	3.92	3.93	3.91	4.21	3.91

[a]1 = 'very low ability to provide attribute'; 5 = 'very high ability'.

Table 4.8 Relative attractiveness of five destinations for recreational and educational travel

Destination	Total attractiveness scores	
	Recreation	Education
Hawaii	225.07	219.12
Australia	222.37	220.94
Greece	217.79	220.51
France	215.47	218.17
China	205.62	212.17

Conjoint measurement

Description

Consider the following choices for a vacation:

- Two weeks in a rented cottage at a beach during the summer. The cottage costs $800 per week. The only recreational activities available are those you can provide yourself. Residents of neighbouring beach cottages are predominantly families with two or three children.

  OR

- One week in a ski lodge in the mountains during the winter. A room in the lodge costs $600 per week. Recreational activities include skiing, skating, sauna, bar, games room and indoor pool. The other people in the lodge are predominantly single and in their late twenties and early thirties.

Which would you choose?

The choice you make between these two reflects your preferences for vacations in certain seasons, the type and range of recreational activities, and the type of other people you are likely to meet. Different people make different choices, and those who select the same vacation may do so for different reasons. The selection of an alternative implies the implicit weighting of each variable (time of year, location, type of accommodation, length of stay, activities, cost, and the type of other guests). The combination of weighted variables is then compared, usually implicitly, to determine which alternative will be likely to give the greatest satisfaction.

This conceptualization of how a choice is made is similar to that behind expectancy-value models. In practice, of course, people do not normally explicitly assign weights; they are conscious only of having made a choice. Because the choice may be the only reliable phenomenon available to the researcher, there may be the opportunity to take the observed choices and to deduce from them (1) the weights implicitly assigned to each variable and (2) how the variables are combined in the consumer's mind to arrive at the ultimate expressions of interest.

Shepard (1957, 1962), Luce and Tukey (1964), and Kruskal (1965) developed a method known as conjoint measurement which allows a researcher to begin with simple rank orders, such as the expressed order of preferences for a series of vacation packages, and to analyse these choices to determine the weights the individual consumer implicitly assigns to each quality.

An important feature of conjoint measurement is that it presents the choice problem to respondents in a format that is more realistic than that used in expectancy-value models. Respondents are given a profile of several commodity choices similar to those introduced at the beginning of this section which they are then asked to rank. This avoids the sometimes unrealistic task of having the respondent assign ratio scale weights to a series of disconnected product attributes.

The profiles presented to the respondents consist of several specific variables, such as the number of stops on an airline flight, and some specified level for that variable, such as 'non-stop'. Each profile contains the same list of variables but different combinations of levels. Conjoint measurement produces a series of coefficients, 'part-worths', from the initial rankings. A separate part-worth is calculated for each level of each attribute. These values represent the relative importance of each attribute and its associated levels in terms of its contributions to the overall attractiveness of any particular combination. In other words, you can add part-worths for a series of different attributes' levels to obtain a score that is directly proportional to the relative attractiveness of that specific combination of attributes. Part-worths may be calculated for an individual or for a group of individuals. The analysis of an individual's preferences (as revealed by his part-worths) is more 'precise' than the analysis of the preferences of a group, because a consumer's preferences are, by definition, an individual phenomenon. However, marketers are rarely interested in the choice of process of specific individuals. They are more interested in the preferences of 'average' individuals or of entire market segments.

Conjoint measurement is based on several assumptions. The product or service must be capable of being described as a number of objective attributes. The selection of that product involves a certain degree of risk (e.g. the spending of a noticeable sum of money). The choice of the product is closely correlated with a combination of objective attributes. These attributes are combined in such a way that specific levels of attributes may be traded-off to determine overall product preference. In other words, poor 'performance' on one attribute can be offset by excellent performance on one or more other attributes. The various combinations of attributes that are described through conjoint measurement are realistic and believable. And, finally, the perceptions of attribute combinations (i.e. the hypothetical products) are reasonably consistent among respondents.

The technique also has a number of limitations. Conjoint modelling is not only based on objective attributes, limitations in respondent endurance mean that the total number of attributes and their associated levels must also be limited. Perhaps five attributes with two to four levels each is the maximum that can be handled in any particular analysis.

Most conjoint models do not allow for interaction effects. For example, the combination of excellent food and excellent service may contribute more to the overall attractiveness of a restaurant than the simple sum of both attributes. The extent to which interaction effects influence the validity of conjoint results is still debated. Interaction effects do exist in human behaviour, but several empirical studies have suggested that their impact on the accuracy of conjoint measurement is minimal. Johnson (1973, 1974, 1976), Green and Devita (1973), Green and Wind (1975), and Green and Srinivasan (1978) have found that a simple additive model, without interaction effects, provides a level of accuracy in conjoint measurement that is comparable to that of more complicated interaction models.

Another limitation is that results are relevant only for specified attribute levels. If one attribute is price, and the specified levels are £100, £200 and £300, one cannot reliably interpolate to obtain a part-worth for £150 nor extrapolate to obtain a part-worth for £400.

Conjoint measurement requires the use of objective product attributes, but those attributes need not be quantitative. Brand names, colours, styles and other qualitative conditions may be used if they can be described in objective terms (which may include drawings or photographs). The choice of attributes and their levels is an important task in the development of a conjoint model. You must identify the most salient features in a product profile. Further, these features should be those you can influence. If, for example, a certain brand name must be associated with a product, there may be little point in using different names as part of the product profile. (Although in some circumstances you may want to experiment with different brand names to determine your company's competitive edge or image in the market-place in the context of a specific product.)

Since the introduction of conjoint measurement to market research by Green and Rao in 1971, hundreds and perhaps thousands of applications have appeared in the marketing literature. Its use in tourism is much more limited, although applications are increasing.

Tourism analysts may still find it difficult to find commercially available software for conjoint measurement; however, there are a number of alternatives to 'brand name' conjoint packages. The basic algorithm for most conjoint analyses is monotonic analysis of variance (MONANOVA), which is widely available. However, if even this is a problem, you can consider using a general linear model, such as dummy variable multiple regression. Cohen (1968) and Green and Srinivasan (1978), have demonstrated the functional equivalence of different forms of analysis of variance, including MONANOVA, with general linear models. Empirical tests indicate that differences in numerical results begin to appear, if at all, in only the third decimal place. This is usually not significant in marketing work. Such findings mean that most common multiple regression packages can produce conjoint results comparable to those obtained with more specialized MONANOVA. Dummy variable regression, besides being more widely available, also has the advantage of providing significance tests and a measure of the level of explained variance that MONANOVA does not provide.

Procedure

1 Develop a sampling framework appropriate for your problem.

2 Identify a set of objective destination or product attributes and associated levels for each attribute. As noted, these do not need to be quantitative, but they must be described clearly and precisely. This includes expressing the attributes in a series of discrete levels or qualities. The attributes must reflect the most important attributes from the perspective of the tourist's decision. If the research is being done for management or planning purposes, it may also be useful to concentrate on those attributes that can be manipulated by the manager or planner.

   A balance must be struck with respect to the number of attributes in the product profile. Three attributes are usually a minimum; to have more than five or six raises difficulties. Very complex descriptions with numerous attributes and attribute levels quickly confuse respondents. The number of profiles required to statistically handle multiple attributes rises quickly as the number of attributes increases.

3 Prepare a set of destination or product profiles composed of different levels of the attributes. These profiles provide the basic information on which conjoint measurement operates, so they must be developed with care. This step is not only one of the most important, it is one of the most difficult. If you have only three attributes with two levels each, it is easy to develop profiles showing all possible combinations. However, as the number of attributes and levels increases, the possible combinations increase exponentially. For example, a set of five attributes of four levels produces $5^4 = 625$ combinations. This is an impossible number for your respondents to compare.

   Fortunately, alternative research designs exist in the form of

Latin squares, Latin–Graeco squares, balanced incomplete designs, and partially balanced incomplete block designs. These designs allow you to consider only a small number of the total number of combinations that would otherwise need to be examined. The trick here is in the choice of combinations. Consider the selection of alternative airline flights for a trip between Toronto and Vancouver. The profile consists of three levels of price, three levels of connections, and three levels of service. A 3 × 3 × 3 Latin square of the following format could be used to design the profiles:

	Level of service		
	Full meal and **complementary bar**	Full meal and **charge for bar**	Snack and **charge for bar**
Non-stop	$600	$900	$1200
One stop in Winnipeg	$900	$1200	$600
One stop in Winnipeg with plane change	$1200	$600	$900

The distinguishing feature of this type of design is that each level of the third attribute, price, occurs once and only once in each row and column. With this design you can limit your survey to nine profiles; with this design you would be forced to use $3^3 = 27$ profiles, a number too large for most respondents to compare efficiently.

More complex designs are possible for problems involving more than three attributes and for attributes having different numbers of levels. The design of these formats is a specialized skill, requiring training in the design of multi-factorial experiments or the advice of a statistician. Software programs are also available to help with this task. Some basic references you may wish to consult for more information on multi-factorial designs are Placket and Burman (1946), Bose and Bush (1952), Cochran and Cox (1957), Addelman (1962) and Winer (1971). Green (1974) provides a helpful, non-technical review of some research designs developed specifically for conjoint measurement.

As a rule, 20 to 25 profiles are the most a respondent can reliably rank. Sometimes larger numbers can be accommodated through the use of the two-stage ranking procedure described earlier.

4 Once the profiles have been developed, they are presented on cards or other appropriate format such as an interactive computer program to respondents. Respondents are asked to rank the choices

given to them, from most preferred (typically scored '1') to least preferred (typically scored *n*, where *n* is the number of choices). The data for analysis also include the specific profiles for each alternative. These are usually coded as a series of dichotomous variables, where each variable is a specific level of each attribute. If a profile includes a particular level of a given attribute, it is coded '1'; if that level is not part of the profile, it is coded '0'.

5 The data are analysed using conjoint measurement. As noted, the most common algorithm is MONANOVA (Kruskal 1965). The rank is the dependent variable and the levels of each attribute in the profile (coded as '1' or '0', i.e. 'yes' or 'no') are the independent variables. The analysis may be done for an individual, for an 'average' individual (defined by the mean responses over a number of individuals), or for a group of respondents. Marketers will typically find that groups or segments of respondents often provide the most useful marketing information.

You must be alert to the danger of over-specifying your model when using dichotomous coding for attribute levels. An example will make this point clearer. In the illustration below, one of the variables is the quality of food served in a restaurant. Three levels are specified: average, above average and excellent. Each of these levels could be coded as a '1' or '0', indicating whether the food associated with a particular restaurant profile was average or not, above average or not, and excellent or not. Although three levels are noted, only two variables are required to completely specify the quality of food. You take one level arbitrarily as the reference level. Let us use the middle quality, 'above average', as an example. You simply code the levels of 'average' and 'excellent' as dichotomous variables – but *not* 'above average'. The logic is illustrated by the following.

Food quality is one of three possibilities. If it is 'excellent', you code the dichotomous variable 'excellent food quality' as '1' (for 'yes') and the dichotomous variable, 'average food quality' as '0' (for 'no'). Conversely, if it is 'average', you code 'excellent food quality' as '0' and 'average food quality' as '1'. If the food quality is 'above average', you simply code 'excellent food quality' as '0' and 'average food quality' as '0'. No additional variable for 'above average food quality' is needed. This strategy reduces the number of variables used in your analysis and it automatically sets the part-worth value of your reference level as 0.00. The part-worths for the other levels are defined in terms of your reference level.

6 One measure of the validity of the conjoint measurement function is Kruskal's stress test. This measure, a standard part of the output of most MONANOVA programs, is a type of residual sum of squares that describes the goodness of fit between the observed rank order data and the ranks predicted by the analysis. No test of significance is available for Kruskal's stress test, so Kruskal (1964) has offered the following guidelines:

Stress (%)	Goodness of fit
20	Poor
10	Fair
5	Good
2.5	Excellent
0	Perfect

7 The part-worths (or coefficients from a dummy regression) are of special interest. They indicate the relative importance of each level of each attribute in terms of its contribution to the overall worth of a product. You may examine these values to make conclusions about how alterations in a product or destination's attributes would affect its perceived attractiveness and thus its probable market share. They allow you to construct and assess the relative attractiveness of products with attribute combinations that were not included in the original design – as long as the specific attribute levels are the same as those specified in the original model.

The part-worths can also be generalized to indicate the relative importance of each attribute with respect to other attributes, without reference to the individual levels. To do this:

(a) Subtract the lowest part-worth for an attribute from the highest part-worth to determine the total range of part-worths for that attribute.
(b) Sum the ranges to get a measure of the total variation in utility (overall part-worth) of all destinations or products.
(c) Divide the range for each attribute by the total variation obtained in the previous step. The result indicates the relative importance (as a percentage) of each attribute with respect to the total utility of a product.

Example

June and Smith (1987) undertook a conjoint measurement of the influence of social context on the selection of restaurant for a meal. The four contexts were:

1 an intimate dinner with a friend or spouse;
2 dinner with a group of friends to celebrate a birthday;
3 lunch with business associates; and
4 dinner with the family.

Five restaurant attributes and associated levels were also defined, as listed in Table 4.9. Because this number of attributes and levels produces 324 possible combinations, an orthogonal design was employed to reduce the total number of combinations required for respondents to rank. This particular design produced 18 profiles and is summarized in

Table 4.9 Restaurant attributes and associated values

Attribute	**Level**	**Values**
Price	a	<$10 for entrée
	b	$10 – $15 for entrée
	c	>$15 for entrée
Service	a	Inattentive servers
	b	Moderately attentive servers
	c	Very attentive servers
Atmosphere	a	Little privacy
	b	Moderate privacy
	c	Substantial privacy
Food quality	a	Average quality
	b	Above average quality
	c	Excellent
Liquor licence	a	Not licensed
	b	Licensed

Table 4.10 Orthogonal design for restaurant profile evaluation

Profile	**Price**	**Service**	**Atmosphere**	**Liquor Licence**	**Food quality**
1	a	a	a	a	a
2	a	b	b	a	b
3	a	c	c	a	c
4	b	a	b	a	c
5	c	a	c	a	b
6	b	b	c	a	a
7	c	c	b	a	a
8	b	c	a	a	b
9	c	b	a	a	c
10	a	b	b	b	a
11	a	c	c	b	b
12	a	a	a	b	c
13	b	b	c	b	c
14	c	b	a	b	b
15	b	c	a	b	a
16	c	a	c	b	a
17	b	a	b	b	b
18	c	c	b	b	c

Table 4.10. A series of cards was prepared with one of the 18 different profiles on each card. Profile 9, for example, contained the following description:

Price: More than $15
Service: Moderately attentive servers
Atmosphere: Little privacy
Food: Excellent quality
Liquor: Not licensed

The cards in each deck were randomly sorted, and the four decks (one for each context) presented in random order to a panel of 50 respondents.

The rankings provided by the respondents for each context were analysed using a conjoint measurement program. The rank (1 through 18, where 1 = 'most preferred') was the dependent variable, and the individual levels were the independent variables. The middle level for attributes with three levels was taken as the reference (i.e. not coded as a dichotomous variable), and 'not licensed' was taken as the reference for the liquor licence attribute. The resulting part-worths are presented in Table 4.11. The 'stress' associated with each model was under 4 per cent, indicating a good fit. Note that because the ranks begin with '1' as the most preferred and end with '18' as the least preferred (an inverse scale), smaller or negative part-worths indicate more desirable attribute levels. A price under $10 for a birthday party, with its part-worth of –0.13, for example, is more desirable than a price over $15, with its part-worth of 2.57.

Table 4.11 Part-worths for restaurant attributes

Attribute	**Context 1 (intimate dinner)**	**Context 2 (birthday dinner)**	**Context 3 (business lunch)**	**Context 4 (family meal)**
PRICE				
<$10	–0.74	–0.13	–0.07	–0.48
$10 – $15	0.00	0.00	0.00	0.00
>$15	0.78	2.57	3.16	3.89
SERVICE				
Very attentive	–1.50	–1.44	–1.68	–1.99
Moderately attentive	0.00	0.00	0.00	0.00
Not attentive	3.80	4.29	5.20	4.79
ATMOSPHERE				
Privacy	–1.29	–1.10	–0.97	–1.02
Moderate privacy	0.00	0.00	0.00	0.00
Little privacy	0.22	0.64	0.28	0.54
LIQUOR LICENCE				
Licensed	–6.21	–6.24	–2.68	–3.21
Not licensed	0.00	0.00	0.00	0.00
FOOD QUALITY				
Average	1.70	1.22	1.32	1.80
Above average	0.00	0.00	0.00	0.00
Excellent	–0.87	–0.30	–0.16	–0.09

A preference function for a restaurant can be determined by specifying the pertinent attribute levels and then summing the part-worths associated with those levels. In the context of intimate dining, restaurants with entrées less that $10, very attentive service, with a private atmosphere, licensed, and average-quality food would have the following preference function:

$$0.74 - 1.50 - 1.29 - 6.21 + 1.70 = -6.56$$

A restaurant with entrées costing over $15, inattentive service, very little privacy, no liquor licence, and excellent food would have the function:

$$0.78 + 3.80 + 0.22 + 0.00 - 0.87 = 3.93$$

The first restaurant, with its large negative score, is more preferred than the second restaurant with its relatively large positive score. Similar calculations may be done for any combination of attributes, whether or not they match any of the original 18 profiles.

The individual part-worths can guide management decisions on product development. For example, consider a restaurateur concerned with compensating for a lack of privacy for intimate dining in the interior layout of his restaurant. If he cannot adjust the interior design at an acceptable cost, he faces a loss of utility of 0.51 units in comparison to an identical restaurant offering significant privacy (0.22 – (–1.29) = 1.51). This could virtually be offset by improving service from being only moderately attentive (if that were the case) to being very attentive – an increase of 1.50 units of utility.

It should be noted that the units of utility do not have any physical meaning. They are simply coefficients that have been estimated at values that allow the original ranks to be reproduced as closely as possible. The significance of part-worths is in their relative size, not absolute magnitudes.

An alternative way of examining the importance of part-worths is to determine the proportion of total product attractiveness attributable to each characteristic in the profile. In the case of a restaurant serving business lunches, how important is service in comparison to price or other attributes? This question may be answered by following step 7 in the procedure. The relative importance of each attribute for each context in June and Smith (1987) is summarized in Table 4.12. Service and the availability of liquor are generally the most important qualities, accounting for more than 50 per cent of the total attractiveness of a restaurant whether for business or one of the other three social contexts.

Summary

Although there are many models available for analysing consumer choice behaviour, the two presented here – expectancy-value and conjoint measurement – illustrate many of the important issues in

Table 4.12 Relative contribution of each attribute to restaurant attractiveness

Attribute	Intimate dinner	Birthday	Business lunch	Family dinner
Price	0.018	0.151	0.207	0.245
Service	0.324	0.320	0.446	0.381
Atmosphere	0.092	0.097	0.080	0.088
Liquor	0.379	0.348	0.172	0.180
Food	0.157	0.085	0.095	0.106
TOTAL	1.000	1.000	1.000	1.000

consumer research. They are also two of the more important models in their own right. Both are based on the assumptions that the choice process is ultimately rational and that rationality can be mathematically analysed. Empirical testing of consumer choices indicates that these assumptions have some basis in fact. This is, of course, a fortunate finding. But the objective modelling of rationality still poses unique challenges for tourism analysts. For example, many tourists return to the same location year after year because of habit, or perhaps a family cottage. This is rational from the perspective of the people doing it, and it certainly represents predictable behaviour, but the problem is a trivial one for analysts. In contrast, other people seek new places for vacations each year. They desire novelty, uniqueness and perhaps a touch of adventure. Their choice, too, is rational given their values. Rather than being a trivial problem, this aversion to doing the same thing two years in a row makes for a difficult modelling exercise. In still other cases, the choice of a tourism destination is influenced by the preferences or expectations of people whose values are normally not included in a tourist-choice model. Thousands of people spend their holidays each year in places innocent of any tourist appeal because their families or friends want them to visit.

All these locational decisions can probably be explained in retrospect on a case-by-case basis; but it would be difficult to develop a predictive model that can accommodate equally those who decide by habit, those who seek constant novelty, those whose decisions are made by others, and those whose choices are based on a linear weighting of objective product attributes. And of course, the same person may employ each of these criteria at different times.

Both models presented in this chapter are designed to explain preferences or behavioural intentions, rather than actual behaviour. The use of stated preferences allows the analyst to measure responses to a wider range of alternatives than would normally be possible if he limited himself to actual behaviour. Preferences can also be obtained for combinations of destination attributes that may not exist in the real world, but that may still offer valuable theoretical insights. On the other hand, statements of preferences are often made without any reference to reality or the need to commit oneself to following through on the statement of preference. The correlation between what people say and what they do is notoriously low. Inclusion of some description of the context of a purchase decision, as in the case of the June and Smith example or as described by Hu and Ritchie (1993) can help to rectify this weakness (Kakkar and Lutz 1975), although it does not eliminate it. The use of overt behaviour may be a better indication of 'real' preferences, but that behaviour may reflect variables other than preferences, such as the availability of alternatives.

An important difference between the two models concerns how the weights of variables or attributes are determined. The expectancy-value example illustrates the self-explicated approach. The analyst asks a series of respondents to assign their own weights to the attributes. This is simpler than the statistical derivation of weights, as with conjoint measurement, but the results may be unreliable. The self-explicated approach assumes that potential customers are able to consider product attributes separately and consistently as they assign weights to each. The large body of empirical research based on Fishbein models suggests there is merit in this assumption, but the approach is still quite removed from how consumers really choose among products. The use of profiles and an analytic or decompositional approach (where choices are 'decomposed' to arrive at weights), represented by conjoint measurement, appears to reflect more closely the actual decision-making process. As a result, this latter approach may give generally more valid results.

No single choice model is appropriate for all circumstances. Ultimately you will need to sort out the relative merits and costs associated with the various models and their modifications, and then match them to the problem you are studying. If possible, the best approach is to experiment with several different models to determine which provides the most accurate, reliable and useful results.

Further reading

Bello D C and **Etzel M J** 1985 The role of novelty in pleasure travel experiences. *Journal of Travel Research* **24**(1): 20–4.

Card J A and **Kestel C** 1988 Motivational factors and demographic characteristics of travellers to and from Germany. *Society and Leisure* **11**: 49–58.

Cattin P and **Wittink D R** Commercial use of conjoint measurement: a survey. *Journal of Marketing* **46**: 44–53.

Chon K S 1989 Understanding recreational travellers' motivation, attitude, and satisfaction. *Revue de Tourisme* **44**: 3–6.

Cosper R and **Kinsley B L** 1984 An application of conjoint measurement to leisure research: cultural preferences in Canada. *Journal of Leisure Research* **16**: 224–33.

Dann G 1981 Tourism motivation: an appraisal. *Annals of Tourism Research* **8**: 187–219.

Gitelson R and **Kerstetter D** 1990 The relationship between sociodemographic variables, benefits sought, and subsequent vacation behaviour: a case study. *Journal of Travel Research* **28**(3): 24–9.

Hoynega K B and **Hoyenga K T** 1984 *Motivational explanations of behavior*. Brooks/Cole Publishing, Belmont, California.

Mansfield Y 1992 From motivation to actual travel. *Annals of Tourism Research* **19**: 399–419.

Nichols C M and **Snepenger D J** 1988 Family decision-making and tourism behaviour and attitudes. *Journal of Travel Research* **26**(4): 2–6.

Pearce P L 1982 *The social psychology of tourist behavior*. Pergamon Press, Oxford.

Pearce P L and **Stringer P F** 1991 Psychology and tourism. *Annals of Tourism Research* **18**: 136–54.

Plog S C 1974 Why destination areas rise and fall in popularity. *The Cornell Hotel and Restaurant Administration Quarterly* **14**(4): 55–8.

Smith S L J 1990 A test of Plog's allocentric/psychocentric model: evidence from seven nations. *Journal of Travel Research* **28**(4): 43–5.

CHAPTER 5

Segmenting the tourism market

Introduction

Diversity among people and places is a fact of life. Variety adds colour and excitement to all aspects of life, but it also confounds easy understanding and simple solutions to problems. A challenge in tourism analysis as in other realms of social science is to find some workable compromise between ignoring the great heterogeneity of the world and being overwhelmed by that same heterogeneity. One method of doing this is segmentation.

The potential for segmentation to bring some degree of order at an acceptable cost into marketing studies was first stated by Smith in 1956. His basic argument was that groups of consumers could sometimes be defined in such a way that their purchasing behaviour would be relatively homogeneous. If an entrepreneur could identify these segments, he might be able to design products and advertising messages to increase sales over what would be expected if the product or promotion were designed for the general population. Since Smith's pioneering article, tens of thousands of marketing studies have been completed that have used and improved concepts of segmentation.

The logic of segmentation resembles that of regionalization. The analyst's task is to define groups of consumers that are relatively similar with respect to some internal criteria and that are relatively different from other groups. The definition of groups is a classification exercise and, as we shall see in Chapter 8, many of the issues involved in regionalization have their counterparts in segmentation. Both processes may be either agglomerative or deglomerative; both are done for some practical purpose beyond just the definition of groups; and methodological issues concerning the definition and number of groups are central to both processes.

In practice, it is not always possible or necessary to define market segments. A population will exhibit one of three different patterns – only one of which supports meaningful market segmentation. First, everyone may be so similar that, in effect, everyone belongs to the same segment. In such circumstances, segmentation of the population into multiple groups is impractical. Alternatively, everyone may have unique characteristics. Concentrations of people sharing similar qualities do not exist, at least not in sufficient numbers to permit the identification of marketable segments. Finally, one or more concentrations of similar consumers do exist that are also relatively distinct from each other. Only this latter pattern permits segmentation.

This chapter begins with an examination of some of the critical issues associated with segmentation as an analytic and a marketing tool. Two general segmentation procedures are then described. Finally, some guidelines for the use of segmentation in tourism research are provided. The literature on segmentation is vast. A few references that can help you get into tourism segmentation are provided at the end of the chapter.

Issues in segmentation

Types and use of segmentation

If you begin to work your way through the segmentation literature, you will soon become aware that not only are there many different types of segmentation methods, there are a great many ways of classifying them. Jefferson and Lickorish (1988: 74) identify four common to tourism marketing: socio-demographics, socio-economics, travel motivations and psychographics. Vanhove (1989: 564) sees five: geographic (origin and degree of urbanization of the origin), demographics, socio-economic status, personality or psychographics, and buyer behaviour (benefits sought, brand loyalty, and marketing factor sensitivity). Middleton (1988: 70) classifies tourism segmentation into six categories: purpose of travel, buyer needs and motivations, buyer characteristics, demographic/ economic/geographic characteristics, psychographics and price.

While the outlines of these authors (and many more not cited here) are informative, a simple two-part classification will be sufficient for our purposes. Segmentation methods in tourism may be functionally described as being either **a priori** or **analytic**. The characteristics and differences of these two approaches will be discussed in detail in a subsequent section.

Segmentation is a multivariate technique, but the common distinction between dependent and independent variables does not apply in segmentation work. The task is not to predict or explain but to describe. As a result, the challenge is to find variables that work well as descriptors for the problem at hand. Table 5.1 contains a list of some of the more common descriptors. In addition to those listed, analysts have employed variables such as the degree of novelty-seeking (Snepenger 1986), length of stay (Uysal and McDonald 1989), and destination region (Opperman 1993). The distinction between business and pleasure travel is also a common basis for a priori segmentation, although these two

Table 5.1 Descriptors often used for segmentation

Social class
Occupational status
Ethnic background
Other demographic variables (age, sex, marital status, number of children, etc.)
Motives for buying
Personality characteristics
Psychographic characteristics
Geographic characteristics of residence
Price sensitivity
Brand loyalty
Frequency or volume of purchasing
Cash versus credit purchaser
Product use patterns
Images or perceptions of product

categories are often too imprecise to be of practical use for marketing. For example, Jefferson and Lickorish (1988: 77) argue that business travellers should be further subdivided into five sub-groups: (1) independent business travellers, (2) conference delegates, (3) visitors to and exhibitors at trade shows, (4) 'incentive' travellers, and (5) travellers attending training programmes.

Segmentation is undertaken for different reasons. Although most segmentation research is conducted for marketing problems, it is also a tool for the study of consumer behaviour. In the case of marketing, segmentation may be used to increase total sales, to improve the cost-effectiveness of advertising, to improve net profits, or to increase market share. Segmentation can help achieve these goals in several ways. Given an appropriately defined set of descriptors, segmentation research can provide information on:

1. the reasons different groups of people buy a product or visit a destination;
2. how big these groups are;
3. the spending patterns of these groups;
4. their loyalty to brand names or destinations;
5. their sensitivity to price;
6. how they respond to various advertising, pricing and distribution strategies;
7. how to design an advertising message or new product to generate sales in a specific market;
8. which advertising channel will most effectively reach the target market;
9. whether a new product should be introduced; or
10. whether an existing product should be redesigned, re-positioned or discontinued.

Academic researchers may be interested in these questions from a conceptual perspective. For example, segmentation can define life-style characteristics for the study of subpopulations in performing arts audiences, visitors to festivals, or families that take vacations. Segmentation can help provide insights into the motivations and other relevant characteristics of people who have particular political orientations, such as being pro-tourism or anti-tourism development. As noted before, segmentation does not explain or predict, but it can define groups who share common sets of values and thus assist in the analysis of social forces by identifying homogeneous groups on which to test models.

Running through all these seasons for segmenting people are several assumptions. The assumptions are, in effect, a summary of the beliefs shared by segmentation researchers. First, of course, is the belief that people differ from each other in measurable, comprehensible ways. This is balanced by the belief that the differences are largely differences in degree and that some people are more alike than others. Further, these differences are related in some ways to other aspects of people's lives, especially their market behaviour. Next, these differences can be objectively measured so that relatively homogeneous and meaningful groups of people may be defined. The use of segmentation implies a belief in the possibility of balance and compromise. As noted previously, you will often find yourself pulled between two extremes: the temptation to oversimplify by ignoring all differences among consumers and the temptation to give up looking for any generalities by seeing everyone as unique. Practical segmentation is the search for a workable compromise.

Special questions

Unit of analysis

Many social scientists study the individual. Even when population aggregates are studied, their characteristics are usually conceived in terms of multiples of the individual. However, much consumer behaviour reflects group decision-making, frequently that of a household. The choice of a vacation, for example, may involve negotiations (explicit and implicit) among the members of the household. A wife and husband will come to an agreement about whether to vacation together and, if so, where they shall go. If they have children living at home, the preferences of the children may be sought or at least given tacit weight by their parents as they evaluate alternatives, budgets, timing and previous experiences. More formal types of negotiations come into play when groups of unrelated adults work out a mutual vacation, such as a seniors' club planning a motor coach tour.

There are several reasons why the analysis of group decisions is rare in tourism. First is intellectual inertia. There are few successful models available for simulating group decision-making. Lacking good examples about how such research can be done, most tourism analysts are unlikely to try it themselves.

If you are willing to try, three questions must be answered. The first concerns the choice of the relevant group making the decision, e.g. all family members or just certain ones? Answering this question alone can be a formidable task. Next, you must decide how to define the decision reached by the group. There are usually multiple aspects of any decision regarding a travel experience. Choice must be made about destination, mode of transportation, timing of the trip, accommodation and activities. Different combinations of individuals may be responsible for different aspects of the travel plan.

Once this has been worked out, there is the problem of specifying the variables that must be considered when trying to model the group decision. Do you consider characteristics of each individual, characteristics averaged over all individuals, group characteristics, or some other combination? The continuing lack of experience by researchers in answering these questions means that, in practice, most segmentation studies will focus for the foreseeable future on the individual.

Working definitions

This is a perennial research issue. Every variable to be used as a segmentation descriptor must be operationally defined. If you define segments on the basis of 'heavy half' or 'light half' consumption patterns, are the halves defined in terms of frequency of purchase, frequency of use, value of total investment, total purchases made in a specified period of time, or percentage of all purchases of some type of good that is a certain brand? Should you use actual values or some form of scaling? Should the data be based on recall or on diary information? Such questions must be worked out for every variable you will be using.

Reliability

Reliability is the measurement quality that indicates the degree to which the same information will be collected every time identical procedures are implemented under identical conditions or that the same conclusion will be reached every time the same set of data is analysed with the same technique. Although reliability is desired in segmentation research, it is often not assessed. Most researchers just assume that their data and conclusions are reliable as long as they seem reasonable.

Reliability is of special concern with analytic segments. A simple way to illustrate the problem is to ask whether the segments derived through the use of factor-cluster analysis (described below) would have been derived if a different sample from the same population had been drawn. A useful technique is to replicate the analytic segmentation on randomly drawn halves of the sample and to retain only those solutions that are identical for both halves.

Certain types of data tend to be more reliable than others. Objective characteristics such as the age or sex of survey respondents are usually more reliable than subjective variables such as motivations or beliefs. As a result, simple a priori segments based on objective variables are more likely to be reliable – in the strict statistical sense – than segments derived from multi-stage techniques.

Validity

As with reliability, the issue of validity concerns both original data and segment structure. In the case of original data, do your variables measure what they are designed to measure? If you want to measure 'adventuresomeness', does a question about the degree of advanced planning a respondent puts into a vacation tell you anything about this quality?

With regards to structure, do the segments you have defined exist in the larger population or are they just artefacts of your analysis? In most cases, it is difficult to determine whether the data are truly valid. Possible methods include correlating variables selected for segmentation with other variables that have been accepted by other researchers as valid measures of the concepts you are attempting to measure. A high degree of correlation does not guarantee validity, of course. It may mean that both measures are invalid in the same way; still, a high degree of correlation provides tentative evidence of validity.

The issue of validity also covers the concept of 'discreteness' – whether the segment can be meaningfully isolated from other groups using an appropriate segmentation method.

Stability

Market segments may be reliable and valid, but if they are not stable they may still be useless for practical applications. Stability, in this context, means at least one of the following: (1) the characteristics that define a segment remain constant over a reasonable period of time; (2) the individuals who constitute a segment tend to remain associated with that segment; and (3) the size of the segment remains relatively constant. The most important type of stability is the first. If the existence and character of segments is relatively constant, a manager can develop various strategies to improve market performance with reasonable assurance that those strategies will not quickly become dated as the segments disappear or change unrecognizably. A closely related concern is, of course, whether the size of the segment is sufficiently large to justify developing a special marketing strategy to reach it. The other two forms of stability are less critical from a marketing perspective, although they may still be of interest. For example, a sales manager might be especially interested in knowing which segments show potential for growth and which are likely to remain stagnant or decline. Individual membership in segments as well as membership change may be of theoretical interest, but generally the identities of specific individuals in various segments are unimportant to marketing applications – as long as an acceptable number of consumers is present in the segment.

As with reliability and validity, stability is an oft-neglected topic in segmentation research. There are few guidelines concerning how long a period one might hope for stability in segments – six months, six years, or what? Bass (1977) and Calantone and Sawyer (1978) are two early studies that examined stability in segmentation that illustrate some of the questions and methods that can be used to assess stability.

Homogeneity

Market segments are characterized as relatively homogeneous groups of individuals, yet this ideal is not necessarily matched by reality. A lack of homogeneity on important characteristics is a special risk with a priori segmentation methods (to be described later in this chapter). In the case of a priori segmentation, one or more segments are defined on the basis of a particular characteristic, such as whether segment members have visited a specified destination. Once the general population has been divided into groups, the segments are typically are 'analysed' by evaluation of the means of demographic variables such as sex, age and income. The intent is to identify a variety of variables that distinguish the two segments from each other. Often, however, the variability in demographic variables within each group is so great that there is no significant difference between the segments. Despite their popularity, demographic variables are often poor descriptors for variations in travel preferences and patterns.

Generalizability

The final question is the generalizability of the results. This is largely a function of the sample. Segmentation is not done simply to describe a sample but to provide information that may be generalizable to a larger population. Whether you conduct your own survey, rely on a consulting company, or tap a secondary data source, you need to ensure the representativeness of your sample. Advice concerning adequate sample sizes, reliable sampling designs and sources of potential bias were covered in Chapter 3.

We now turn to two general strategies for segmentation. The first is a priori segmentation.

A priori segmentation

Description

A priori segmentation is a procedure in which the analyst selects at the outset the basis for defining the segments. The basis may be an intuitive belief of the researcher about how people should be grouped, but segmentation is more commonly tied to the use of one objective variable the researcher believes is critical for understanding travellers' behaviour. 'Heavy-half/light-half' is one such variable. It utilizes some measure of consumption such as the number of days spent on vacation or the value of package tours purchased. Individuals or travelling parties are then ranked and bisected at the median. Those who spent or consumed less than the median are the 'light-half'; those who spent more are the 'heavy-half'. A business will often wish to develop marketing strategies or product designs that have a special appeal to the heavy-half – those consumers who account for the greatest volume of sales.

A priori segmentation may be based on other variables besides heavy-half/light-half comparisons. In the context of tourism you might find purpose of travel, mode of travel, use of travel agents, use of package tours, accommodation types, destinations, distance travelled, length of time spent planning the trip, and duration of the trip to be useful descriptors. The choice should be made on the basis of whatever descriptor is of the greatest relevance to the marketing issue prompting the segmentation. Although we examine heavy-half/light-half segmentation here, the general logic is similar for other types of a priori segmentation.

Procedure

1. Develop a sampling procedure to obtain data representative of the population you wish to study. Heavy-half/light-half methodology examines only those people who have actually purchased a product, so the sampling design will reflect the subpopulation of consumers rather than the general population.
2. Select the variable on which the segmentation is to be made (the segmentation base). This will be some measure of participation, purchasing or usage behaviour such as the number of trips made. Operationally define the variable to be examined in each segment. This may be a socio-economic characteristic or benefits sought from the travel experience.
3. After collecting data from the sample population, order the observations with reference to the segmentation base. Identify the median value to divide the sample into two halves.
4. Calculate the percentage split between heavy-half and light-half consumers for nominal and ordinal characteristics. For example, you might want to calculate the ratio of males and females in each half. For variables that have interval or ratio scale properties, such as age and income, calculate the means for each half. Appropriate significance tests can be used to assess the observed differences.
5. Identify those variables for which there are significant differences between the two halves and which might be used to guide the promotion or design of vacation packages and other tourism commodities. Finally, interpret the results to develop specific recommendations for promotional and product development strategies.

Example

An illustration of this type of segmentation can be drawn from the 1991 Canadian Pleasure Travel Market Survey, sponsored by the US Travel and Tourism Administration, the Secretaria de Tourism (Mexico), and the Las Vegas Convention and Visitor Authority. This survey consisted of two phases: (1) a telephone survey of 5850 Canadians who had either travelled outside Canada in the three years prior to the survey or intended to travel outside Canada within the next two years and who agreed to participate in the second phase; and (2) a self-completed mail questionnaire sent to these individuals. The mail survey yielded a 60 per cent response rate (n = 3506).

One of the questions asked in the survey was the number of trips taken for pleasure within the three years prior to the survey and that lasted three nights or longer. This question provides an indication of the frequency with which the sampled household takes vacations. Other questions concerned destinations, trip timing, sources of information, activities, satisfaction levels, benefits sought from travel, product preferences, demographics, and media habits.

For the purposes of this example, we will focus on segmenting the market for 'small town tourism' – an important tourism product in many parts of Canada. Those respondents who indicated that small towns were a 'very important' tourism product were identified. This yielded a sub-sample of 1025 respondents. The 1025 respondents were then ranked according to the number of pleasure trips lasting three nights or more they had taken within the last three years. Values ranged from 0 to 60 trips, with a median of four (52 per cent of the population took four trips lasting three nights or less). Those individuals who had taken four trips or fewer within the last three years were classified as belonging to the light-half; those taking five trips or more were classified as the heavy-half. A total of 535 respondents had taken four trips or fewer; 490 had taken five trips or more over the last three years.

Once the travellers were assigned to the two segments, we can cross-tabulate segment membership with a variety of other variables. Tables 5.2 and 5.3 summarize some of these patterns.

The heavy-half travellers interested in small towns are less likely to

Table 5.2 Selected socio-economic characteristics of Canadian travellers

Household socio-economic characteristics	**Light-half (n = 535) (%)**	**Heavy-half (n = 490) (%)**
CHILDREN UNDER 18[a]		
Yes	62.4	37.6
No. of people over 18		
1	24.2	22.9
2	52.7	54.8
3 or more	23.1	22.3

Table 5.2 *continued*

Household socio-economic characteristics	**Light-half (n = 535) (%)**	**Heavy-half (n = 490) (%)**
MARITAL STATUS		
Single	29.0	25.2
Married	47.3	52.6
Living together	7.4	7.3
Divorced or separated	15.5	14.1
Other	0.7	0.8
AGE[a]		
18–24	14.9	8.0
25–34	30.1	28.7
35–44	21.3	26.3
45–54	14.9	14.8
55–64	10.0	11.1
65+	8.9	11.1
RESPONDENT'S OCCUPATION[a]		
Self-employed	22.7	20.3
Owner/manager	10.4	13.8
Professional	23.9	30.1
Clerical	11.6	7.2
Sales	6.3	6.6
Services	14.7	11.6
Primary industry	1.0	0.9
Skilled blue-collar	5.7	4.1
Unskilled	1.4	0.7
Student	9.0	5.9
Unemployed	0.4	0.7
Homemaker	6.9	5.7
Retired	6.7	11.4
RESPONDENT'S EDUCATION[a]		
Grade school	4.1	2.2
Some high school	8.9	5.5
High school diploma	19.6	15.2
College	24.9	17.6
Some university	17.0	19.2
University graduate	25.6	40.4
HOUSEHOLD INCOME[a]		
Under $20 000	15.6	6.9
$20 000–39,999	32.2	26.3
$40 000–59,999	24.5	27.1
$60 000–79,999	16.2	16.7
$80 000–99,999	6.4	11.5
$100,000+	5.1	11.5

Source: author's calculations.

[a]Significant differences between the two segments (based on chi-square; 0.05 level of probability).

Table 5.3 Magazines read by Canadian travellers

Magazine	Light-half (n = 535) (%)	Heavy-half (n = 490) (%)
Atlantic Monthly	2.2	2.0
Canadian Geographic	11.6	12.8
Canadian Living	20.4	24.9
Equinox	8.4	9.6
Harpers	1.9	1.8
Harrowsmith	4.5	5.9
Hockey News	3.6	2.7
Life	6.2	9.8
Macleans	23.0	27.3
National Geographic	26.4	31.4
Newsweek	6.5	8.8
Report on Business	4.9	10.2
Saturday Night	5.0	4.9
Sports Illustrated	5.0	6.3
Time	18.1	21.6
L'Actualitie	12.9	8.6
BC/ALTA/SASK/MAN Business	3.1	4.3
Flare	5.2	5.9
Chatelaine	30.5	24.5

Source: author's calculations.

Table 5.4 Selected product preferences of Canadian travellers[a]

Product	Light-half (n = 535) (% citing product as 'very important')	Heavy-half (n = 490) (% citing product as 'very important')
Local cuisine	37.7	42.9
Nightlife	13.2	8.2
Bed and breakfasts	19.9	14.0
Local crafts	25.7	33.1
Outdoor activities	18.6	24.4
Inexpensive restaurants	33.5	26.9
Public transportation	44.7	35.0
Resorts	19.6	12.0
Warm welcome	54.8	50.1
Environmental quality	44.3	37.1
Overnight cruises	13.8	8.2

Source: author's data.

[a] Only products cited by significantly different percentages of the two segments are shown.

have children under the age of 18, are slightly more likely to be married, and tend to be older. They are more likely than the light-half to be professionals, managers, or business owners, and are also more likely to be retired (Table 5.2). They are well educated and have significantly higher incomes. The percentage of either segment regularly reading any of the magazines shown in Table 5.3 is too small to yield statistically significant differences. However, it appears that a higher percentage of heavy-half travellers tend to read business, news and geographic magazines than do light-half travellers. Finally, heavy-half travellers are more interested in local cuisine and crafts and less interested in economizing on food, accommodation and local transportation than the light-half (Table 5.4).

The results of this a priori segmentation provide a general picture of the differences between heavy- and light-half segments for the small-town tourism market. Clearly, the heavy-half is of more interest to destination marketing organizations and businesses involved in promoting small towns to tourists. The heavy-half comprises not only more frequent travellers (by definition), but they have higher levels of income and are significantly more interested in two important products that small towns offer: local cuisine and local crafts. This segmentation also identifies – with some uncertainty due to the small sample size – which national magazines are more likely to be read by the target segment, and thus may be of interest to advertisers promoting small town products or freelance travel writers seeking outlets for their stories.

There are two main limitations with this particular segmentation procedure. First, of course, is the limited sample size. Although we began with over 1000 respondents, the number quickly shrank as we began to segment and then to divide into more precise categories. For example, readership for the heavy-half segment for many of the magazines listed in Table 5.4 was only about 50 or 60 people.

The second limitation is related to the fact that the analysis utilized a secondary data set. Although there are many benefits associated with using secondary data sets (including cost-savings), the analyst is limited to the sampling design, sample size, and the content of the original questionnaire. For example, our segments were based on the total number of overnight (three nights or longer) pleasure trips made within the last three years, regardless of the type of destination visited. It is not possible with the data set used to pull out those individuals who actually visited a small town; we were limited to identifying potential segments on the basis of expressed interest in small towns, not actual visits to them. None the less, the example illustrates how a simple a priori segmentation procedure can yield useful insights for marketing of a specific tourism product.

Factor–cluster segmentation

Description

Factor–cluster segmentation is a more complex method of defining segments. It requires familiarity with factor analysis and clustering algorithms. Unlike a priori segmentation, in which the analyst specifies the number and identity of the segments, factor–cluster segmentation produces segments analytically. The analyst still has some degree of control over the formation of segments through selection of variables and the particular algorithms used, but this control is much less than with a priori segmentation.

Factor–cluster segmentation, as the name implies, is a two-step procedure. The first step defines characteristics of the segments through factor analysis of a large number of descriptive variables. These characteristics are then used to cluster individuals into statistically homogeneous segments. This 'automatic' creation of homogeneous clusters is one of the advantages of factor–cluster segmentation over a priori segmentation.

On the other hand, the stability and reliability of factor-analytic segments is less certain than those derived by a priori segments. The use of split-half replications can verify whether the segments are reasonably reliable or if they are merely statistical artefacts. This type of replication, however, indicates nothing about stability over time or the generalizability of the results to other populations.

As we have noted, the purpose of segmentation is not just to define groups; the groups must be useful for some other purpose. One of the criteria to be used when evaluating segments should be whether they are 'reachable' for marketing purposes. In other words, is there any characteristic of the groups that would allow a marketer to communicate effectively and efficiently with them? Such characteristics include media habits (e.g. newspaper or magazine readership, television-viewing habits), geographical concentration (which might permit a direct mail campaign), or purchasing habits (e.g. individuals who make frequent calls from foreign countries to their home or office telephone number using a telephone calling card are likely to be frequent travellers and thus might be good prospects for other tourism product promotions). On the other hand, if there are no characteristics that permit the marketing message to be delivered more efficiently than a general, population-wide campaign, there may be no point in pursuing a segmentation strategy.

Procedure

Before describing the steps associated with this method, there may be value in reviewing briefly the two procedures on which it is based: factor analysis and cluster analysis. Factor analysis is a general term that refers to traditional factor analysis and principal components analysis. The technical and conceptual distinctions will be noted later, but for most purposes 'factor analysis' can be interpreted to include both methods.

Factor analysis is a procedure that identifies a hidden structure in a set of variables. This structure is composed of a number of statistically independent factors (combinations of variables). For example, if you were to study a set of socio-economic and travel characteristics of a population, you would be likely to find a number of significant correlations between variables. Frequency of travel and income are likely to be positively correlated. Strong preferences for wilderness camping, national parks, and opportunities to view wildlife are likely to be inversely correlated with an interest in nightlife, gambling and expensive restaurants. Factor analysis is a statistical procedure that can be used to identify and measure these relationships quantitatively and efficiently.

Factor analysis begins with the construction of a correlation matrix in which the values of each sampled individual on each variable are compared to their values on all other variables. The form of the matrix is a square with the rows and columns representing the variables. The correlations between variables range from –1.0 (perfect inverse correlation) to 1.0 (perfect direct correlation). Most values are not very close to either extreme, indicating some degree of imperfect correlation. The main diagonal of the correlation matrix (the values running from the upper left-hand corner to the lower right-hand corner) is usually a vector of 1.0s because each variable is perfectly correlated with itself. This diagonal is replaced in factor analysis by communality estimates, which are estimates of the correlation between each variable and the set of all other variables. If communality estimates are not available, you can retain the values of 1.0 in the diagonal as a preliminary estimate. The replacement of 1.0s in the main diagonal by estimated communalities is a feature that distinguishes it from principal components analysis (which retains the 1.0s). Principal components analysis produces final communality estimates that can be examined upon completion of the analysis.

After the correlation matrix is computed, factor analysis examines the pattern of correlations to find the best combination of variables that will summarize the pattern. A new set of variables, called factors, is defined. Each factor is a set of the original variables multiplied by weights, called loadings, that represents correlations between the original variables and the newly defined factor. Each factor is statistically independent of every other factor. As many factors as original variables are produced, but only a portion of these are meaningful. You must use some guideline to determine which are worth keeping for further analysis. In the case of traditional factor analysis, you specify a priori the number of factors to be extracted. If there is no theoretical basis to estimate the number of factors, statistical guidelines may be used. The use of theory as opposed to statistics is another characteristic separating factor analysis from principal components analysis.

The most common statistical guideline is the eigenvalue. An eigenvalue is a measure of the explanatory power of each factor relative to the set of original variables. The first factor explains much more of the variance in the data than the original variables did individually; subsequent factors are less powerful or meaningful. As a result, the first factor will have an eigenvalue above 1.0, indicating its greater relative explanatory power. Each additional factor will have a lower value, indicating their lower power. All those factors with eigenvalues above

1.0 explain more of the overall variance than the 'average' original variable; factors with an eigenvalue below 1.0 are not as meaningful as the original variables. Typically, you will work with only those factors whose eigenvalues are greater than 1.0.

After the factors have been selected, you try to name them. Factor names are selected on the basis of the loadings produced for each factor. High loadings (near –1.0 or +1.0) indicate a high degree of correlation between certain original variables and the factor. These variables are examined to see if they suggest the identity of the new factor.

The initial factor solution often produces many high loadings on the first factor, with successively fewer factors on subsequent factors. The loadings also tend to be spread between –1.0 and +1.0 with numerous loadings of only moderate size. A clearer pattern – with relatively fewer loadings on the first factor and all loadings falling nearer –1.0 and +1.0 or near 0.0 – can be obtained through a procedure known as 'rotation'. The most common form of rotation is varimax rotation. This forces the loading to approach –1.0, +1.0 or 0.0 as closely as possible on each factor while retaining the same level of explained variance obtained from the original factor solution. This forced dichotomization of loadings often makes interpretation of the factor structure easier.

The next step in factor analysis is to calculate factor scores for each individual for each of the new factors. To calculate a factor score, you multiply the loading of each variable on a factor by the individual's original value for that variable. This is repeated for all variables in the factor for that individual. These are then summed to give a preliminary score. The process is repeated for all factors for that same individual. This is repeated for all other individuals. Finally, all scores are standardized to a mean of 0.0 and a standard deviation of 1.0. The factor scores are recorded in a factor score matrix where the rows are the individuals and the columns are the factors.

Because each factor is independent of each other factor, they can be interpreted as defining a multi-dimensional space. For example, two orthogonal (independent) axes define a two-dimensional space; three axes define a three-dimensional space. Although impossible to visualize, four, five or more factors define a four, five or higher dimensional space. The set of factor scores for each individual locates that individual in space, just as latitude and longitude locate a traveller on the face of the earth. The more similar two individuals' scores are, the closer they will be together in this multi-dimensional space. Factor scores thus are measures of similarity. Groups of relatively similar people – segments – can be defined by identifying those individuals with similar factor scores. This is done through cluster analysis.

Several different types of cluster analysis are used in the social sciences. The basic type used in factor–cluster segmentation is hierarchical clustering. Hierarchical clustering produces successively larger clusters by combining smaller clusters. An individual belongs to only one cluster, but the number of clusters depends on how inclusive you wish to make the clusters. Clusters are derived from factor scores by measuring the distance between each individual in the multidimensional space defined by the factor structure. This process involves the use of a generalized Pythagorean theorem. It can be summarized by the following steps:

1 Calculate the distances between all pairs of points, using the Pythagorean theorem generalized to *n* dimensions.
2 Identify the smallest distance. Replace that pair of points with a new point midway between them.
3 Recalculate distances between all pairs of points, using the new point, but excluding the two points it replaces.
4 Continue to some termination solution.

This procedure begins by considering every individual as a cluster of one and can continue until everyone is grouped into a single cluster. These two extremes are rarely useful solutions. Some compromise balancing numerous highly homogeneous clusters with very few, highly general clusters is normally desired. One technique for finding a good solution is to plot the error sum of squares or other measure of the increasing variance in each cluster (based on statistics produced by the clustering program). When this measure is plotted against the decreasing number of clusters, you may find a sudden increase in the level of variance. The clustering solution just before this jump may be a good choice. Often, this solution occurs near the point where the number of clusters equals the number of factors.

With this background in mind, we can now examine how factor analysis and clustering may be used to define market segments.

1 Develop a sampling design to obtain data representative of your study population. The data may include both tourists and non-tourists and may consist of socio-economic variables, attitudinal data, buying behaviour and other personal characteristics. Because factor analysis is based on regression analysis, the variables should be measured on an interval scale or as dichotomous (yes/no) variables. Factor analysis is based on the assumption that each variable is relevant or common to each individual, but that their values vary. In practice this means that you should avoid variables for which 90 per cent or more (or 10 per cent or fewer) have the same value (such as often occurs for some activity participation variables). Be sure to attach an identification number to each respondent's data record but do not use this identifier as a variable in the factor analysis.
2 Organize the data into a matrix with the rows representing respondents and the column representing variables. Reduce this matrix through factor analysis. The conventions of varimax rotation and an eigenvalue cut-off of 1.0 are usually appropriate. Check the communality estimates for each variable and eliminate any variables that have low communalities. The definition of 'low' is a matter of judgement, but many researchers use a threshold of 0.4 or 0.5 to retain a variable. Low communalities may indicate a low

degree of variance for a given variable; for example, an activity in which either virtually no one or nearly everyone participates.

3 Repeat the factor analysis if any variables have been eliminated. You may find it desirable to experiment with different solutions based on different numbers of factors.
4 Identify each factor in your tentative solution by examining the pattern of factor loadings. Your solution should not only be statistically valid but should also make conceptual sense. After you have decided upon a solution, obtain the factor scores for each individual.
5 Cluster the individual respondents using a hierarchical clustering program.
6 Once the clusters have been formed, begin to characterize each one. Sort through each cluster to obtain the identification numbers of each respondent assigned to each cluster. Determine the average factor scores for those individuals. Large factor scores (greater than +1.0 or –1.0) indicate factors that may be important in determining the identity of each cluster. You might want to calculate *F*-ratios and *t*-tests for each factor in a cluster. The *F*-ratio is a measure of the variance in each factor; the *t*-test is a measure of the difference between the clusters on each factor. Ideally, you want to find factors that have high mean scores, low *F*-ratios and high *t*-test scores.
7 Finally, once all clusters have been formed and identified, note their relative sizes. Interpret the results and make a recommendation for application to the marketing problem at hand.

Example

This type of segmentation has been employed in the series of long-haul pleasure travel market studies conducted by Tourism Canada and the US Travel and Tourism Administration by Market Facts of Canada. The analysis of the United Kingdom pleasure travel market provides a typical example (Market Facts of Canada 1989).

A total of 1209 personal in-home interviews were conducted in England, Scotland and Wales in the summer of 1989. Respondents had to be 18 years of age or older and either have taken a trip of four nights or longer outside of Europe and the Mediterranean in the three years prior to the survey or be planning to take such a trip within the next two years. The survey included an extensive array of questions on travel patterns, travel attitudes, motivations, benefits sought, sources of information, booking patterns, media habits, and other personal data. Data relating to benefits sought consisted of the respondent's ranking (on a four-point scale ranging from 'very important' to 'not at all important' of each of 30 potential benefits such as 'getting away from the demands at home' and 'reliving past good times'.

The segmentation yielded three segments: adventure travellers (30 per cent of the total market), getaway travellers (31 per cent), and social safety travellers (39 per cent). The individual segments may be described as follows (Market Facts of Canada 1989: 126–7):

- **Adventure travellers**: These travellers seek thrills and excitement; they are daring and adventuresome. Sports and physical activity are important. They are willing to rough it in their seek to escape from the ordinary.
- **Social safety travellers**: The second segment prefers to visit family and friends or travelling together as a family activity. They enjoy meeting people with similar interests. They tend to be cautious and conservative, and are concerned about personal safety.
- **Get-away travellers**: This segment seeks to escape pressure of home life and their job by having a vacation in which they can relax – do nothing at all. They are not interested in family activities, visiting friends or relatives, or engaging in physical activities.

The consultant also examined cross-tabulations between membership in each segment and demographics, travel behaviour and media habits among other characteristics. Some of the patterns observed included the following:

Adventure travellers

Demographics
More likely than average to be male, younger, single, a student or skilled worker, and university-educated.

Travel behaviour
More likely than average to have visited or be interested in visiting the Far East and Asia, taken a touring trip, travelled solo, used books/library for trip planning, and travelled by plane and train within the US or by boat in Canada.

Media
More likely than average to read the *Sunday Times*.

Social safety travellers

Demographics
More likely than average to be older, divorced/widowed/separated, retired or a homemaker, from northern England or Scotland, less educated, and to have lower incomes.

Travel behaviour
More likely than average to have taken few pleasure trips in the last three years, to be highly interested in visiting Canada, to travel to see friends and relatives, and to travel by a private car in the US and Canada.

Media habits
More likely than average to read the *Daily Express*, *Sunday Express* and *Readers Digest*.

Get-away travellers

Demographics
More likely than average to be female, middle-aged, married, a couple with children, and to have higher incomes.

Travel behaviour
More likely than average to have visited the US in the past three years, spent fewer nights on the trip, visited a resort, combined business and pleasure, visited a theme park, attended a special event, purchased a vacation package that included flight and/or accommodation, and rented a car.

Media habits
No readership habits significantly different than average.

The consultant concluded from these and other findings that the get-away market segment was the most promising segment for the US to target. A key part of the strategy would be to emphasize products that would appeal to the group's desire to escape from their daily routine and to relax. The social safety segment is the market most likely to be interested in Canada; however, they are motivated primarily by visiting friends and relatives, and thus are not easily influenced by marketing campaigns. Adventure travellers express less interest in visiting North American destinations, but may be open to products that emphasize adventure, sport and physical recreation. Canadian wilderness and outdoor opportunities may offer a special appeal to this market.

Summary

Market segmentation can improve the efficiency of advertising aimed at different groups of potential consumers by increasing the appeal of advertisements or products for specific groups. Segmentation can also be a tool for other analytical projects such as the development of forecasting models for different social groups or for the study of the motivations and behaviours of different types of individuals.

There are two types of segmentation methodology: a priori and factor–clustering. The first is based on the arbitrary choice of one or more variables by the analyst to develop the segments. Factor–cluster segmentation produced statistically defined segments. Factor–clustering can produce more insightful results, but it requires more data and statistical ability and may produce less stable segments than simple a priori segmentation based on variables such as sex or age. Although more analytically based, the results of factor–clustering can still be modified by various options and decisions involved in the application of the routines used.

Regardless which method is chosen to define segments, there are several criteria that any good segmentation should satisfy:

1 **Accessibility**. The marketer must be able to reach the segments through existing information channels; ideally the channels should allow the message to reach only the target audience. At a minimum, the channels should reach the target audience at a higher rate than other groups not likely to purchase the product. If the target market cannot be reached within the limits of the available advertising budget, or they cannot be singled out for a focused campaign, there is little point in identifying them.
2 **Size**. The segments must be of a size sufficient to make them economical to reach. In other words, they must be big enough to justify the cost and effort of a directed marketing campaign. This is a special concern with factor–cluster segments. A solution that yields many small segments might make statistical sense, but could be meaningless in practical terms.
3 **Measurability**. The segments must be defined in such a way that you can obtain adequate information about their market behaviour to monitor the effectiveness of a marketing campaign. This is also a concern when segments are being defined in order to be used in forecasting models. The characteristics used for defining the segments must be those for which adequate trend data are available for forecasting.
4 **Appropriateness**. Some products such as wide-bodied passenger jet aircraft will be purchased by relatively few firms or governments. As a result, there may be no special need to segment the market. Other companies may enjoy a monopoly in a geographical region, such as a tourist magazine in a small resort community. With no competition, there can be no competitive advantage gained from segmenting the market. On the other hand, firms might use segmentation procedures to develop specialized products – other types of commercial aircraft or other publications – that cater to different markets they could serve.

Further reading

Cook V J, Jr and **Mindak W A** 1984 A search for constants: the 'heavy user' revisited. *Journal of Consumer Marketing* **1**: 79–81.

Madrigal R and **Kahle L R** 1994 Predicting vacation activity preferences on the basis of value-system segmentation. *Journal of Travel Research* **32**(3): 22–7.

Mills A S, Couturier H and **Snepenger D J** 1986 Segmenting Texas snow skiers. *Journal of Travel Research* **25**(2): 19–23.

Moutinho L 1987 Consumer behaviour in tourism. *European Journal of Marketing* **21**(10): 5–44.

Muller T 1991 Using personal values to define segments in an international tourism market. *International Marketing Review* **8**: 57–70.

Shoemaker S 1989 Segmentation of the senior pleasure travel market. *Journal of Travel Research* **16**(3): 14–21.

Shoemaker S 1994 Segmenting the US travel market according to benefits realized. *Journal of Travel Research* **32**(3): 8–21.
Snepenger D J 1987 Segmenting the vacation market by novelty seeking role. *Journal of Travel Research* **26**(2): 8–14.
Spotts D M and **Mahoney E M** 1991 Segmenting visitors to a destination region based on the volume of their expenditures. *Journal of Travel Research* **24**(4): 24–31.
Woodside A G, Cook V J, Jr and **Mindak W** 1987 Profiling the heavy traveller segment. *Journal of Travel Research* **25**(4): 9–14.

CHAPTER 6

Forecasting tourism demand and market trends

Introduction

This chapter introduces two important concepts for tourism analysis: forecasting and demand. Both are common words, but the relationship between them may not be immediately apparent. Forecasting, of course, refers to making predictions. The majority of forecasts made by tourism analysts arguably concern the demand for tourism commodities. An understanding of the concept of demand and of its various connotations can help deepen your understanding of the practice and problems of forecasting tourism trends. Conversely, an appreciation of the nature of forecasting can shed light on how demand is studied and measured. The chapter begins by describing the nature of forecasting and some of the general issues associated with selecting a forecasting model. The nature of demand is then considered, with special attention given to both the definitions of demand and the forces that cause demand to change. Finally, several forecasting models are described that illustrate some of the more common approaches used in forecasting tourism trends.

The nature of forecasting

Virtually all policy analysis and planning problems require forecasts of future conditions. Estimates of future levels of demand for different commodities, travel volumes, the market share of various destinations or businesses, household incomes, interests rates, changes in consumer intentions and tastes, and many other social and economic variables are vital to managing and planning tourism development. Forecasting can give you an idea of what future conditions may be like if you fail to take corrective action, and it can provide you with an assessment of the possible outcomes of alternative courses of action. The challenges of successful forecasting are more than just the technical difficulties of developing an accurate model. Forecasting models must be developed with a clear understanding of both the nature of the situation for which a forecast is desired and the resources available for making the forecast. Stynes (1983) identifies four factors that should be considered when developing a forecasting model: (1) the organizational environment; (2) the decision-making situation; (3) existing knowledge; and (4) the nature of the phenomenon being studied.

The organizational environment

Each organization has resources, structures, ways of operating, and objectives specific to it. These characteristics influence the types of forecasts the organization requires. An agency that prides itself on being politically neutral and on producing objective, high-quality forecasts will have different standards for forecasting than an organization devoted to lobbying from a pre-determined political position. The availability of resources such as data banks, computers, statisticians and other technical experts will also influence the type of forecast that can be developed. The forecaster needs to be aware of all these organizational aspects in order to design a model that will be accepted and will function effectively within that organization.

The decision-making environment

Some organizations need to make decisions quickly for their immediate future or to respond to governmental policy initiatives; others work with a more distant planning horizon and have a longer time period available for developing their model. The level of precision required for a decision is also important in selecting the appropriate forecasting technology. Generally, the greater the precision required, the more complex the model and the longer the lead time required. Another aspect of the decision-making situation is the level of accuracy required. Whereas precision refers to the amount of detail, accuracy is a measure of the correctness of the forecast. For example, a forecast that demand for air travel between Australia and China will increase by 98.76 per cent is fairly precise, but probably not accurate. On the other hand, simply saying that the demand is likely to increase may well be accurate, but it is not precise. As with precision, greater accuracy usually requires more resources and a longer lead time.

Existing knowledge

Scientific forecasts are based on information about past and current conditions. Some types of forecasting models, such as trend extrapolation or simulation models, require significant amounts of historical data. Other methods, such as the Delphi technique, require less data. The issue of existing knowledge also refers to the forecaster's understanding of theoretical issues associated with the phenomenon being forecast and his familiarity with forecasting technology. A match must be made between the theoretical and technical requirements of the problem and the abilities you are able to bring to the problem.

The nature of the phenomenon being forecast

Certain phenomena show a high degree of stability. The percentage of Canadians taking a vacation has remained virtually unchanged at about 55 per cent since the early 1970s. Other phenomena exhibit dramatic changes from year to year in response to fads, local crises or other forces. The former are, of course, easier to predict than the latter. You will also need to consider whether the phenomenon you are modelling is best studied with a stochastic model (one that predicts probabilities) or a deterministic model (one that predicts absolute numbers). A choice between a linear and a non-linear model will also be guided by the nature of the phenomenon you are forecasting. Knowledge of the forces that have affected the past behaviour of your phenomenon can assist you in selecting the most useful variables and model structure.

The nature of forecasting models

Forecasting models in tourism may be classified as belonging to four types: (1) trend extrapolation models, (2) structural models, (3) simulation models and (4) qualitative models.

Trend extrapolation models, as the name suggests, rely on the extrapolation of an historical series of data into the future. One of the simplest of such models is a plot of data on a graph. The ordinate (the vertical axis) is some measure of demand or market activity, while the abscissa (the horizontal axis) is time, measured, perhaps, in quarters or years. A line may be visually fitted to the data and then extrapolated beyond the observed data to a desired point in the future. Statistical techniques are available to accomplish the same task. These include simple regression models, exponential models, logistic models, quadratic equations and harmonic analysis. Despite the differences in the statistical complexity of each and the shape of the extrapolation curve, the forecaster makes the same assumption in each case: that the observed trend will continue for some reasonable period into the future.

Structural models depend on the identification of the relationship between a measure of tourist demand and one or more causal variables such as price, income, competition or distance. These relationships are usually identified with a multiple regression equation or analysis of variance and either cross-sectional data (e.g. data drawn from different destinations at the same time) or time series data (e.g. data drawn from one destination over a period of time). Once the models have been calibrated, estimates of future values of the causal variables are used to make a forecast of future demand.

Simulation models are a complex set of equations that typically combine trend extrapolation and structural models into a more comprehensive system. Relationships among many variables – including synergistic and dampening effects – are specified through a set of interrelated equations. These models also rely on historical data for calibration. Forecasts are made by specifying expected values for the causal variables and then solving the system of equations to arrive at predicted values of the dependent variables.

The term 'qualitative models' refers to a variety of non-scientific techniques including intuition to anticipate future developments. The best-known qualitative technique may be the Delphi technique. This procedure involves the formal and structured soliciting of expert opinion from a panel of knowledgable individuals concerning a given forecasting problem. The results of their collective opinions are given back to the panel one or more times to encourage them to move towards a consensus of opinion about the nature and timing of future events.

The selection of the most appropriate model involves consideration of the four factors described earlier. The choice often requires trade-offs between accuracy, precision, and constraints of time, money and other resources. A summary of the requirements and characteristics of the four general forecasting model types is provided in Table 6.1. A review of the table illustrates the fact that no single model is best on all criteria. It is significant that two important criteria are not included in this table: accuracy and precision. All models are capable of producing good-quality forecasts if they are properly developed and applied, if adequate data are available, and if the problem being studied conforms closely to the assumptions implicit in the model. The degree to which the development and application of a forecasting model departs from these conditions ultimately determines the quality of the forecast.

The nature of demand

Demand is an ambiguous word with at least five definitions relevant to tourism. The traditional definition is that of neoclassical economics: demand is a schedule of quantities of a good or service that will be consumed at various prices. Higher consumption is usually associated with lower prices, and vice versa. Demand, in this sense, can be described graphically as in Figure 6.1. The downward sloping line, *DD'*, reflects the inverse relationship between price and consumption. Consumption refers to the purchase of some good or service, such as a hotel room; participation in some activity, such as a game of golf; or attendance at an attraction, such as visiting a museum.

Demand is also used to refer to actual consumption. This definition could be represented as a single point on *DD'* in Figure 6.1. Such a point, *X*, is specified by the pairing of a specific price, say P_1, and observed consumption, Q_1. This is arguably the most common use of the word 'demand', but it is of limited utility because it tells us nothing about trends or levels of unmet demand. It is a static definition, useless for forecasting.

A third definition is that of unmet demand, which is sometimes

Table 6.1 Summary of requirements and characteristics of forecasting models

	Trend extrapolation	**Structural**	**Simulation**	**Qualitative**
Technical expertise required	Low to medium	Medium to high	High	Low to medium
Type of conceptual knowledge or data required	Time series data	Cross-sectional data plus causal relationships	Time series and cross-sectional data, causal relationships, and change processes	Expert and experiential
Required data precision	Medium to high	High	High	Low
Appropriate forecast horizon	Short	Short to medium	Long	Long
Time required for forecast	Short	Short to medium	Long	Medium to long
Type of problem best suited for	Simple, stable or cyclic	Moderately complex with several variables and known, stable relationship	Complex with known and quantifiable relationships and some feedback effects	Complex with known but qualitative relationships and elements of uncertainty

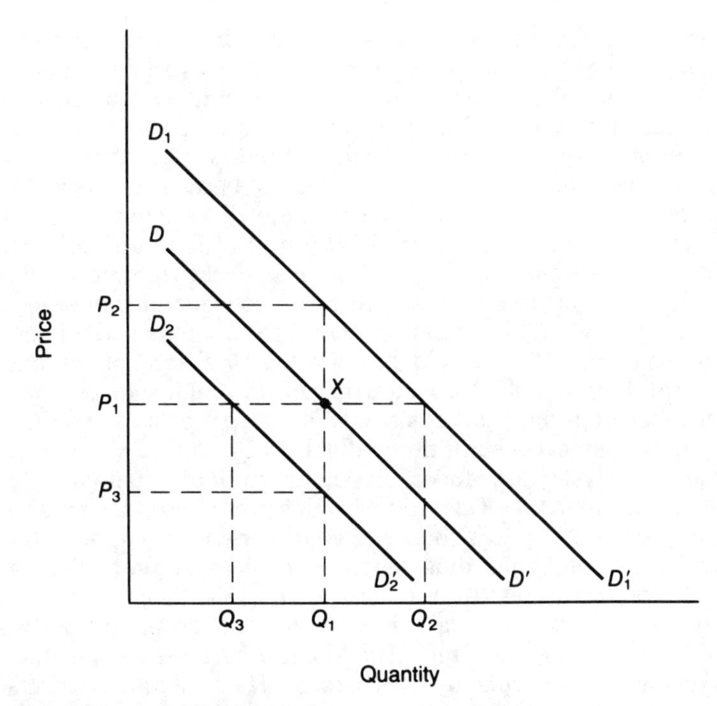

Figure 6.1 The concept of demand

called latent demand. Latent demand is a measure of the difference between the potential level of consumption and the observed level. The difference may be due to a supply shortage, excessively high prices, scheduling problems, or other barriers. Latent demand is of special interest to tourism planners because it represents the potential for market expansion. However, it can be difficult to measure reliably unless you can specify the relationship between price and quantity, which brings you back to the neoclassical definition.

Demand also refers directly to a forecast of future consumption. This concept of demand is closely related to the neoclassical definition, but there are important distinctions. Demand as future participation is seen as a function of many variables, not just price. It thus implies knowledge of the anticipated mix and values of other variables that affect demand. Demand as future consumption is thus simultaneously broader and narrower than the neoclassical definition. It include more variables, but the focus is on estimating one single value, not a schedule of values.

Demand shifters

The neoclassical conception of demand defines demand as a function of only a single variable: price. However, many other variables influence how much of a commodity an individual will consume. These variables are known as demand shifters. Demand shifters include consumer characteristics such as age, education, tastes, previous experience with the product, as well as advertising, product innovation, government policy and new technology. The reason such variables are called demand shifters may be seen in Figure 6.1. Consider *DD'* as the demand for rooms at a given hotel. If the hotel adds a recreational complex, shuttle services to a nearby airport, or adds a 'frequent sleeper' promotion, the demand will increase. That is, because of greater value, consumers will be willing to either pay more (P_2) for the same number of visits or use the hotel more frequently (Q_2) at the same price. Either condition may be represented by shifting *DD'* to the right, to $D_1D'_1$.

If the hotel begins to deteriorate through poor maintenance, lessened service or unfavourable publicity because of hotel problems, the demand may drop. This would be reflected in a shift of the demand curve to the left, to $D_2D'_2$. Consumers will be willing to purchase the same number of rooms as before only if the price drops to P_3. If the prices stay the same, consumption will fall to Q_3.

Demand analysis may focus on either an individual or a group. Modelling the demand of an individual for any product tends to be more complex and have a higher degree of variance than modelling the demand of a group and thus tends to be less accurate than group forecasts. The main reason for this is that the behaviour of large numbers of individuals in a group tend to average out individual idiosyncrasies. Young and Smith (1979) describe the effects of the level of aggregation on demand forecasting. Their work confirms the experience of many others who have noted that the reliability and accuracy of models increases as the level of aggregation increases. While this is desirable to a point, the most accurate models are often obtained at the most general levels of analysis – analysis so generalized and based on such highly aggregated data that the results have little value for planning or policy analysis. As with the other issues surrounding the selection of a forecasting model, you may have to make a trade-off between a highly accurate, highly aggregated but less relevant model and one that has a lower level of aggregation and thus potentially greater usefulness but with lower accuracy and reliability.

Elasticity

The slope of *DD'* (Figure 6.1) indicates the degree to which consumption changes given a change in price. Such change is called 'elasticity'. Graphically, a line with a steep slope indicates that a large change in price has relatively little effect on consumption; a fairly flat line indicates that even a modest change in price can cause a large change in consumption. Quantitatively, elasticity is the ratio between the observed percentage change in consumption and a 1 per cent change in price. A commodity whose consumption changes at the same rate as price (e.g. a 1 per cent increase in price results in a 1 per cent decrease in consumption) has unitary elasticity. If consumption changes at a rate lower than that of price – reflected in a steep line – the commodity is described as being relatively inelastic.

Two characteristics of commodities influence their degree of elasticity.

Commodities that are necessities tend to have inelastic demand. Food staples such as bread and salt, modern necessities such as gasoline or telephone service, and life-supporting goods such as some prescription drugs show relatively little short-term variation in consumption due to price changes. In contrast, luxury goods, which include many tourism commodities, tend to be elastic. Goods for which there are few substitutes also tend to be inelastic whereas those for which we have many substitutes tend to be elastic. Thus one can find intense price competition among gasoline stations or beer retailers (in those jurisdictions where gasoline or beer are not government- or industry-controlled monopolies), soft-drink bottlers, and many other types of goods where one brand is an acceptable substitute for another. Note that in the case of gasoline, the demand for gasoline as a generic product tends to be inelastic in the short run, whereas the demand for gasoline from specific refineries can be highly elastic. Consumer responses to price changes in the long run may exhibit different strategies from simple changes in the volume of purchasing. For example gasoline price increases have not, in the long run, reduced the demand for gasoline, *per se*, but have encouraged motorists to buy more fuel-efficient automobiles.

Elasticity may also be examined from the perspective of income. This perspective is of special interest in tourism analysis because of the close relationship between the ability to pay for tourism expenditures (reflected in income) and the willingness to pay for them (reflected in demand). If we plot income on the ordinate of a graph against quantity consumed of most tourism products on the abscissa, we typically obtain a curve with a positive slope (see Figure 6.2). In other words, higher levels of income are usually associated with higher levels of consumption. The degree of association, reflected by the slope of the line, is the income elasticity. Commodities that are purchased at only slightly higher levels as incomes rise are relatively inelastic; those whose

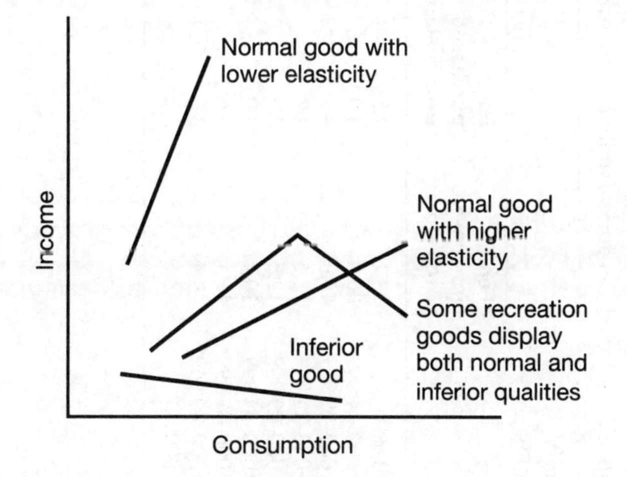

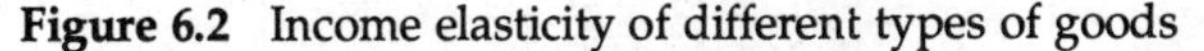

Figure 6.2 Income elasticity of different types of goods

Table 6.2 Selected Canadian family expenditure data testing Engel's Law

Family income	Sector (%)						
	Food purchased from stores	Food purchased from restaurants	Housing	Clothing	Air transportation	Recreation	Tobacco and alcohol products
<$10 000	23.1	4.5	43.3	6.8	0.9	4.1	5.9
$10 000–14 999	20.0	4.0	30.2	6.1	0.6	4.0	5.4
$15 000–19 999	15.0	4.9	23.7	7.3	1.0	5.1	4.9
$20 000–24 999	13.3	4.0	21.5	6.0	0.8	4.2	4.2
$25 000–29 999	12.5	3.9	20.3	6.2	0.5	4.6	3.9
$30 000–34 999	10.4	4.0	18.0	6.4	0.7	4.5	3.4
$35 000–39 999	11.4	3.6	17.6	6.2	0.6	4.7	3.0
$40 000–49 999	9.4	3.8	15.6	5.7	0.7	4.8	2.7
$50 000+	6.6	3.2	11.4	5.4	0.7	4.1	1.9

consumption increases at higher rates than income increases are highly elastic. Goods that show some degree of positive elasticity (consumption rises when incomes rise) are called normal goods. The purchase of some commodities may actually decrease as incomes rise; these are known as inferior goods. Inferior goods are often associated with social class. For example, participation in tent camping as a vacation tends to decline with increasing incomes (after peaking at lower middle income levels).

Ernst Engle, a German statistician, was among the first analysts to formalize observations about the relationship between incomes and consumption. Engle observed that as incomes in his society rose: (1) the percentage spent on food fell; (2) the percentage spent on lodging and clothing remained about the same; and (3) the percentage spent on all other goods rose. Current census information allows us to test the applicability of Engle's Law, as his observations have become known, to contemporary society. Consider the data in Table 6.2, family expenditure data for Canada (Statistics Canada 1984). As Engle observed, the percentage of income spent on food shows a clear decline with increasing income. The pattern is especially strong with food purchased from stores, rather than food purchased from restaurants. In contrast with Engle's observations, the percentage spent on housing also decreases, as does the percentage spent on clothing. Expenditure on air transportation varies relatively little across income categories. Commodities that are recreational in nature also tend to increase to middle incomes and then fall. Finally, the percentage of income spent on alcohol and tobacco products drops with increasing incomes. It should be noted that Table 6.2 does not include all expenditures. Higher income Canadians generally must allocate a much higher percentage of their incomes on taxes; they also tend to put higher percentages of their incomes into tax shelters, such as retirement savings plans.

There is one other aspect of demand important in tourism analysis: consumer surplus. Consumer surplus refers to the value received by a consumer above what he pays for a commodity. It is the basis for inferring demand curves for commodities that are not distributed throughout the market-place and thus do not carry prices set by the market. Consumer surplus and its relationship to demand and resource valuation is examined in greater detail in Chapter 9.

Trend extrapolation: simple regression analysis

Description

One of the simplest but most useful methods of forecasting is trend extrapolation through simple regression. As the name suggests, the basic logic of trend extrapolation is that some past trend, such as visitation rates to a particular destination, will continue into the near future.

Simple regression is a method for correlating two variables against each other – in this case, time and some measure of tourism demand. To use simple regression, you must work with two interval scale variables. The form of a simple regression model is:

$$Y = a + bX \qquad [6.1]$$

where: Y = dependent variable reflecting a measure of tourism demand;
a,b = coefficients to be statistically calibrated;
X = independent variable associated with changes in tourism demand.

Y is usually a variable such as the number of visitors at an attraction, the number of international border crossings, total receipts, or the number of scheduled airline flights. X can be any variable that influences demand, such as price, income, advertising budgets and so on. However, in the case of trend extrapolation, X is some measure of time, usually quarters or years. If changes in demand are fairly stable over time, a reasonably accurate forecasting model may be developed by correlating demand with time.

Regardless of the independent variable selected, the process of making a forecast is the same. The coefficients in equation [6.1] are estimated using a procedure known as least-squares estimation (described below) using historical data. Once the model has been calibrated, you substitute an expected future value of the independent variable (usually obtained from an independent forecast) into equation [6.1] and solve for Y. This new value of Y is the forecast value of future demand.

The central problem in simple regression is the calibration of a linear function that best summarizes the trends in a set of data. The role of least-squares estimation can be described graphically. If a set of data describing two variables, say hotel occupancy and advertising expenditures, is plotted on a graph (Figure 6.3), least-squares estimation specifies the position of a line that best fits the scatter of points. The best fit is the line that minimizes the sum of the squares of the vertical distance between each point and the line. In other words, least-squares identifies the line that best approximates the overall trend in the scatter of points. This line may be interpreted as the plot of predicted values of the dependent variable given any particular value of the independent variable.

Two pieces of information are needed to define the position of the line: the point where the line intersects the ordinate and the slope of the line. The intersection of the line with the ordinate is represented by a in equation [6.1]; the slope is given by b. The equations for calculating a and b are given below. A more extended discussion of the theory and applications of regression analysis may be found in Draper and Smith (1966) or similar texts.

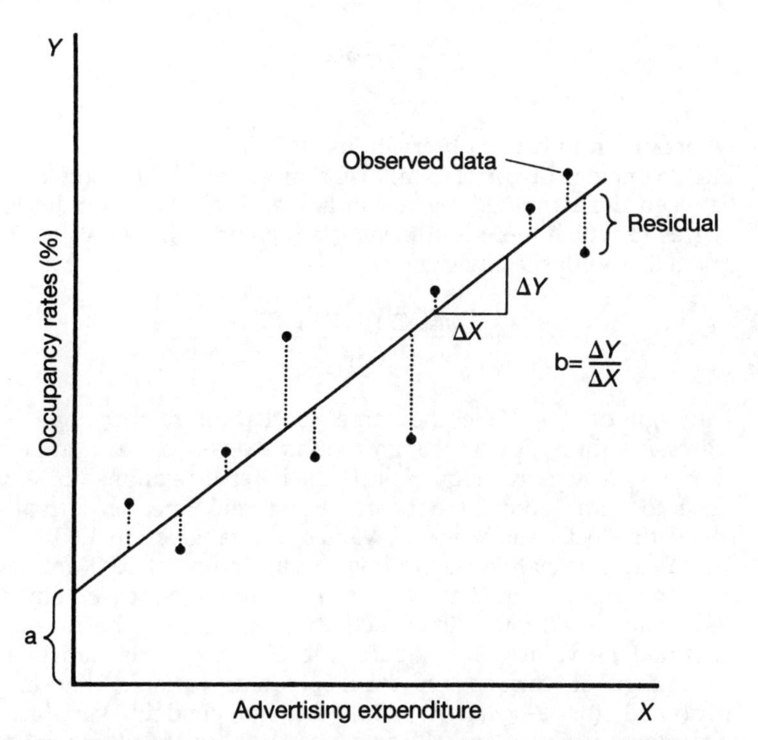

Figure 6.3 Regression line relating hotel occupancy rates with advertising expenditure

Procedure

1. Select an appropriate dependent variable and an independent variable measuring the passage of time. Designate the dependent variable as Y and the independent variable as X. Select appropriate units of analysis and collect data. A minimum of 10 to 15 observations is normally desirable.
2. Prepare a table similar to Table 6.3. The first column, X, contains values of the independent variable; the second column, Y, lists values of the dependent variable. The third column is the product of X and Y. The fourth and fifth columns are X^2 and Y^2 respectively. Calculate the sums of each column.
3. Calculate b with the equation:

$$b = \frac{n(\Sigma XY) - (\Sigma X)(\Sigma Y)}{n(\Sigma X^2) - (\Sigma X)^2} \quad [6.2]$$

4. Once you have a value for b, the value of a is obtained from:

$$a = \frac{\Sigma Y - b(\Sigma X)}{n} \qquad [6.3]$$

where: n = number of observations.

5 The coefficient of correlation, r, is a measure of the goodness of fit between the estimated regression line and the data. It indicates the degree to which there is a linear relationship between X and Y. It is calculated with the equation:

$$r = \frac{n(\Sigma XY) - (\Sigma X)(\Sigma Y)}{\sqrt{(n(\Sigma X^2) - (\Sigma X)^2)}\sqrt{(n(\Sigma Y^2) - (\Sigma Y)^2)}} \qquad [6.4]$$

The sign of r will be the same as that of b, the slope of the regression line. A positive sign reflects a direct correlation between X and Y; a negative sign reflects an inverse relationship. A value close to zero indicates a nearly horizontal line, or virtually no correlation between X and Y. Values of r range from 1.00 to –1.00. These extreme values as well as the midpoint of 0.00 are easy to interpret, but intermediate values are more frequently obtained and less easily interpreted. One method for assessing the meaning of intermediate values is to square r, to obtain r^2. This statistic is the percentage of variance in the dependent variable that may be attributed (or associated with) the independent variable. The extreme values of r^2 are 0.00 and 1.00. A value of 0.80 indicates that the independent variable 'explains' 80 per cent of the variation in the dependent variable. A model with an r^2 of 0.80 is four times as strong as one that produces an r^2 of 0.20.

The word, 'explains', is in quotation marks in the previous paragraph because the explanation is statistical – not necessarily causal. In other words, you are simply observing a certain degree of relationship between two variables – which may or may not represent a causal connection between the two. A common misuse of r^2 is to interpret a high r^2 as proof of a causal relationship. Whereas a low r^2 is evidence of a lack of a causal connection between X and Y, a high r^2 does not necessarily prove that such a connection exists. It is merely suggestive evidence that could be coincidental. The coefficient, r^2, is a measure of the strength of the **linear** hypothesis implied by equation [6.1]. Two variables may be strongly related through a curvilinear function, but r^2 (as we have defined it here) would fail to measure this because it is predicated on the assumption that the relationship is linear (see Figure 6.4).

6 Once the regression equation has been calibrated, select a value for the independent variable reflecting the anticipated future condition. Substitute this value for X in equation [6.1] and solve for Y.

If the values of X are in years, a short cut is possible to simplify the calculations of a and b. The trick is to replace the values of X as years with values that will cancel out when summed. If the number of years is even, replace the years with the values: . . . –3, –2, –1, 1,

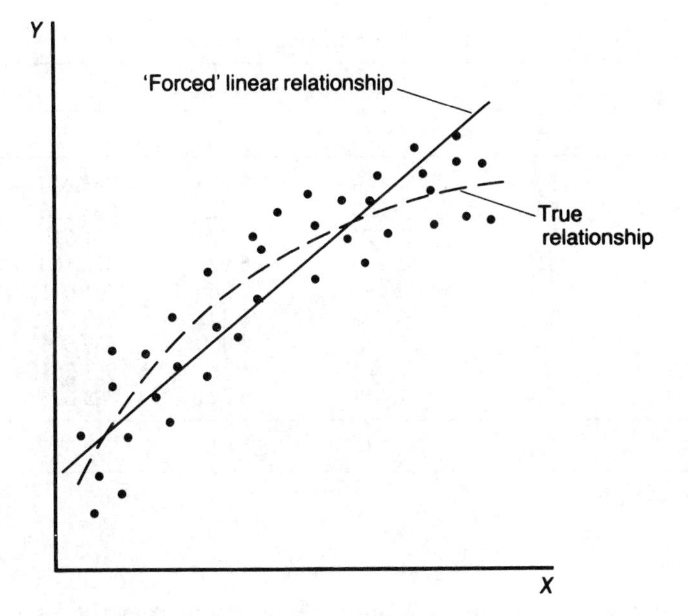

Figure 6.4 A curvilinear relationship

2, 3 . . . If the number is odd, use: . . . –3, –2, –1, 0, 1, 2, 3 . . . The equations for a, b, and r become:

$$a = \frac{\Sigma Y}{n} \qquad [6.5]$$

$$b = \frac{\Sigma XY}{\Sigma X^2} \qquad [6.6]$$

$$r = \frac{n(\Sigma XY)}{(\sqrt{n(\Sigma X^2)})\,(\sqrt{n(\Sigma Y^2) - (\Sigma Y)^2})} \qquad [6.7]$$

A forecast is then made by extending the series of codes to the future year. If your data extended from 1985 to 1994, these years would be recorded as:

1985 = –5	1990 = 1
1986 = –4	1991 = 2
1987 = –3	1992 = 3
1988 = –2	1993 = 4
1989 = –1	1994 = 5

A forecast to 1998 would involve recoding 1998 as 9. The rest of the forecast would follow the procedure described previously.

Table 6.3 Forecast of world-wide air passenger travel

Year	Millions of passengers			
X	Y	XY	X^2	Y^2
74	515	38 110	5 476	265 225
75	534	40 050	5 625	285 156
76	576	43 776	5 776	331 776
77	610	47 970	5 929	372 100
78	679	52 962	6 084	461 041
79	754	59 566	6 241	568 516
80	748	59 840	6 400	559 504
81	752	60 912	6 561	565 504
82	764	62 648	6 724	583 696
Σ = 702	5 932	464 834	54 816	3 992 518

Example

An illustration of the use of simple regression may be seen in the following model of world-wide air passenger travel. Table 6.3 is a listing of total world-wide airline passenger counts for 1974 through 1982, inclusive (American Express 1985). The value of b may be calculated using equation [6.2]:

$$b = \frac{n(\Sigma XY) - (\Sigma X)(\Sigma Y)}{n(\Sigma X^2) - (\Sigma X)^2}$$

$$= \frac{9\,(464\,834) - (702)\,(5932)}{9\,(54\,816) - (702)^2}$$

$$= 35.6$$

This positive value indicates a positive relationship between the number of air passengers and years, or, more precisely, increasing air travel over time.

The intercept, a, may be determined from equation [6.3]:

$$a = \frac{\Sigma Y - b(\Sigma X)}{n}$$

$$= \frac{5932 - 35.6\,(702)}{9}$$

$$= -2117.7$$

Finally, r may be calculated as:

$$r = \frac{n(\Sigma XY) - (\Sigma X)(\Sigma Y)}{(\sqrt{(n(\Sigma X^2) - (\Sigma X)^2})(\sqrt{n(\Sigma Y^2) - (\Sigma Y)^2}))}$$

$$= \frac{9\,(464\,834) - (702)\,(5932)}{\sqrt{(9\,(54\,816) - (492\,804))}\,(\sqrt{(9\,(3\,992\,518) - (35\,188\,624))})}$$

$$= 0.962$$

This value is positive, as expected given a positive value for b. The value also indicates a strong linear relationship. If we calculate r^2, we obtain 92.5 per cent, which is quite strong. Given such results, we can feel comfortable in using the model to forecast future airline travel. The forecast model is obtained by substituting the estimate parameter values into equation [6.1]:

Passengers = –2117.7 + 35.6 (year)

To make a forecast to 1984, we substitute 84 for 'year' and solve:

Passengers = –2117.7 + 35.6 (84)

= 873 (million)

Data for 1984 permit a check of the accuracy of the forecast. The actual volume was 832 million. This represents an error of about 4.9 per cent, which is acceptable for most tourism marketing forecasts.

Gravity model

Description

The gravity model is a well-known structural forecasting model. As its name suggests, the gravity model is based on an analogy to Newton's law of gravitation. Newton's law states that the gravitational attraction between any two bodies in the universe is directly proportional to their masses and inversely proportional to the square of the distance between them:

$$I_{ij} = \frac{GM_iM_j}{D_{ij}{}^2} \qquad [6.8]$$

where: I_{ij} = gravitational attraction between two bodies, i and j;
G = gravitational constant;
$M_{i,j}$ = masses of i and j;
D_{ij} = distance between the centres of gravity of i and j.

This rather simple formula has been the inspiration for a growing body of travel and interaction models in the social sciences. Interaction in this context refers to any form of interchange between groups. This may be financial flows, telephone calls, mail volumes, marriages, visits or any of literally hundreds of other variables. The masses of the social groups may be represented in terms of population, relative wealth, retail floor space, destination attractiveness and many other variables. Distance is usually measured in terms of physical separation, but travel time or indices of social distance (class or distinctions) can also be used.

Crampon (1966) was the first to demonstrate explicitly the usefulness of the gravity model in tourism research. He introduced the form of the social gravity model that has become the standard in travel research:

$$T_{ij} = \frac{GP_iA_j}{D_{ij}{}^a} \qquad [6.9]$$

where: T_{ij} = a measure of tourist travel between origin i and destination j;
G, a = statistically estimated coefficients;
P_i = a measure of the population size, wealth, or propensity to travel at origin i;
A_j = attractiveness or capacity of destination j;
D_{ij} = distance between i and j.

As with other structural models, equation [6.9] must be calibrated with historical data. If this calibration is done successfully, not only do you have a forecasting model, but the coefficients G and a may have some intrinsic interest. The value of a, for example, reflects the relative strength of distance as a deterrent to travel. The larger the estimated value of a, the greater the effect of distance in reducing the number of trips. The value of G is less easily interpreted. In Newton's model, G is a universal constant – one of the four universal constants that shape the structure of the universe. For tourism, G is a proportionality constant that adjusts the magnitude of the other variables so that they explain as closely as possible the observed level of tourism activity, T_{ij}. The relative values of G in different modelling situations might contain some meaning of use to tourism analysts, but the subject has not been explored.

The most important reason for developing a gravity model is not to replicate observed travel patterns or to examine the magnitudes of a and G, but to make forecasts. Given estimates of future values of P_i and A_j, and assuming a, G and D_{ij} remain constant, you can predict future levels of tourism trips.

The gravity model is used to forecast trips between a single origin and a single destination within a specified time period. If you want to forecast trips for multiple origins and/or multiple destinations, you can calibrate a separate model for each origin–destination pair. You should, however, be aware of two concerns with this approach. First, simply calculating trips for an increasing number of origins and destinations implies that there is no competition among the destinations. In other words, you are assuming implicitly that the number of trips generated by a specified set of origins can increase without limit as the number of destinations increases. This situation is described as being unconstrained and is, in most situations, unrealistic. Second, the number of calibrations increases quickly with more origins and destinations. A set of forecasts for the ten Canadian provinces and two territories, the 50 states of the USA, and the District of Columbia would require (63 origins × 63 destinations =) 3969 separate equations.

Before we turn to the procedures section, we should consider further some of the limitations in gravity models as well as some of the modifications that have been proposed to overcome them.

One weakness in the basic gravity model has already been mentioned: equation [6.9] is unconstrained. There is no upper limit to the number of trips the equation can forecast. If, for example, you were calibrating a model to predict travel to Cancun, Mexico, and you were using the number of hotel rooms as the attraction component, your model would predict a doubling of the number of visitors every time you doubled the number of hotel rooms. Unlimited visitation rates do not occur. There is some upper limit to the number of trips a given destination can expect to receive or that a given population can make in one year. The solution to this problem is to develop a constrained gravity model in which a realistic upper limit is identified. Constrained gravity models usually consist of two components. The first estimates the total number of trips that can be generated under specified conditions; the second allocates those trips to alternative destinations. A common form of the trip-generation component of a constrained gravity model is a simple regression equation relating income, population size, or mobility to the total number of trips expected. A multiple regression equation might be developed to combine several independent variables into a more accurate and comprehensive model. The expected number of trips is then fed into the trip distribution component to allocate the trips among competing sites. An example of a trip-distribution model is described later in this chapter, under the section on the probabilistic travel model.

A long-standing criticism of gravity models has been that they have no theoretical basis. This criticism was historically correct but is irrelevant and no longer valid. Stewart (1948) and Zipf (1946), who independently developed the concept of the gravity model, based their formulations explicitly on an analogy to Newton's law of gravitation. Although their work had no theoretical basis, the original gravity models and subsequent refinements have been as successful or more successful than models derived directly from theory in empirically predicting travel flows. Further, Niedercorn and Bechdoldt (1966) were able to demonstrate that the gravity model can be derived from existing economic theory. They demonstrated that the gravity model is a logical and theoretically sound solution for the problem of maximizing individual satisfaction subject to time and budget constraints.

One reason that the gravity model continues to be a popular

structural forecasting tool is that it allows for substantial refinement and modification. The history of forecasting tourism and recreational travel is, to a great extent, the history of the development of gravity models. One part of the gravity model that has undergone extensive development and experimentation is the distance variable. This experimentation stems from another weakness of most gravity models: a tendency to over-predict very short trips and to under-predict very long trips (Martin *et al.* 1961). One of the first researchers to address this problem in recreation and tourism was Whitehead (1965). He suggested adding a second distance term to allow for feedback between the distance travelled and the response to that distance (willingness to travel further):

$$T_{ij} = G\Sigma\left(\frac{P_i P_j e^{-\lambda D_{ij}}}{D_{ij}^b}\right) \qquad [6.10]$$

where: $P_{i,j}$ = population of origin and destination zones, respectively;
e = natural logarithm base;
λ = coefficient to estimated;
other symbols are as defined previously.

Wolfe (1972), while agreeing with Whitehead on the problem, suggested a different strategy for overcoming the difficulty. Drawing again from physics for an analogy, Wolfe proposed incorporating an 'inertial' term in the gravity model. This term would have the effect of lowering the predicted number of short trips to reflect the 'inertia' of starting up, while increasing the number of long trips predicted, reflecting the effect of 'momentum'. Wolfe's model is of the form:

$$T_{ij} = G\left(\frac{P_i A_j}{D_{ij}^b}\right)\left[D_{ij}^{\,[(\log D_{ij}/m)/n]}\right] \qquad [6.11]$$

where: m, n = coefficients to be estimated;
other variables are as defined previously.

A still more elaborate modification of the distance variable was proposed by Edwards and Dennis (1976). Their approach was to define a distance measure more generalized than physical distance alone:

$$C_{ij} = \left[\frac{[(X_1)(X_2)(X_3)] + (X_4)}{X_5}\right] X_6 \qquad [6.12]$$

where: C_{ij} = cost of travel between i and j;
X_1 = cost of gasoline per litre;
X_2 = litres consumed per kilometre;
X_3 = average kilometres travelled per hour;
X_4 = value of leisure (travel) time, set at 25 per cent of hourly wage;

X_5 = number of people per car;
X_6 = total travel time.

This variable was then incorporated into the model:

$$T_{ij} = P_i A_j \exp(-\lambda C_{ij}) \quad [6.13]$$

Edwards and Dennis next proposed a modification of the recreational 'pull' or attractiveness component, A_j:

$$S_{ij} = A_j \exp(-\lambda C_{ij}) \quad [6.14]$$

where: S_{ij} = pull of destination on travellers at origin i.

The total number of trips generated between all origins and all destinations is then defined as:

$$T_{ij} = P_i S_i^{\ b} \quad [6.15]$$

where: S_i = ΣS_{ij}; or if expressed as a constrained gravity model:

$$T_{ij} = T_i \left(\frac{S_{ij}}{S_i} \right)$$

where: T_{ij} = total number of trips to be generated at origin i.

Combining equations [6.12], [6.14] and [6.15] yields:

$$T_{ij} = P_i S_i^{\ b} A_j \exp(-\lambda C_{ij}) \quad [6.16]$$

The calibration of equation [6.16] involves statistical estimation procedures beyond the scope of this book. None the less, Edwards and Dennis's work illustrates how the simple gravity model has become the seed for increasingly sophisticated and powerful forecasting models, each building on earlier versions. Additional examples of the development of gravity models in tourism and recreational travel forecasting can be found in Smith (1983).

Procedure

1 Specify the origin–destination pairs and a relevant time period for data collection. Collect data on: (1) the total number of trips from the origin to each destination; (2) the population of the origin; (3) the capacity or attractiveness of the destination; and (4) some measure of the distance between origin and destination.
2 Define the per capita trip rate for the origin's population:

$$k = \frac{\Sigma T_{ij}}{P_i} \quad [6.17]$$

where: k = per capita trip rate;
ΣT_{ij} = number of trips to all destinations by residents of i;
P_i = population of i.

3 Calculate the total attractiveness of all destinations by summing the attractiveness measure of each destination:

$$A = \Sigma A_j \quad [6.18]$$

where: A = aggregate attractiveness of all destinations;
A_j = individual attractiveness of each destination.

4 Calculate the expected number of trips for all travellers from the origin to each destination under the assumption that distance has no effect:

$$V_{ij} = \frac{kP_iA_j}{A} \quad [6.19]$$

where: V_{ij} = expected number of trips.

5 Calculate the effect of distance on the expected number of trips by dividing the actual number of trips by the expected number:

$$T_{ij}/V_{ij}$$

where: T_{ij} = trips from each single origin, i, to the destination being modelled.

6 Obtain a measure of the distance between each origin and each destination and do the following regression:

$$\log \frac{T_{ij}}{V_{ij}} = \log a + b\,(\log D_{ij}) \quad [6.20]$$

7 Remove logs, substitute using equation [6.19], and let $G = \alpha k/A$ (where α = antilog a) to obtain:

$$T_{ij} = \frac{GP_iA_j}{D_{ij}^b} \quad [6.21]$$

8 To make a forecast, substitute predicted values for P_i and A_j for the values used in calibrating the model. Solve for predicted T_{ij}. This procedure may be used for either future travel patterns between an existing origin and destination, or the model may be calibrated using an existing origin and destination pair and then applied to a different but similar pair.

Table 6.4 Ontario vacation travel data used for calibrating a gravity model

	Trips T_{ij} (000s)	Distance D_{ij} (km)	Population A_j (000s)	V_{ij} (000s)	T_{ij}/V_{ij}	Predicted T_{ij} (000s)
Newfoundland	[a]	2100	579	338.48	0.01[b]	18
Prince Edward Island	67	1400	124	72.49	0.92	8
Nova Scotia	90	1300	853	498.66	0.18	61
New Brunswick	85	1200	708	413.90	0.20	58
Quebec	865	500	6 314	3 691.16	0.23	2 222
Ontario	12 542	200	8 547	4 996.58	2.48	13 939
Manitoba	185	1400	1 029	601.55	0.30	65
Saskatchewan	[a]	2100	971	561.65	0.01[b]	31
Alberta	50	2500	2 083	1 217.72	0.04	50
British Columbia	58	3100	2 642	1 554.51	0.04	44

[a] Data not released due to high variance.
[b] 0.01 used arbitrarily to avoid missing cases; actual value is 0.01 or less.
$P_i = 8547$.
$A = \Sigma A_j = 23\,850$.
$k = \Sigma T_{ij}/P_i = 1.63$.

Example

The following example illustrates the development of a gravity model to travel by residents of Ontario to other Canadian provinces. The initial step was to collect data on travel volumes between Ontario and the nine other provinces as well as intra-provincial travel for a recent year. These cross-sectional data, shown in Table 6.4, were obtained from Statistics Canada. Note that the number of trips to Newfoundland and Saskatchewan was too small to be released. Statistics Canada, as many government agencies, has policies prohibiting the release of data if the sampling variance exceeds some specified maximum. The purpose of this policy is to prevent the release of data whose variance is so great that they are virtually useless for scientific purposes. In the case of Statistics Canada, the maximum variance is 33.3 per cent; Newfoundland and Saskatchewan's data had variances greater than this due to the small number of observations on which the trip estimates were based.

Great circle distances between the centre of population of Ontario and every other province (D_{ij}) were obtained. Distance for vacation travel within Ontario was estimated at an average of 200 km one-way. This information is also shown in Table 6.4. The attractiveness of the destinations, A_j, was assumed to be proportional to their population, reflecting the fact that the single most popular motivation for travel in Canada is to visit friends and relatives. Population estimates were obtained and are presented in Table 6.4. The total attractiveness, A, is the sum of these.

Per capita travel, k, was calculated at 1.63 using equation [6.17]. Following the steps in the procedure, V_{ij} was estimated; the ratio, T_{ij}/V_{ij}

was also obtained. Note that a value of 0.01 was used arbitrarily for Newfoundland and Saskatchewan, necessitated by the fact that T_{ij} is unknown. The actual value of T_{ij} is small, so the estimate of 0.01 is not likely to be too far in error.

Table 6.5 Calibration results for Ontario vacation travel gravity model

Dependent variable: log $[T_{ij}/V_{ij}]$
Intercept (a) = 4.2876; antilog (a) = α = 19 390.99
b for log (D_{ij}) = −1.67
r^2 = 0.99
Standard error (as % of mean Y) = 24.3

$$G = \frac{\alpha k}{A} = \frac{(19\,390.99)\,(1.63)}{23\,850} = 1.32$$

T_{ij}/V_{ij} was regressed against D_{ij}. The results are summarized in Table 6.5. Following the substitution in equation [6.21], and after obtaining the value of G, the following gravity model was estimated:

$$T_{ij} = \frac{1.32\,(P_i)\,(A_j)}{D_{ij}^{\;1.67}} \qquad [6.22]$$

where: T_{ij} = predicted numbers of vacation trips between Ontario and province j;
P_i = population of Ontario;
A_j = population of j;
D_{ij} = great circle distance between Ontario and j.

With this equation, it is possible to predict *ex post facto* (i.e. to attempt to replicate the original observations) the number of vacation trips from Ontario to each province. This prediction (or replication) will provide us with a measure of how accurately our model reproduces the original data. The predicted results are shown in Table 6.4. If we regress predicted trips against observed trips, we obtain an r^2 of 98.9 per cent. This is a very high correlation coefficient – much higher than might be expected given the data in Table 6.4. The reason for the high r^2 is the high values associated with Ontario's intra-provincial data. These values, approximately 10 times higher than those of Quebec and 100 times higher than those of other provinces, distort the value of r^2. A more meaningful comparison of the predicted and observed values may be obtained by examining the root-mean-square error as a percentage of the mean observed number of trips. The root-mean-square error is a statistic produced by most regression software. If we divide the root-mean-square error by the mean of our dependent variable, we obtain an estimate of the standard error of estimate as a percentage of the mean number of trips. For our example, the root-mean-square error was 500.11; the mean number of trips was 2055.88. The ratio, 500.11/2055.88 gives a standard error of estimate of 0.243 or 24.3 per cent. This is the expected mean error of this model's forecasts. Whether this level of accuracy is acceptable is for the analyst to decide based on the context of the forecasting situation. For the purposes of comparison, it is worth noting that a standard error of estimate of 20 to 25 per cent is common for tourism and recreational travel gravity models.

A forecast of future travel may be made by using predicted values of P_i and A_j instead of the observed values. If we predict that the population of Ontario is likely to rise to 8.8 million in some particular year, and that of Alberta will become 2.4 million, we can predict the travel from Ontario to Alberta as:

$$T_{ij} = \frac{(1.32)\,(8800)\,(2400)}{2500^{1.67}}$$

$$= 58.98$$

or about 60 000 trips. Note that the values of P_i and A_j are recorded in thousands and that the model was calibrated using these figures reported in thousands.

This example highlights several issues. The first concerns the measure of distance. The use of great circle distances is convenient, but more sophisticated measures could be developed. For example, surface travel from Ontario to Prince Edward Island or Newfoundland requires the use of sea-going ferries. Some additional weight might be added, therefore, to travel to these provinces to reflect the added inconvenience and cost of ferry travel. Quebec is a francophone province and Ontario is anglophone. The cultural differences between the two provinces make Quebec an attractive destination for many Ontario residents. On the other hand, the open hostility of some Quebecois to anglophones and the rise of a strong separatist movement in that province have made Quebec more 'distant' to many anglophones. The position of the Great Lakes means that automobile travel from southern Ontario to the western provinces must take a circuitous route. The great circle distance from Ontario's centre of population to that of Manitoba is 1400 km, but the road distance is 2000 km.

The pattern of predicted versus observed trips, clearly illustrates the tendency of the gravity model to over-predict short trips (Ontario and Quebec) and under-predict long trips (British Columbia and the Maritimes). Modifications such as those proposed by Wolfe (1972) may help correct this problem.

The use of population as the measure of attractiveness is, of course, simplistic. More comprehensive measures including accommodation, events, attractions and climate would probably add greater accuracy to the estimates of provincial attractiveness. Chapter 8 includes a method that can be used to estimate the attractiveness of tourism regions using a variety of characteristics; such an aggregate measure could be developed for each province.

Finally, this model is an unconstrained model. The potential problems associated with this type of model have already been described. We have also referred to the concept of a constrained gravity model. One of the two components of a constrained gravity model is a trip-distribution model. A common form of a trip-distribution model is the probabilistic travel model, which we now examine.

Probabilistic travel model

Description

The probabilistic travel model is another example of a structural forecasting model. This particular model differs from the unconstrained gravity model described in the previous section in that forecasts are expressed in terms of probabilities or percentages of total trips rather than as numbers of trips. This model can be combined with a trip-generation model to develop a constrained gravity model.

The model described here is a derivation of a consumer choice model originally developed by Luce (1959) and applied to recreational travel by Wennergren and Nielsen (1968). This model is based on the argument that the probability of a consumer selecting a particular product such as a tourist destination is directly proportional to the 'utility' of that product with respect to all alternative products. An important advantage of this model for tourism is that it allows the analyst to avoid the unrealistic assumption that a tourist will always go to only the most 'desirable' destination (however 'desirable' is defined), ignoring all other destinations. The model 'permits' the same traveller to go to different destinations. Specifically, the model assigns a probability to each destination expressing the odds that the average traveller will select that particular option. Because the probabilities add up to 1.00 for all destinations in a set of competing destinations, the probabilities may also be interpreted as the expected market share of each tourism product or destination – if the definition of utility has adequately included all relevant variables. Recall our discussion in Chapter 4 about the potential problems in inferring market share for different destinations when those destinations differ greatly in size.

As just implied, the central issue associated with the use of this model is the definition of utility. Utility reflects more than just attractiveness; it also includes the effects of cost, capacity, and access limitations. Although many characteristics affect utility, you are limited – in this model – to those that can be measured on an interval scale. Wennergren and Nielsen, in their analysis of boaters' choices of reservoirs, used reservoir surface and travel distance. Although the authors recognized that other variables may influence a boater's choice of a reservoir, these two variables were believed to be the most important and the most reliably measurable. Their results – a model that explained 80 per cent of the variance in boater attendance – suggests their reasoning has merit.

Before we examine the procedures to be followed in developing a probabilistic travel model, it will be helpful to identify several assumptions implicit in such models. First, the model is based on the assumption that travellers from any origin are homogeneous in their tastes, their willingness to travel, and in their perceptions of utility. Or to put this differently, it is assumed that the average traveller in any region is an adequate surrogate for all travellers from that region. This assumption may be relaxed by developing separate models for different types of travellers, such as different income strata. The model structure, though, remains the same; you simply calculate a larger number of separate probabilities.

The model is also based on the assumption of equal (not necessarily perfect) knowledge. All destinations in the set being considered by travellers are assumed to be equally familiar to the potential travellers. This, too, may be unrealistic. The assumption might be relaxed if you are able to estimate weights that can be used to adjust probabilities to reflect different knowledge levels. This may be a case of 'easier said than done'. For some travellers, lack of knowledge may translate into a greater desire to visit – because of the desire to explore new destinations – while for others, lack of knowledge may translate into a lower probability of visiting.

The importance of finding a valid measure of utility is obvious in this procedure. The challenge involves not only the identification of the relevant variables but also the proper specification of how they are to be combined. A common assumption is that variables are combined multiplicatively, with exponents of 1.0. Other structures or variable transformations are possible. There is empirical evidence to suggest that distance, when used as part of a gravity model, should carry an exponent somewhere between 1.0 and 2.0.

Finally, it will be helpful to describe a test of the explanatory power of this model. The accuracy of the calibrated model may be estimated using the formula:

$$r^2 = 1 - \frac{-\Sigma(P_{ij} - \hat{P}_{ij})^2}{\Sigma(P_{ij} - \overline{P}_{ij})^2} \qquad [6.23]$$

where: r^2 = the coefficient of determination;
P_{ij} = actual percentage of trips made by travellers from i to j;
$\overline{P}_{ij}$ = average percentage of trips made by travellers from i to all destinations;
$\hat{P}_{ij}$ = predicted percentage of trips made by travellers from i to j.

The interpretation of r^2 in equation [6.23] is the same as r^2 calculated in simple regression.

Procedure

1 Develop a quantitative measure of destination utility, incorporating both positive and negative qualities relevant to the travel system you are studying. The definition of utility must include not just the variables that affect utility, but also how these variables are to be

combined. Wennergren and Nielsen (1968) suggested that their variables, reservoir size and distance between origin and destination, should be combined multiplicatively:

$$U_i = \frac{S_j}{D_{ij}} \quad [6.24]$$

where: U_j = utility of reservoir j;
S_j = surface areas of j in acres;
D_{ij} = distance from i to j in miles.

2 After deciding on an appropriate measure of utility, collect the data necessary to calculate utility for every destination. Record the data and calculate total utility by summing the individual utility measures: ΣU_j.
3 Determine the probability of a traveller choosing any particular destination by dividing the utility of that destination by the total utility of all destinations:

$$P_{ij} = \frac{U_j}{\Sigma U_j} \quad [6.25]$$

where: P_{ij} = probability that a traveller from i will select j.

Example

Assume we are interested in predicting the probabilities that travellers from a city in southern Ontario, Canada, will choose each of three resorts along the American coast of the Gulf of Mexico. In this example, let the distance from the origin to each destination be equal. Further assume that the amenities and quality of service in the three resorts are closely correlated with their price. Finally, assume that the major difference in the resorts, besides cost, is their capacity. The utility of each resort can thus be defined as:

$$U_i = \frac{C_j}{R_j} \quad [6.26]$$

where: U_j = utility of resort j;
C_j = number of rooms at resort j;
R_j = average daily room charge at j.

The probability that the average traveller from i will choose any particular resort is:

$$P_{ij} = \frac{C_j/R_j}{\Sigma(C_j/R_j)} \quad [6.27]$$

Table 6.6 contains the capacity and room rates for three hypothetical resorts. The utility of the resort identified as '1' is equal to the capacity of the resort divided by the average room rate: 250/90 = 2.78. The same calculation is performed for the other resorts, and a total utility is derived by summing the individual utilities: 8.36. The probability that a traveller from *i* will select the first resort is 2.78/8.36 = 33.2 per cent. The probabilities of selecting other resorts may be calculated in the same way.

Table 6.6 Hypothetical resort data for calibration of a probabilistic travel model

Resort	Daily room rate R_j (\$)	Number of rooms C_j	$U_j = C_j/R_j$
1	90	250	2.78
2	100	300	3.00
3	120	310	2.58
		$\Sigma(Cj/Rj)$ =	8.36

$$P_{i1} = \frac{2.78}{8.36} = 33\%$$

$$P_{i2} = \frac{3.00}{8.36} = 36\%$$

$$P_{i3} = \frac{2.58}{8.36} = 30\%$$

If desired, one can also develop a multiple regression model to predict the number of trips that travellers from origin *i* are likely to generate to all three resorts, thus producing a two-component model that is equivalent to a constrained gravity model.

Delphi technique

Description

The Delphi technique is one of the best known and sometimes more controversial forecasting methods. It was pioneered by the RAND Corporation in the early 1950s (Dalkey and Helmer 1963) as a method for forecasting events when historical data are unavailable or when the forecast requires significant levels of subjective judgement.

The technique depends on a panel of experts, assembled by the analyst, who respond to several rounds of carefully constructed questionnaires. The questionnaires are designed to move the panel to a consensus on the identity, probability and timing of future events. The panel may be assembled face-to-face, through telephone conference calls or a computer network, but the use of mail questionnaires is probably the most common. A major advantage of the mail format is that it avoids the potential biasing effects of peer or committee pressure and other psychological influences on the respondents' answers. Most panels consist of 40 to 50 experts, although Brockhoff (1975) successfully used a panel of only four members in a study of computer technology, while Shafer *et al.* (1974) worked with a panel of 904 respondents in their study of future recreational environments.

Delphi, like other forecasting models, begins with a question about the future. This question often concerns qualitative trends or the emergence of discoveries or other unprecedented events that cannot be studied with conventional extrapolation, structural or simulation tools. Delphi is often selected, therefore, as the forecasting tool of last resort. This feature is perhaps Delphi's greatest attraction, but it does have several other strengths.

The method brings together, in a controlled format, experts with diverse abilities who can challenge one another's assumptions and arguments, and complement each other's strengths. Delphi is relatively simple to conduct, with no particular need for computer or statistical skills. This quality, though, is deceptive. Some researchers naïvely approach Delphi, thinking it to be 'quick and dirty'. In practice, it can take from nine months to two years to recruit a panel, administer several rounds of questionnaires, and reach a consensus. The administrative costs and complexities of Delphi can be significant, and grow rapidly with the size of the panel and the length of the project.

The success or failure of a Delphi forecast depends on the qualifications of the experts and the skill of the researcher in designing and administering the questionnaires. Personal but unintentional biases on the part of the researcher can be incorporated into the wording of the questionnaire or the analysis of the results. The stability of the panel is also important. If the exercise lasts a year or more, some panel members are likely to drop out. If too many leave, the validity of any consensus is suspect.

The predictions made by experts are typically expressed in percentage probabilities. Such probabilities, however, are subjective judgements and do not have any of the properties that true, quantitative probabilities exhibit. The use of numerical probabilities can impart an air of unwarranted precision and scientific rigour to what are ultimately personal opinions. The use of Delphi normally does not permit various events to be related to each other in any systematic fashion. In other words, if forecasts are made about ten different events that will influence the future of tourism, Delphi treats these events as discrete and independent, with no interaction among them. This is not necessarily a valid assumption. Some panel members may attempt to incorporate such interactions, but there is no way to ensure that all do in a consistent fashion.

The most successful Delphi forecasts concern technological or scientific developments. Experience with applying Delphi to social interactions, human values, political changes, economic growth, and the dynamics of the market-place have been much less successful.

Given these limitations, it should be apparent that Delphi is not foolproof. It may be the method of last resort when it comes to forecasting, but Delphi cannot be counted on to produce successful forecasts. None the less, the method may have value in helping tourism planners and policy analysts anticipate possible future trends when no other forecasting tools are available.

Procedure

1 Define the forecasting problem and assemble the panel of experts. Problem definition is a crucial step in any forecasting problem, and especially Delphi. A period of time devoted to reading, thinking, and identifying possible panel members are important parts of this initial step. The panel should include individuals from a wide range of backgrounds to ensure representativeness and comprehensiveness of insights. For most tourism applications, panel members should come from both the private and public sector as well as from universities, consulting companies, and perhaps from related areas such as conservation districts, visitor and convention bureau, investment firms and advertising agencies. The size of the panel is determined by the number of available experts, the complexity of the problem, and the resources for managing the panel; many panels have 40 to 50 members.
2 Develop and distribute the first-round questionnaire. This first questionnaire is designed to introduce the general area of study and to invite the panel to identify possible future events, the probability that they will occur, and/or the likely date of their occurrence. One method for framing questions is to ask panel members to identify those events that can conceivably occur within some specified time period, say 20 years. They are then asked to indicate the date by which they believe the event has at least a 50 per cent chance of occurring.
3 Once the first-round questionnaires have been returned, the results are tabulated and summarized. The summary includes the median date (the date that divides the range of forecasts in half). The two middle quartile ranges are also frequently noted – the range around the median date that encompasses 50 per cent of total forecasts.
4 Summary statistics from the first round are presented in the second-round questionnaire, which is then mailed to the same panel of experts. Copies of their first-round questionnaire may also be provided for reference. Each panel member is then asked whether he wishes to change his forecast in light of the group's statistics. Those whose personal forecasts fell outside the two middle quartiles are also invited to explain the rationale for their predictions if they chose not to alter their original positions.
5 The results of the second round are tabulated and summarized. These results now include both new predictions as well as comments from some members about why they do not agree with the emerging consensus.
6 Summary results and comments from the second-round questionnaire are incorporated into a third round. The instructions with the third-round questionnaire are similar to those in the

second round. The major difference is the addition of arguments for dissenting forecasts.

7 Once the results of the third-round questionnaire are in, you need to decide whether a fourth round is desirable to refine the consensus of your respondents. One of the criteria you may use in making this decision is whether there appears to be much chance of the respondents making further alterations in their positions. The spread in the forecasts for any particular event can also indicate the need for further rounds. If most forecasts are already tightly clustered around the median date after the first two rounds, there is little point in prolonging the exercise. On the other hand, if the predictions show a high degree of variance, a third and even a fourth round, with special emphasis placed on explanations of differing viewpoints, might help move members to a consensus.

Upon completion of the final round, the results are summarized. These results include median dates, inter-quartile ranges, and the identity of those events for which no consensus was reached.

Example

Shafer *et al.* (1974) used the Delphi technique to identify future developments likely to occur in the US that would influence park and recreation management. A panel of 904 experts including recreation and park managers, biological scientists, demographers and environmental technologists was formed (the panel eventually shrank to 405 as members dropped out). They were asked in the first round to list those events they believed had a 50 per cent probability of occurring by the year 2000 and to estimate the most probable date by which the event would occur.

After circulating the results of the first round in the second-round questionnaire, the authors discovered that the consensus of the panel regarding certain events was that they were likely to occur after the year 2000. As a result, they dropped 2000 as a cut-off date and retained all events and dates. Two more rounds of questionnaires were distributed, using the basic procedure described above. The complete results are summarized in the authors' report, but a summary of some of the tourism-related predictions illustrates the type of results produced. These are presented without comment, other than to note that they were made by an American panel in 1973–74.

By 1980

1 Computers will be used to advise recreationists where to go.
2 Interpretive material on flora and fauna as well as historic sites will be available at a majority of public sites.

By 1985

1 Tax credits will be created for private landowners to protect scenic resources.
2 Cable television will be available at a majority of campgrounds.

3 Use of wilderness will be restricted.
4 Special fishing areas will be established in urban areas for disabled, elderly and youth.

By 1990
1 Year-round skiing on artificial surfaces.
2 Salt-water fishermen will be required to have federal fishing licences.
3 National campground reservation system for public parks.
4 Public schools will open year-round with staggered vacations.
5 Most homes will have video-tape systems.

By 2000
1 Eight hundred kilometres will be considered a reasonable one-way distance for weekend pleasure travel.
2 Average retirement age will be 50.
3 Middle-class American families will take vacations on other continents as commonly as they take vacations in the US in the 1970s.
4 Electric power or other non-polluting engines will replace internal combustion engines in recreational vehicles.
5 Travel in large parks will be limited to low-impact mass transit, e.g. tramways, air transport and underground rapid transit.

By 2020
1 Man-made islands will be created solely for tourism and recreation.
2 Most metropolitan areas will provide adequate outdoor recreation so the majority of their residents do not feel the need to go to the country for recreation.

By 2030
1 Most middle-income Americans will own vacation homes.

After 2050
1 First park on the moon.
2 Fees at public recreation areas will be set to cover capital and maintenance costs.
3 Self-contained underwater resorts.
4 Average life-span will be 100 years.

Summary

Forecasting demand is an important task in tourism analysis. The process of forecasting in tourism is a bit like a game of golf. A golfer may have to approach the green along a path that may have a sharp dog-leg and several bunkers or other hazards. The approach is taken in a series of steps, with a different club for different shots. In a similar fashion, a forecaster rarely has a clear and straight shot at the ultimate goal. He must approach his goal, an accurate prediction, in a series of steps, avoiding traps while attempting to improve the accuracy of his forecasts by using different tools at each step.

There are a range of tools available to the forecaster, including both qualitative and quantitative methods. We have examined some of the basic techniques in this chapter and noted strengths and weaknesses of each. Two of these models, trend extrapolation and the gravity model, are based directly on statistical analysis of past behaviour. Forecasts are made by assuming that historical patterns will continue. Such an assumption is not unreasonable for short-term and middle-term forecasts, but it is increasingly dubious for the longer term.

A third model, the probabilistic travel model, also utilizes empirical data. It is different from the first two in that the data do not directly relate to observed tourist demand but to measures of presumed utility. As a result, the predictions are not expressed in terms of demand *per se*, but rather in terms of probabilities or market shares.

The last model, Delphi, relies heavily on expert opinion and the debatable assumption that a group consensus derived from an exchange of those opinions will eventually produce reliable forecasts.

Great progress has been made in the last 25 years in tourism and recreation forecasting. Although forecasting is still as much an art as a science, continued progress is likely in both technical aspects of forecasting and in the application of forecasting to management and planning goals.

Further reading

Bull A 1991 *The economics of travel and tourism*. Pitman, Melbourne.

Calantone R J, **di Benedetto C J** and **Bojanic D** 1987 A comprehensive review of the tourism forecasting literature. *Journal of Travel Research* **26**(2): **28**–39.

Crouch G I 1992 Effect of income and price on international tourism. *Annals of Tourism Research* **19**: 643–64.

Davies B and **Mangan J** 1992 Family expenditure on hotels and holidays. *Annals of Tourism Research* **19**: 691–9.

Green H, **Hunter C** and **Moore B** 1990 Assessing the environmental impact of tourism development: use of the Delphi technique. *Tourism Management* **11**: 111–20.

Linstone A H and **Turoff M** (eds) 1975 *The Delphi method: techniques and applications*. Addison-Wesley Publishing, Reading, MA.

Morley C L 1992 A microeconomic theory of international tourism demand. *Annals of Tourism Research* **19**: 250–67.

Salanick J R, **Wenger W** and **Helfer E** 1971 The construction of Delphi event statements. *Technological Forecasting and Social Change* **3**: 65–73.

Smeral E, **Witt S F** and **Witt C A** 1992 Econometric forecasts: tourism trends to 2000. *Annals of Tourism Research* **19**: 450–66.

Smith S L J 1984 A method for estimating the distance equivalence of international boundaries. *Journal of Travel Research* **22**(3): 37–9.

Wheeler B, **Hart T** and **Whysall P** 1990 Application of the Delphi technique: a reply to Green, Hunter, and Moore. *Tourism Management* **11**: 121–2.
Witt S F and **Witt C A** 1992 *Modeling and forecasting demand in tourism.* Academic Press, London.
Witt C A, **Witt S F** and **Wilson N** 1994 Forecasting international tourist flows. *Annals of Tourism Research* **21**: 612–28.

CHAPTER 7

Selecting a site for business development

Introduction

An adage in business development states, 'The three most important factors to consider when starting a business are location, location and location.' The emphasis on location is strong because changing a site once a business has been established is difficult and expensive, and because the process of site selection can significantly influence a business's profitability, perhaps even its survivability.

Despite the expressed importance of location, there is little published research on site-selection principles and procedures in tourism. Some of the larger hotel, resort and restaurant franchises allegedly have sophisticated locational models, but these are proprietary. They are not made publicly available to the academic community because of the legitimate concern that competitors could benefit from knowledge of the site-location strategies of these firms.

The lack of published site-selection research in tourism, while unfortunate from a scholarly perspective, should not be surprising. As Grether (1983) notes, academic researchers interested in marketing and entrepreneurship have largely ignored spatial and locational questions in their analyses. Since tourism research, even in universities, has been dominated by the marketing perspective, it is not unexpected that tourism research should have many of the same shortcomings as market research. To be fair, Grether also notes signs of a growing recognition of the need for locational research. This growth is fed by the need to understand and manage all the forces that influence a business's operations and performance, including factors such as location that directly influence a customer's decision about which business to patronize.

Fortunately for marketing and tourism analysts, they do not need to start from scratch in developing locational guidelines. There is already a large body of literature on the economics of location and the geography of business. This work may be found in the journals and books of applied geography and, to a degree, in regional science and economic geography.

At the risk of over-simplifying a complex corpus of research, this work may be classified into two categories: (1) locational theory and (2) site-selection guidelines. The former tends to be nomothetic, formal, abstract and mathematical. The latter tend to be idiographic, applied, empirical and sometimes based on intuition as much as on objective procedures. As the title of this chapter suggests, our interest is in the second category. The reason for this is that this book emphasizes practical research techniques rather than the development of theory. However, both perspectives are important for tourism research in the long run, and we will consider briefly the contributions of locational theory before turning to site-selection procedures.

Location theory

Classical location theory began with the work of von Thünen (1885) on a highly simplified model involving an isolated market-place surrounded by an isotropic plane. Von Thünen developed a theory that predicted a concentric pattern of crops around the market centre. His model focused on the comparative value of crops, the cost of transportation of each type of crop, their bulk and their perishability. Land closest to the market would be used for highly perishable, valuable or bulky items such as truck crops, milk from stall-fed cows, and eggs from laying hens. The next zone was devoted to forest because of the contemporary importance of wood for fuel. This was followed by more durable and less bulky crops such as potatoes. Then would come fields for cereal grains, pasture-land (for animals that could be driven to market), and finally wilderness tracts for hunting and trapping. Although modern agricultural economics, food production and storage technology, and transportation have created very different patterns, von Thünen's emphasis on the central role of transportation in determining land uses was seminal. His model and the models of those who followed his lead (e.g. Weber 1928, 1931; Hoover 1948) are commonly recognized as the beginnings of the 'transportation cost school' of locational theory because of their emphasis on the importance of transportation in determining the location of economic activities. Their work also became the inspiration for several tourism development models such as those of Yokeno (1974), Vickerman (1975), and Miossec (1977).

Work by some economists in the 1930s shifted the focus from rural land-use patterns to urban development and industrial lands. One of the most important of these early modifications of von Thünen's work was the replacement of his central market with a localized resource used by some industry (Ohlin 1935). This line of enquiry grew in complexity and scope, leading ultimately to formal statements about the interrelationships of urban areas, industries and economic regions developed by Christaller (1933) and Lösch (1944). Their work is known as 'central place theory' or as the 'locational interdependence school'. The first name comes from Christaller's focus on urban areas as 'central places' or concentrations of economic activity and political administration, while the second title is derived from Lösch's attention to the interrelationships among the development of economic regions, and the nexus of urban areas and industrial concentrations.

Central place work continued to be a major force in economic geography and regional science into the 1970s. Among the contributions this work made to the geography of business activity has been the formalization of the concepts of:

- **threshold populations** – the minimum number of people needed to support a given industry or retail activity;
- **hinterlands** – the geographic area containing the threshold population; the actual size varies with the type of commodity under question and the surrounding population density;
- **hierarchies** – the pattern of a few large cities offering many goods and services and more, smaller cities and towns offering fewer goods and services.

Despite its importance to regional science, central place theory has had little to contribute to tourism research. In fact, when Christaller (1964) looked at the spatial patterns of tourism, he observed that the pattern seemed to be the result of processes that were in direct opposition to those that formed central places. However, his argument that resorts and other tourism developments were 'peripheral places' reflected an inaccurate understanding of the nature of tourism.

Despite some successes of central place theory, regional scientists grew dissatisfied with it. Their major concern was that central place theory permitted a large number of different, independent and conflicting models to develop. Each appeared to work under certain circumstances, but failed under others. The dissatisfaction with this growing number of contradictory *ad hoc* models led to the emergence of the 'generalized market school'. This school explicitly attempted to build on the strong points of earlier models while avoiding their weaknesses.

Two major approaches have been employed to develop a more powerful synthesis of economic concepts desired by the generalized market school. One of these was pioneered by Isard (1956). He examined all available models to identify common themes or components, assuming that each model had some truth in it. By identifying common features and unique strengths, he attempted to combine them into one comprehensive model, without making any new detailed structural analysis or major innovations.

A contemporary, Greenhut (1956), argued that one could not overcome the weaknesses in existing models simply by combining parts of them. He believed it was necessary to adopt a new perspective on the problem of locational theory and site selection. One of the major contributions that Greenhut made was his emphasis on the role of risk and uncertainty in locational decision-making. Greenhut rejected the common assumption that firms would act as would-be spatial monopolists (an assumption common to most previous locational theorists) and assumed, instead, that they would behave as spatial oligopolists. In other words, Greenhut based his model on the belief that each firm would look at several other firms in the immediate vicinity, but with a conscious recognition that no firm's owners can know for certain what the other owners are planning. Greenhut's model is still abstract, but it is more realistic than the models proposed by other location theorists. His emphasis on risk and uncertainty also helped prepare the way for the application of simulation and game theory to locational development (e.g. Baligh and Richartz 1967; Barcun and Jeming 1973; Rao 1981).

Locational theory has developed a useful base of concepts that support the scientific analysis of industrial, settlement and urban development patterns. However, such theories are of limited utility to practical problems because too many variables and issues are assumed away. And often those issues that remain require mathematical tools beyond the abilities of most planners and business people. None the less, location theorists have developed several concepts that are useful for site-selection work in business, including tourism:

1 Location is both theoretically and practically important in determining the size and success of a firm.
2 The choice of a good location involves trade-offs among transportation costs, production costs, resource availability, labour availability, market accessibility and land costs.
3 Certain types of businesses do well if they avoid locating close to competitors; others benefit from such closeness; still others are indifferent.
4 Population size and the number and location of competing firms can limit the potential for new business growth in predictable ways.
5 Businesses that require the shipping of heavy or bulky items required for production or that are closely tied to particular resources will tend to locate close to those resources. Businesses that produce heavy or bulky products will tend to locate close to their market.

These general observations provide some of the conceptual foundation for the more practical site-selection guidelines we will now examine.

Checklist method

Description

The checklist method is a simple, systematic consideration of the key characteristics of a number of pre-selected sites. This list may be developed on the basis of intuition, expert opinion, or from a listing of available real estate. This particular method allows the developer to select either an optimal site or to rank order available sites. The checklist is probably the most common tool used by business planners. Its advantages include low cost, simplicity, flexibility and general objectivity. It can be applied equally well to different types of businesses and to virtually any scale of development, from the selection of a major new international resort to the best corner for an ice-cream vendor's wagon.

The ease with which the checklist method can be employed, however, introduces risks. Although the specific site characteristics to be evaluated are identified objectively, the evaluation of each site in terms of these characteristics depends on the skill, insight and experience of the evaluator. Further, while comparisons of each site on each individual criterion are simple, the method does not provide any form of calculation to compare sites on multiple criteria. For example, a business planner might note that one site performs very well on two relatively unimportant characteristics but only moderately well on an important quality while an alternative site performs very well on the important quality but only weakly on the two lesser important qualities. How does the planner choose? The checklist method does not provide any advice.

Finally, the checklist method does not provide any quantitative estimates of market share, return on investment, or net profitability. Such estimates might be derived as part of the site-selection process, but they involve techniques beyond the checklist method.

Procedure

1 Identify the important locational criteria for the type of business being considered. These criteria should be divided into 'critical' and 'desirable' attributes. Critical attributes are those whose absence (or possibly presence) would eliminate a site from further consideration regardless of other qualities. For example, a local ordinance against serving alcoholic beverages would immediately eliminate a location as a potential place for a bar. Critical attributes act as filters to screen out undesirable locations, reducing the number of alternatives that have to be considered in detail.

   Desirable attributes are those whose presence, in some degree, enhances the suitability of a location. As with critical attributes, these will vary from business to business. Desirable attributes may be expressed either negatively or positively. With respect to adjacent land uses, for example, the particular issue may be a concern for either the presence of other businesses that would enhance the potential success of a business or the desire to avoid selecting a site near incompatible uses.

2 Once the list of critical and desirable attributes has been developed, each attribute must be operationally defined. This definition should be expressed in such a way that the planner or consultant can

clearly understand what is meant by the attribute and what levels or features are associated with different degrees of desirability. For example, if labour supply is an issue, what types of skills are sought? How many job applicants are desired? How far can one expect potential employees to travel? Is there potential for hiring unemployed workers and retraining them?

If desired, an operational definition can be expressed as an ordinal scale. 'Neighbourhood cleanliness', for example, could be ranked on a scale from '1'(very clean) to '5'(very dirty). Other types of qualitative attributes, however, may not be meaningfully expressed as an ordinal scale, such as zoning or deed restrictions. These types of qualities have to be carefully assessed in detail to compare sites.

3 Draw up a list of potential locations and identify appropriate data sources. This list may be based on informed judgement, a systematic consideration of all possible locations (clearly not feasible if the number of businesses runs into scores or hundreds), or locations on which the developer may hold options.
4 Prepare a worksheet listing all attributes. Each location might be evaluated on one or more sheets if lengthy verbal descriptions are necessary for some of the attributes. The critical attributes should be placed at the top of the list.
5 Begin completing the worksheets for each location by examining the critical attributes. Complete the worksheets only for those locations that meet the critical requirements.
6 Compare the locations. Resist the temptation to reduce all evaluations to a standard ordinal scale so that the scores could apparently be added up to obtain an overall ranking. While this may be appropriate in some cases, it is based on the assumption that all attributes are measurable on comparable, valid and interval-level scales. This assumption is not often met. The evaluation of the checklists may be done best in consultation with experts who can assist in the evaluation of criteria associated with engineering, real estate and legal matters.

Example

A detailed example is not appropriate because the method is largely qualitative and idiosyncratic – depending on field inspection and professional judgement. However, an example of a typical checklist will be informative. The following is drawn from the Kitchener–Waterloo Area Visitor and Convention Bureau's (Ontario, Canada) guidelines for selecting staffed seasonal visitor information kiosks (Table 7.1). These kiosks are usually located in or near hotels, restaurants and tourist attractions. Each location is assessed on each factor using a four-point scale, ranging from '0' (unacceptable) to '3' (excellent). These assessments are then weighted by the importance of the factor ('1' – desirable but not essential; '2' – important; '3' – vitally important). Any location that receives an 'unacceptable' rating on any 'vitally important'

Table 7.1 Checklist for staffed, year-round visitor information centres

Site factor[a]	Factor importance[b]
Major access point to the area	3
High visibility traffic route	3
Visibility from roadway and exterior signage potential	3
Consistent hours of operation	3
Hours of operation match peak enquiry hours	2
Availability of clean washrooms and additional facilities/services	3
Free parking	3
Convenient parking	3
Geographic location	3
Interior space available	2
Compliance with all VCB policies	2
Availability of backup personnel on-site	2
Accessibility when un-manned	2

[a] Assessed on a four-point scale ranging from 0 (unacceptable) to 3 (excellent).
[b] 3 = vitally important; 2 = important; 1 = desirable.

factor or two 'unacceptable' ratings on two 'important' factors is immediately dropped from further assessment.

Analogue method

Description

The analogue method was pioneered by William Applebaum, a site-selection expert for the Kroger Company (a US-based grocery store chain) in the early 1930s. It has found extensive application in other companies (Rogers and Green 1978; Thompson 1982) and has been adopted by the US Army Corps of Engineers for predicting visitation rates at Corps recreation sites (US Army Corps of Engineers 1974). The model lacks any theoretical base and, like many other site-selection tools, is highly dependent on the skill of the individual analyst. In the hands of an experienced and competent researcher, the method is quite useful.

The analogue method guides site selection by producing a forecast of visits or sales by comparing a proposed site(s) with an existing site(s) that is similar in relevant characteristics. Central to the use of the analogue method is the analogue database – a series of profiles of existing sites to be used for comparison and forecasting. Because of the importance of this database, the analogue method is usually limited to those situations in which a prior investment in data collection has been made. With sufficient lead time, however, one can develop a series of profiles that can then be used to assess a set of alternatives. Each analogue site profile describes a successful site. The basic logic of the method is that if a proposed site closely resembles an existing successful site, the proposed site has a good chance of success as well. Further, the level of success (measured in terms of revenues or visitors) is also likely to be similar.

Because the analogue method compares potential sites on the basis of how well they compare to known sites, the method depends on: (1) the development of precise and valid profiles of existing sites or businesses; (2) the potential for finding close matches between the proposed site and existing sites; and (3) the ability of the analyst to 'adjust' for discrepancies between the analogues and the proposed sites to make more accurate forecasts of potential sales or attendance.

Procedure

1. The first and often the major task is the development of the database – the file of existing site profiles. Each profile should focus on characteristics relevant to the particular business being studied. Although some variation will be necessary for different types of businesses, the following are common variables:

   (a) Location, size, and other basic information about the firm.
   (b) A graph relating per capita attendance or sales by census tract, neighbourhood, county, or other geographical unit of distance. This plot may be a simple graph showing the change in sales over increasing distance, or it may be a plot combined with a regression equation statistically summarizing the relationship between business volumes and distance.
   (c) Population of the geographical units used in item (b).
   (d) A socio-economic profile of the population in the market area of the firm, typically that area surrounding the firm that encompasses 65 to 80 per cent of all business. This profile, however, may be irrelevant if the market area extends hundreds or thousands of kilometres, as in the case of international resorts.
   (e) Descriptions of other relevant site characteristics such as the surrounding transportation network, local land uses, available parking, and significant deed restrictions or zoning regulations.

   To use the profiles for comparing alternative sites, follow the next series of steps.
2. Identify potential sites for development. This list may be drawn from an inventory of available real estate or from informed opinion combined with a preliminary review of properties.
3. Select the analogue site that most closely resembles the proposed site.
4. Delineate the probable market area of the proposed site.
5. Obtain estimates of the population in the census tracts or other geographical units surrounding the proposed site within the market area.

Table 7.2 Demographic profile for six FSAs in Waterloo, Ontario

			% by age				% with income greater than:					Distance to
Forward sortation area	**Population**	**% married**	**<25**	**25–44**	**45–64**	**>64**	**$15k**	**$25k**	**$35k**	**$50k**	**$75k**	**site (km)**
N2J	10 250	51	21	34	32	13	26	26	12	4	2	1.7
N2K	4 950	71	19	52	24	5	33	33	18	6	3	3.9
N2L	16 900	64	21	46	24	9	30	30	17	7	3	0.6
N2T	1 475	82	13	68	17	2	47	47	27	11	4	2.1
N2V	1 325	80	18	68	12	2	33	33	11	0	0	4.1

Table 7.3 Demographic profile for the alternative site

			% by age				% with income greater than:					Distance from centroid to proposed site (km)
Forward sortation area	**Population**	**% married**	**<25**	**25–44**	**45–64**	**>64**	**$15k**	**$25k**	**$35k**	**$50k**	**$75k**	
1	12 300	46	23	33	31	13	55	30	19	7	1	0.9
2	15 125	52	20	59	16	5	51	29	12	5	0	1.5
3	7 050	63	21	36	28	15	52	26	23	13	4	1.9
4	1 200	85	10	46	26	18	56	40	17	12	2	2.8
5	6 550	55	23	60	11	6	60	41	12	4	1	3.6
6	10 000	71	19	58	19	4	59	28	14	8	3	4.1

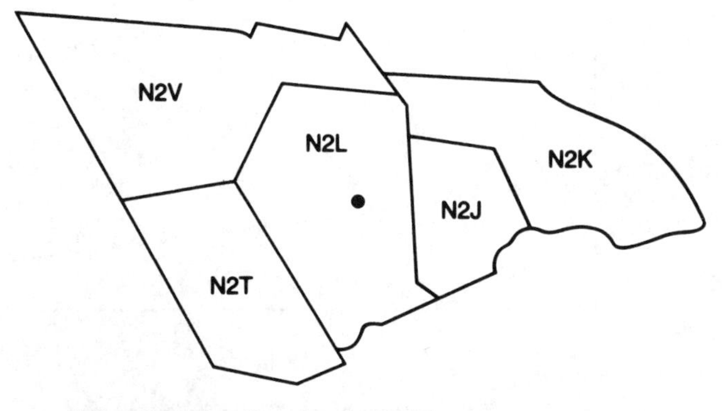

Figure 7.1 Location of analogue restaurant and surrounding market area: Waterloo, Ontario

6 Using the per capita sales (or use) curve from the analogue site and the associated population figures from the proposed site, estimate expected sales for each geographical unit in the market area.
7 Sum the estimated sales figures to arrive at a total. Select the preferred site on the basis of the largest estimated total or other appropriate criterion.

Example

The following method demonstrates the basic logic used in the analogue method. This example considers only one analogue; actual applications will consider two or more. The steps in the calculation of each analogue would be identical, however. This particular example concerns the selection of a site for a restaurant offering table service, entrées averaging $7 to $12 for the evening meal, some finger foods, bar service in an attached lounge, a nostalgia theme in the décor and menu, and catering primarily to a young, moderately up-scale clientele. This particular restaurant (for the analogue) is located in Waterloo, Ontario. A customer survey revealed that over 80 per cent of the customers come from five 'forward sortation areas' (FSAs), which are postal code regions defined by the first three digits of Canada's six-digit postal codes. These FSAs were all within 5 km of the restaurant (see Figure 7.1). Census files for the FSAs provided the demographic profile summarized in Table 7.2. A plot of per capita sales (drawn from survey data and the population of each FSA) against the distance between the population centroid of the FSA and the restaurant is given in Figure 7.2. A regression line was estimated using least-squares regression, as described in Chapter 6.

One of the key features associated with the analogue operation in Waterloo was the proximity of two universities to the restaurant. A majority of the customers were university students, staff and faculty. This allowed the analyst to narrow the search to other Ontario

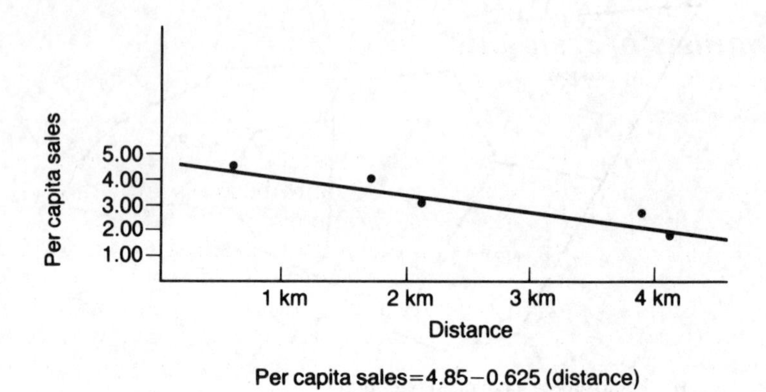

Figure 7.2 Distance versus per capita sales for analogue restaurant

communities that had universities as likely communities for expansion of the restaurant chain. A review of the census profiles of FSAs in the vicinity of restaurants in these communities led to the identification of one especially promising location. The profile of the FSAs from that community are provided in Table 7.3. Although the profiles do not match exactly, the analogue and the proposed site are similar in having a relatively high concentration of younger residents, singles, and mid- to upper-income households. Use of the plot and the regression equation in combination with data from the proposed community resulted in estimated per capita sales and total sales figures (Table 7.4). These forecasts, combined with a review of the existing competition in the city, led the investor to begin negotiations for a site in the selected community.

Table 7.4 Calculation of expected sales for proposed location

Forward sortation area	Population	Distance	Estimated per capita sales ($)	Estimated total sales ($)
1	12 300	0.9	4.3	52 890
2	15 125	1.5	3.9	58 987
3	7 050	1.9	3.7	26 085
4	1 200	2.8	3.1	3 720
5	6 550	3.6	2.6	17 030
6	10 000	4.1	2.3	23 000
			Total:	181 712

Analysis of residuals

Description

Analysis of residuals is a quantitative method for estimating the relative potential of several locations to support business growth. Like the checklist method, it permits examination of several locations simultaneously on the basis of previously specified attributes. Unlike the checklist method, analysis of residuals is applicable only to regional-scale location choice. It cannot be used to select specific sites. Its strength is in its ability to help you narrow the alternatives to be considered in greater detail, perhaps through application of the checklist or analogue methods.

The analysis of residuals method is based, in part, on the notion of economic equilibrium – the idea that the number of businesses a region can support is influenced by a range of social and economic conditions and that the number of businesses slowly changes until an equilibrium is reached. The slowness of change is due to the uncertainty that entrepreneurs face as they start a new business (or cut their losses and close an existing business) and to the fact that the equilibrium will change over time. In fact, it is possible that conditions may change so quickly that the number of businesses will never be in equilibrium with the environment. The study of equilibrium conditions might ideally be conducted through longitudinal research whereby the changing numbers of firms are examined over time as they adjust to changing conditions. However, the time and budget requirements for a longitudinal design are usually too great for most business decisions. What is needed is a practical, efficient method for providing an estimate of the probable number of firms a region could support in at least a short-term equilibrium. Such an alternative is a latitudinal or cross-sectional study. Within limits, one can argue that the economic development patterns in a large number of economically and politically linked regions (such as counties in a province or state) measured at the same time reflect different stages of economic development. In other words, one assume that regions with different levels of development represent different stages in the economic evolution of the average region.

If you use a cross-sectional design to estimate the expected numbers of some type of tourism enterprise, a comparison with the actual number will indicate the degree to which a region is over- or under-developed with respect to that type of enterprise. A basic assumption in this method is thus that the existence of residuals – any difference between the expected and observed number of businesses – is the result of economic disequilibrium. In practice, residuals may be due to a variety of factors. One of the more probable sources of a mismatch is the mis-specification of the predictive model. Whether this is likely to be the explanation in any particular case can be assessed by examining the level of explained variance, R^2, in the model. A relatively high R^2 may indicate that model mis-specification is not likely to be a serious problem.

Omission of important variables is another potential source of residuals. Again, a review of explained variance and the level of model error can give an indication of the potential seriousness of this source of residuals. Review of the model by other researchers may also provide a check against the omission of causal variables. This type of peer review can be useful in determining if there is a source of systematic error in your work. A more difficult situation would be the omission of one or two variables that could play an important role in explaining levels of business development in a few counties, but which might be of negligible importance in the majority of regions. For example, particular zoning restrictions or local variations in venture capital might significantly affect local patterns. It should be recalled, though, that analysis of residuals is intended to be only the first step on a lengthy site-selection process. The omission of one or more important variables at this early stage of research does not necessarily mean that these variables would never be considered at a subsequent stage of research. In the case of zoning restrictions or the availability of capital, for example, these matters would clearly come to the fore before a final development decision could be made.

The dependent variable normally used in the analysis of residuals is the number of businesses. Such a measure, however, ignores the variations in the size and profitability of firms. It also implies that the characteristics of the typical firm in each county are the same – only their number and market conditions vary. Of equal or greater relevance is the fact that the model makes no allowance for clever and aggressive (or incompetent and bungling) management and marketing. A good location is no guarantee of success; a bad location is not an economic kiss of death.

Lastly, this site-selection model assumes that the critical social and economic environment and the primary origin of customers is the local region. For some types of tourism businesses, this will not be true. To the degree that the number of businesses is influenced by conditions in distant regions, the analyst must reflect those conditions in his model or risk producing misleading results.

Procedure

1. Define a set of regions relevant to the locational problem you are interested in. Regions that conform to existing political boundaries such as counties are usually the most practical because most data sources are organized by political units.
2. Define a set of independent economic and social variables related to the success of the tourist business being studied. Collect data for each of these variables in each of the regions identified. Also determine the number of firms located in each region.
3. If data for a large number of variables have been collected and if some of these variables appear to be correlated with each other,

you can reduce the set through factor analysis. This will produce a small set of statistically independent variables.

4 Using multiple regression, regress the dependent variable (the actual number of businesses) against the independent variables. Calculate the expected number of firms expected in each region.
5 Compare the estimated and observed number of firms to obtain residuals (observed – estimated).
6 Small residuals are to be expected (no social science model is perfect). A tolerance interval needs to be defined to indicate whether any residual is 'small' and not worth comment or is 'large' and thus meaningful. A convenient confidence interval is plus or minus one standard deviation. Any region with a residual within one standard deviation of the mean (which will be zero for a least-squares estimation) might safely be ignored, i.e. it can be interpreted as indicating a region near equilibrium.
7 Identify any regions that have unusually large residuals, perhaps greater than plus or minus three standard deviations. These extreme values are outliers that can significantly distort the results of the regression analysis (remember the effect of the very large intra-provincial travel in Ontario and the unusually high r^2 described in the previous chapter). Temporarily remove these from your data set.
8 Once any outliers have been removed, rerun the regression analysis and calculate new residuals. Also calculate the expected number of firms and the residuals for those regions that were removed in the last step because they were identified as outliers. This may be done by using your new regression model and substituting the appropriate values of the independent variables, and then solving for the dependent variable. Note any regions that now have residuals greater than plus or minus one standard deviation. Any region that has significantly fewer businesses than predicted, indicated by a negative residual greater than –1.0, may be interpreted as having the potential for expansion of that type of enterprise. A region with significantly more businesses than predicted will have a positive residual in excess of 1.0, and may be interpreted as having more businesses than local conditions can support. All other regions are near equilibrium and are unlikely to support an expansion in the number of businesses or to force many businesses out of operation.

Example

An example of this type of analysis can be found in Smith and Thomas (1983). In a study of the potential for expansion in the number of golf courses in southern Ontario, Smith and Thomas began by identifying 34 demographic and social variables they believed were related to the demand for golf courses (Table 7.5). Several of these variables were highly correlated with each other, so the analysts used factor analysis to reduce the list of 34 variables to five independent factors (Table 7.6). The

Table 7.5 Independent variables used to predict number of golf courses

Variable label	Age (years)
001 Total number of males:	0–14
002	15–24
003	25–34
004	35–49
005	50–64
006	65+
007 Total number of females:	0–14
008	15–24
009	25–34
010	35–49
011	50–64
012	65+
013 Percentage of males between:	0–14
014	15–24
015	25–34
016	35–49
017	50–64
018	65+
019 Percentage of females between:	0–14
020	15–24
021	25–34
022	35–49
023	50–64
024	65+
025 Number of tax returns showing taxable incomes over $30 000	
026 Urban/rural population ratio	
027 Total personal disposable income	
028 Number of licensed restaurants	
029 Number of licensed private clubs	
030 Population	
031 Population density	
032 Number of households	
033 Percentage of households employed in management or administration	
034 Percentage of labour force employed in farming, forestry or mining	

Source: Smith and Thomas (1983).

factor scores for each county were regressed against the number of golf courses in each county. The first regression resulted in the identification of seven counties out of 43 that were outliers. These were removed from the data set and the regression repeated. The second multiple regression produced the following equation:

$$Y = 6.573 + 1.443 (X_1) + 1.223(X_2) + 0.513 (X_3) + 0.727(X_4) - 0.175(X_5) \quad [7.1]$$

where: Y = number of golf courses;
X_n = factor scores.

The R^2 for this equation was 91.6 per cent, which suggests that the model structure is adequate and the most important variables have been included. Residuals were calculated and are presented in Table 7.7.

Table 7.6 Principal factor analysis loadings (highest loadings only)

Variable label[a]	Factor 1	Factor 2	Factor 3	Factor 4	Factor 5
001	0.974				
002	0.981				
003	0.983				
004	0.983				
005	0.989				
006	0.986				
007	–0.974				
008	–0.983				
009	–0.972				
010	0.986				
011	–0.988				
012	–0.909				
013			–0.848		
014					0.891
015			–0.760		
016			–0.766		
017			–0.900		
018			–0.727		
019			0.802		
020					–0.809
021			0.815		
022			0.888		
023			0.948		
024			0.779		
025		0.902			
026		0.954			
027		0.969			
028		0.973			
029		0.889			
030		0.937			
031		0.973			
032		0.974			
033				0.750	
034				–0.817	

Source: Smith and Thomas (1983).
[a] See Table 7.5 for interpretations of label numbers.

Table 7.7 Number of golf courses (observed and predicted) for southern Ontario

County	Observed	Predicted	Residual	Status[a]
Dundas	1	2.3	–1.3	
Frontenac	5	6.2	–1.2	
Glengarry	0	2.7	–2.7	Under
Grenville	1	3.7	–2.7	Under
Hastings	6	4.5	1.5	
Lanark	4	4.9	–0.9	
Leeds	6	3.3	2.7	Over
Lennox/Addington	1	2.2	–1.2	
Ottawa	21	16.7	4.3	Over
Prescott	2	2.9	–0.9	
Renfrew	1	4.2	–3.2	Under
Russell	1	1.2	–0.2	
Stormont	2	3.9	–1.9	
Brant	6	5.2	0.8	
Dufferin	1	1.1	–0.1	
Haldimand/Norfolk	5	3.5	1.5	
Hamilton/Wentworth	13	12.8	0.2	
Muskoka	8	4.5	3.5	Over
Northumberland	4	4.6	–0.6	
Peel	11	11.8	–0.8	
Peterborough	7	5.7	1.3	
Victoria	3	4.9	–1.9	
Waterloo	13	9.4	3.6	Over
Wellington	5	5.1	–0.1	
York	34	36.2	–2.2	Under
Bruce	5	4.1	0.9	
Elgin	3	4.8	–1.8	
Essex	10	10.4	–0.4	
Grey	5	3.6	1.4	
Huron	5	3.3	1.7	
Kent	6	5.0	1.0	
Middlesex	11	10.6	0.4	
Oxford	4	3.6	0.4	
Perth	3	4.8	–1.8	
Parry Sound	1	2.4	–1.4	
Durham[b]	17	4.7	12.3	Over
Haliburton[b]	1	5.0	–4.0	Under
Halton[b]	16	6.7	9.3	Over
Niagara[b]	21	11.4	8.6	Over
Simcoe[b]	19	7.8	11.2	Over
Ontario[b]	9	25.1	–16.1	Under
Lambton[b]	11	5.4	5.6	Over

Source: Smith and Thomas (1983).

[a] Probable over- or under-development.
[b] Outliers removed during second stage of regression.

Inductive reasoning

Description

Inductive reasoning, in the context of locational research, is more a research strategy than a specific technique. It is, in fact, the 'textbook' approach to empirical scientific investigation, including tourism analysis. As such, it deserves an entire book to itself. Given the relatively poor state of theoretical development in tourism and in site selection, there is value in introducing the topic here.

Simply put, inductive reasoning is generalization. One observes systematically a number of occurrences of some phenomenon and works towards a generalized explanation of the process behind that pattern. Induction is not proof; it can never produce irrefutable laws. It is only the collection of evidence and the making of inferences from that evidence in support of a conclusion. Induction cannot exclude the possibility that any conclusion, however plausible, is false.

Induction is a slow and uncertain way of developing a locational model because it depends ultimately on the intuitive ability of the analyst to recognize a potentially meaningful pattern and to sense the possible cause of that pattern. That intuition may be wrong; even if it is right, the interpretation of the cause of the pattern could be wrong. There is no 'cookbook' method of enquiry or set of inferential statistics that can be easily and directly applied to the inductive search for meaning. This type of enquiry is sometimes described as basic research. In basic research, positive results are not guaranteed. If successful, though, the discoveries can become a part of the fundamental knowledge of social science.

Much of the potential for success in this research design depends on the researcher acquiring as much background knowledge about the phenomenon as possible. You must immerse yourself in the problem, exploring and reading widely on the subject matter to collect ideas, references and insights. Work on inductive model-building will tend to go on for a relatively long time, easily a year and perhaps a decade. This time will be characterized by periods of intense activity of data collection and analysis followed by periods of quiet reflection.

If a researcher has the curiosity, time and necessary skill to undertake this type of research, it can be an absorbing, rewarding, long-term project. If, on the other hand, the motivation for a site-selection project is to solve an immediate problem, some other procedure should be seriously considered.

Procedure

1 Begin with a question. This is not as trivial as it sounds. Induction can lead to a hypothesis only when you have a sense that there may be some meaningful pattern hidden in what, at first glance, appears to be an incoherent set of data. The way that belief is

operationalized is by formulating a research question that defines some sort of goal to direct your research. For site selection, your question will usually be couched in terms relevant to regularities in pattern and process behind existing business locations.

2 Carefully define the units of observation. What types of businesses are you interested in? Are there several categories of the same type of business that should be examined separately? On what scale – neighbourhood, community, county, provincial, national – should data be collected? Should data be collected from many regions at one point in time or should you examine one region over time?

3 Systematically observe and record the observations defined in step 2. One or more maps will be useful as a supplement to tabulations of addresses and other locational characteristics. Some of the descriptive measures in Chapter 9, such as nearest-neighbour analysis or Lorenz curves, may also be useful.

4 Compare the observed locational patterns with the patterns of other phenomena. The location of ski resorts might be compared with patterns of snowfall, topography, access roads, population, and other types of resorts. Urban hotels might be examined in comparison with traffic volumes on major urban arterials, visitor attractions, concentrations of office space, and airports.

5 Using any spatial correlations observed in step 4, other models of locational development, analogies, and any other potential source of ideas or inspiration, develop tentative hypotheses about the processes that give rise to the patterns observed. An important talent in this task is to be able to abstract the patterns enough to get a generalized or 'ideal' pattern without getting caught up in trying to explain every minor detail.

6 Use the hypotheses developed in step 5 to make predictions about the locational patterns to be expected in similar regions. Collect data from those regions and compare the observed pattern with that predicted. Verify the accuracy of the hypotheses and make modifications in them to account for significant discrepancies. Again, a talent is needed for recognizing the difference between insignificant and significant anomalies.

7 After the revised hypotheses have been tested in several different settings, and have been found to be valid, formulate a tentative theory formalizing the hypotheses. The theory should explain the process that generates the pattern. Your theory can then be scrutinized and tested by other researchers, which ultimately will lead to a better understanding of some important aspects of tourism development.

Example

An example of inductive reasoning can be found in Smith (1983). In this study, the author was interested in identifying some of the forces that influenced restaurant location. More precisely, the author was curious about whether certain types of locations were 'better' in the sense that

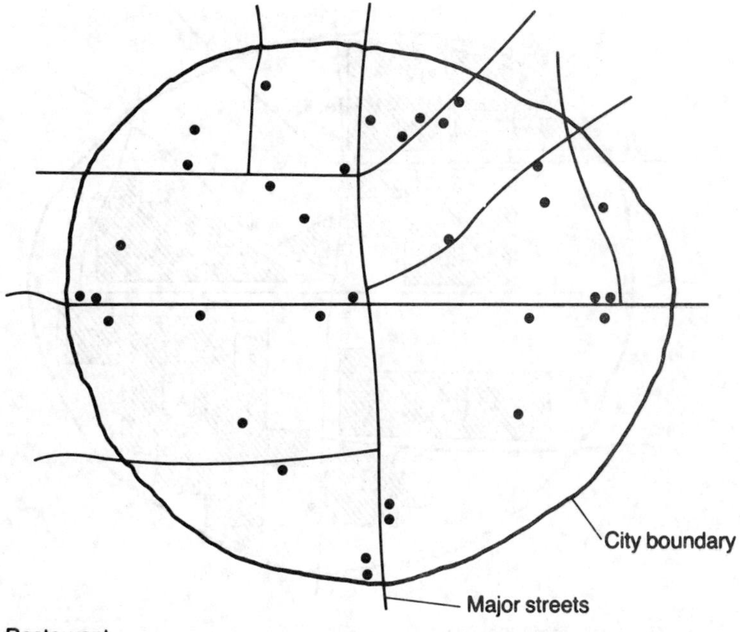

Figure 7.3 Location patterns for several restaurant types combined

restaurants located in such places were more likely to succeed than those located elsewhere. Smith began by mapping every restaurant and related dining establishment (e.g. bars and coffee shops) in a study city. The initial pattern was hard to interpret. There was some clustering, some dispersal, but no consistent pattern. Figure 7.3 is an illustration of the type of pattern observed. When the restaurants were divided into specific types, and each mapped separately, some meaningful patterns began to emerge.

These patterns were compared with land uses, street networks, and other aspects of the urban environment. Several spatial correlations appeared that seemed plausible. For example, pizza parlours tended to locate on moderately busy streets (annual average daily traffic counts of 10 000 to 15 000 vehicles) that delineated major residential areas. Pizza parlours also showed a tendency to disperse from each other, forming a 'necklace' around neighbourhoods (Figure 7.4). Certainly this description does not perfectly describe pizza parlour locations, but the general pattern – reflecting an interaction of spatial competition and the need for high visibility, high traffic counts and easy accessibility – appears valid.

Similar types of analyses were conducted for other types of dining establishments, including ice-cream parlours (which tend to locate inside residential areas in locations with high levels of pedestrian traffic and near or in major shopping centres), fast-food outlets (which

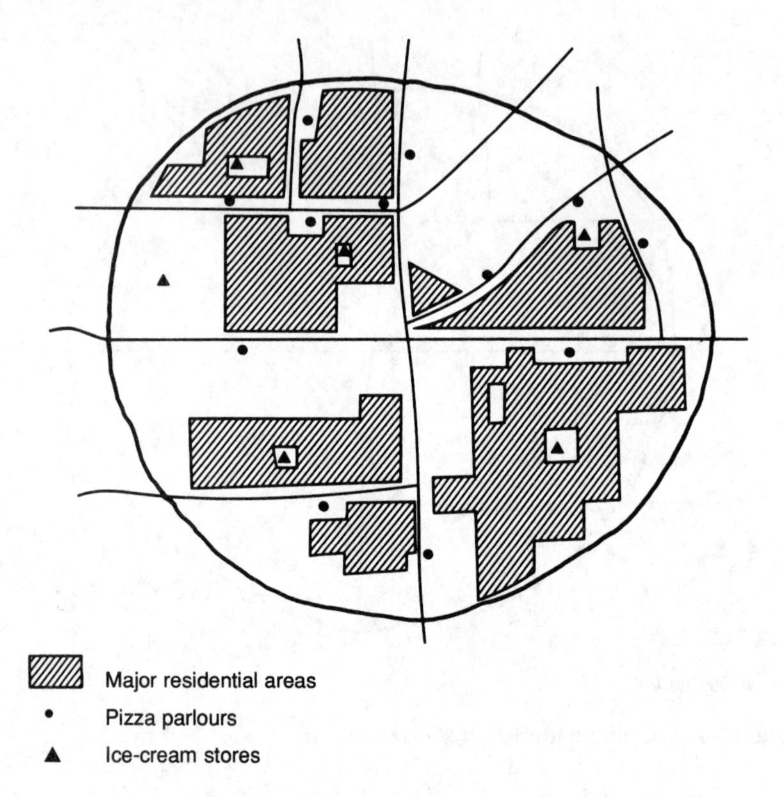

Figure 7.4 Location patterns for pizza parlours and ice-cream stores

agglomerate on heavily travelled streets), and doughnut stores (which locate near fast-food outlets and avoid other doughnut stores). These patterns are summarized in Figures 7.4 and 7.5. They were also formally expressed as a series of testable hypotheses that were examined in the light of data drawn from eight different cities. The results of that testing showed that several of the hypotheses appeared to be fairly accurate while several others needed rewording. This work eventually led to a preliminary articulation of locational principles for urban restaurants.

Summary

The adage that the three most important factors in business are location, location and location over-simplifies the complex problem of understanding the forces that influence business success. Location is certainly not the only decision a business planner must make; it is doubtful whether it is even among the more important – but it is far from negligible.

The question of location in business may be expressed in two ways. You can start with some particular business and then seek a good

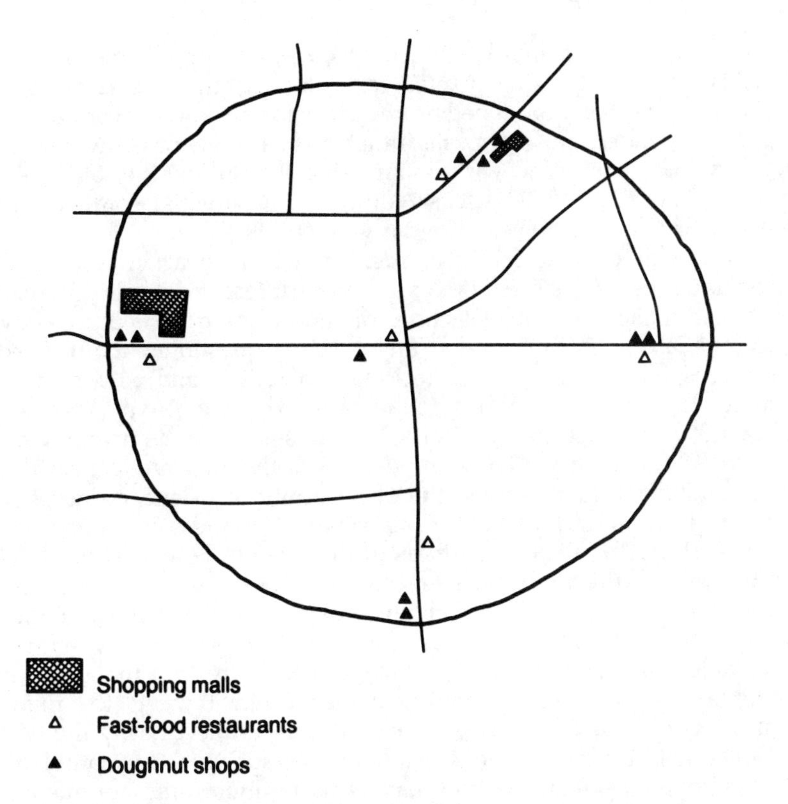

Figure 7.5 Location patterns for fast-food restaurants and doughnut shops

location, or you can begin with a location and then decide which type of business is most suitable for that location. We have concentrated on the first perspective in this chapter. It is the more common of the two, and it is the most relevant here because we are looking only at tourism businesses. With the general type of business already specified, the task is to find the best location.

The methods examined in this chapter were checklist, analogue, analysis of residuals, and inductive reasoning. Although these are among the better-established site-selection tools, other methods have recently been developed. These include the LOCAT model and the analytic hierarchy process (AHP) (Moutinho and Curry 1994). AHP is not a location tool *per se*, but rather a strategy for a decision-maker to assess existing knowledge and data related to any of a number of management issues. LOCAT is similar to the tourism attractiveness index in that it consists of a set of weighted attributes that are combined linearly to 'predict' the best site. LOCAT also includes an attractiveness variable based on the Luce choice theorem, described in Chapter 4.

Our methods and examples in this chapter concern the location of a single business operation. This is arguably the most common context for location decisions. One should note, however, that many location decisions in tourism involve the location of multiple units of a large franchise operation. The criteria for selecting the site can be very different depending on whether the decision-maker is working for a single independent business, is a franchisee with one or a few outlets, or is a franchiser concerned with maximizing the profitability of the entire chain. Zeller *et al.* (1980) discuss some of the potential conflicts in site selection arising from these different viewpoints.

Location models designed to identify which location is best for a particular type of business belong to two different schools. One is the site-selection school. Site-selection methods are designed to provide quick, reliable, useful results for specific questions about specific businesses. The orientation is decidedly practical, and as a result, the models do not often contribute much to the growth of knowledge. Location theory, the second school, is designed to contribute to the growth of knowledge. These models and theories are generalizable, flexible, abstract, and provide the opportunity to extend economic and geographic laws and theories. This type of progress is gained at the expense of the short-run usefulness of the models. Few location theories can be applied directly in the site-selection process of real businesses.

Location choice models, whether based on the site-selection or location theory perspective, can be employed to identify either an acceptable location or an optimal location. In the first case, the researcher is interested in finding a location that meets the minimal requirements for the business to succeed, but not necessarily the best of all locations. From a purely academic perspective, the problem of identifying the optimal location may be more interesting, but one has to ask whether the additional rewards associated with finding the perfect location are worth the greater costs involved in solving that problem. Very often they are not. The distinction between the search for adequacy and perfection is frequently associated with the distinction between site selection (which often leads to an acceptable solution) and location theory (which is more concerned with finding the perfect location). You will need to decide which is more appropriate for your project.

Finally, the locational methods described here do not normally include variables such as purchase price, terms of a lease, deed restrictions, and various financial aspects of site selection. These considerations can be introduced in the checklist method, but they are not always present. This omission should not be taken to imply that researchers are hopelessly naïve about the financial facts of life. The explanation for the omission, when it occurs, is usually a practical one. Much site-selection research demands that the researcher concentrates on selected aspects of potential sites and makes assumptions about other aspects. For example, a planner working for a proposed ski resort might be concerned about soil stability and slope, but will probably ignore soil fertility. He assumes that when the time comes to landscape the resort, someone else can resolve any problems about soil fertility by careful selection of ornamentals or by importing topsoil or soil conditioners. In the case of business location and development, most social scientists are more interested in social and economic variables, assuming that when the time comes for purchase negotiations, financial concerns will be worked out in a thoroughly business-like manner.

So although the ideals of rational comprehensive planning and research might lead one to wish that a site-selection researcher would also be able to make recommendations regarding purchase or lease arrangements for the sites he identifies, a realistic assessment of the business skills of most social scientists would lead the decision-maker to rely on financial advisers. In turn, one should also remember that very few financial and real estate experts have the scientific ability to undertake the original research needed before one even considers making an offer on a site.

Further reading

Akehurst G P 1981 Toward a theory of market potential with reference to hotel and restaurant firms. *Service Industries Review* **1**(1): 18–30.

Beaumont J R 1987 Retail location analysis: some management perspectives. *International Journal of Retailing* **2**(3): 22–35.

Black W C, Ostlung L E and **Westbrook R A** 1985 Spatial demand models in an intrabrand context. *Journal of Marketing* **49**: 106–13.

Curry B and **Moutinho L** 1991 Expert systems for site location decisions. *Journal of Marketing Channels* **1**(1): 23–37.

Green H L 1987 Good site analysts know where to put a new store. *Marketing News* **21**(19): 6–7.

Jones K and **Simmons J** 1990 *The retail environment*. Routledge, London.

Moutinho L and **Curry B** 1994 Modeling site location decisions. In Witt S F and Moutinho L eds) *Tourism marketing and management handbook*, 2nd edn. CAB International, Oxford, pp. 544–9.

Moutinho L and **Paton R** 1991 Site selection analysis in tourism: the LOCAT model. **11**(1): 1–10.

Wrigley N 1988 *Store choice, store location and market analysis*. Routledge, London.

CHAPTER 8

Defining the geographic structure of the industry

Introduction

Tourism is an industry with substantial geographic content. Its very nature involves travel and a sense of place. Tourists leave 'here' and visit 'there'. That may sound trivial, but the ability to describe precisely the difference between 'here' and 'there' is not trivial and often not easy. This task of defining places is a prerequisite for much data collection, planning and marketing effort in tourism. We examine in this chapter some tools used for describing the geographic structure of tourism. These tools are generically labelled as regionalization methods because a common product of their application is the definition of tourism regions. We will begin our review by defining regionalization more fully. We then turn to some of the basic uses of regionalization, different types of regions, regionalization logic, basic principles of regionalization, and some applications of regionalization in tourism. Finally, we examine five methods in detail.

Definitions and goals of regionalization

Regionalization is areal classification. It is the defining of one or more areas on the earth to identify them as distinct entities. The process of definition involves both 'integration' and 'segregation'. Integration here refers to the identification of some internal integrity or homogeneity in the region on the basis of selected features, while segregation refers to the process of distinguishing among regions on the basis of those same features. If the features chosen for defining a system of regions are related to some aspect of tourism development, then each region may be referred to as a 'tourism region'. A few authors, however, have suggested that the phrase 'tourism region' should be used more restrictively. Gunn (1965), for example, has argued that the only regions that should be considered tourism regions are those: (1) located at some distance from potential visitors; (2) seen as potential destinations; (3) that are reasonably accessible to the market; (4) that have some minimum level of economic and social infrastructure that can support tourism development; and (5) that are large enough to contain more than just one community. Gunn's concept of a tourism region is useful for certain purposes; indeed, it is discussed later in this chapter as the basis for 'destination zone identification'. His concept, however, may be too narrow to be a general definition of tourism regions. The basic problem is that Gunn's requirements would often eliminate locations that tourism planners will want to study – locations important for planning and development even though they do not meet all of the definitional criteria proposed by Gunn. For example, his requirement that a region be sufficiently large to incorporate two or more communities eliminates, as a tourism region, numerous places in North America that consist of a single community surrounded by a hinterland which form commercially viable tourism destinations.

For our purposes, therefore, a tourism region is simply a contiguous area that has been explicitly delineated by a researcher, planner or public agency as having relevance for some aspect of tourism planning, development or analysis. The specific purposes for delimiting tourism regions are numerous, but they can be grouped into three broad goals:

1 *Regionalization assists in naming part of the world.* When an analyst or planner wants to talk about some part of the world, whether as an origin, a popular destination, or perhaps as a region that has no potential to ever attract tourists (if any exist), it helps to be able to attach a label to that area. Marketing experts know the value of labels. Names not only help them to organize their work and to improve communication, names can be used to promote destinations. For example, the Ministry of Culture, Tourism and Recreation of the province of Ontario, Canada, has labelled a group of counties in midwestern Ontario as 'Festival Country' because of the large number of festivals and special events held there year-round. 'Midwestern Ontario' would be a more accurate label, but it lacks any appeal that would draw the attention of a tourist browsing through a brochure on Ontario destinations. The points to be made here are two: (1) a person with a job to do usually finds it necessary to name the things he works with, including things like regions; and (2) the choice of names, if done carefully and imaginatively, can make the job of marketing easier.
2 *Regionalization helps to simplify and order knowledge.* No individual can ever hope to know all relevant tourism facts and relationships about every place on earth. Regionalization provides a type of intellectual shorthand that simplifies and orders knowledge about diverse places. Instead of trying to remember all the details about hundreds of Caribbean islands, many of which are similar, it is useful to be able to speak about them as members in a group – the Caribbean region. Using the same logic, tourism planning and development in the Yangtze Delta of China, for instance, is

simplified by our being able to discuss the Yangtze Delta as a region composed of many locales that share essential characteristics. If we could not do this, we would have to treat every place on earth independently; this would greatly increase the cost and lower the efficiency of planning.

3 *Regions permit inductive generalization and predictions to be made.* Regions can be compared with each other to learn more about what relationship[s and characteristics are important for tourism development. Gunn (1979), for example, identified certain structural relationships in tourism regions that he believed were factors that contributed to the success of those regions. If his hypotheses are correct, then one can make predictions about the potential success of other regions as tourism destinations on the basis of whether or not they possess those characteristics.

Before examining types of regions, it will be worthwhile to consider one use that is not an appropriate objective or regionalization: the development of regions, *per se*. The identification of regions is a means, not an end. There is no point in assigning a label to a region if nothing more is implied than a regional name. This does not contradict the first use of regionalization – the assigning of names. It means only that the names must be needed for some larger purpose. Looking at this from another perspective, the caveat against regionalizing for its own sake implies that regions do not really exist as separate entities. If they did, then a perfectly acceptable scholarly objective for tourism analysts would be to identify and name as many tourism regions as possible, just as taxonomic biologists search for new species. Unlike biological species, tourism regions do not exist in themselves; they are created for, and only for, some larger purpose.

Types of regions

Three types of regions are generally recognized: (1) a priori, (2) homogeneous, and (3) functional. The first type is perhaps the most common and is usually the simplest to develop. An a priori region is an area around which someone has drawn a line and assigned a name. Its existence can be thought of as a 'given'. It is not the result of methodical regionalization; rather it is the intuitive beginning of other tasks. The best-known example of a priori regionalization is political division. Most political units are defined on the basis of political advantage, tradition, conquest, history, or convenient natural boundaries such as rivers or coastlines. They are rarely based on a scientific analysis of internal structure. Many tourism regions defined by provincial or national governments or by the tourism industry are also a priori regions. As with political divisions, these are not based on rational analysis, but the intuition or political preferences of a single individual or committee charged with defining the regions. This makes such regions unreliable for planning, marketing and research because they lack a reliable foundation.

A homogeneous region is a region defined by a set of objective, internal similarities. The important issues associated with defining a homogeneous region are the selection of relevant characteristics and specification of the degree of similarity that would cause a locale to be included within the region. Homogeneous regions are the type of region most planners and analysts think of when they hear the term, 'region'. They may also be viewed as the model for a priori regions – but there is a major difference: homogeneous regions are defined on the basis of objective analysis; a priori regions are not.

The third type of region is the functional region. This is an area that has a high degree of internal interaction. For example, people who live in proximity to each other tend to subscribe to the same newspaper. A map of the subscribers to a particular paper will define a region based on the circulation of that paper. The location and extent of this region might be of interest to competing newspapers or to retailers interested in advertising in certain neighbourhoods. Nothing is implied about the internal homogeneity of the people in that region beyond their subscriptions to a certain newspaper. Other functional regions can be defined on the basis of banking flows, spatial patterns of wholesale trade, vacation travel, or support of a sports team.

Types of regionalization logic

We have identified three reasons for regionalization and three types of regions. Coincidentally there are three types of logic used in regionalization: (1) synthetic, (2) analytic, and (3) dichotomous regionalization. Specific methods use one or a combination of these logics. Synthetic regionalization (also known as agglomerative or ascending regionalization) begins with small spatial units, perhaps townships or counties, and proceeds by combining contiguous and similar units. The extent of the synthesis depends on both the inherent similarity of the units on relevant characteristics and on the degree of regional homogeneity desired by the analyst. If a province were composed of 10 000 spatial units, the logic of synthetic regionalization would begin with 10 000 regions of one unit each and could continue until one region of 10 000 units was defined. The initial pattern is one of many regions with a high degree of homogeneity in each while the ultimate pattern is one region with minimal homogeneity. The optimal pattern is usually somewhere in between. This particular form of synthetic regionalization might be more precisely described as areal-based synthetic regionalization.

A second form of synthetic regionalization is point-based regionalization. In this case, data for discrete points located across a broad area of the earth's surface are obtained and then generalized to represent the entire area. The construction of a contour map is an example of this type of process. Regions are defined by delineating all

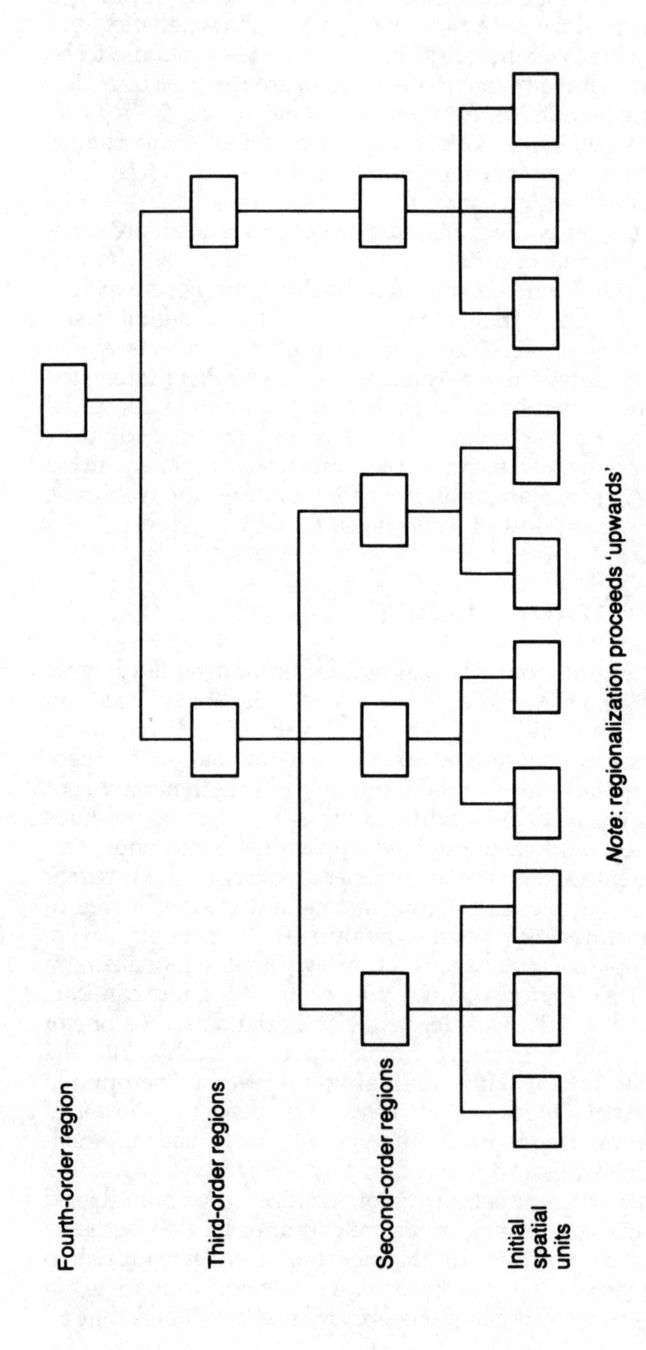

Figure 8.1 Synthetic regionalization

parts of a map where data fall into specified ranges (see, e.g. Dorney 1976).

Figure 8.1 is an illustration of the basic process of synthetic regionalization. Note how a large number of small units are gradually incorporated into a smaller number of increasing general regions. In the case represented by Figure 8.1, ten individual spatial units (called 'first-order' regions because they are the lowest level of regionalization) are grouped into four second-order regions. These are grouped, in turn, into two third-order regions. Finally, the third-order regions are agglomerated into one fourth-order region. The number of orders will vary with the regional system being developed, but in all cases a hierarchy of regions will be developed. The analyst may be interested in the entire hierarchy or in only one level.

Analytic regionalization begins with a relatively large area which is divided into smaller and smaller regions. This procedure is also known as deglomerative or descending regionalization. It is, in effect, the opposite of synthetic regionalization (Figure 8.2). The results of either procedure are the same: a hierarchy of spatial units defined in terms of greater or lesser homogeneity. Tourism analysts have thus far limited their use of analytic regionalization to examining natural resource distributions. For example, analytic regions are found in geomorphological or vegetative hierarchical systems used to describe the natural resources systems that can support outdoor recreation activities in certain areas of the world. One excellent example is the Canada Land Inventory System (1979) (Coombs and Thie 1979). The main reason for the limited use of analytic logic in tourism regions is that the procedure requires a theoretical foundation upon which to base the structure of the hierarchy. Such theory has yet to be developed in many areas of tourism.

The third type of regionalization is dichotomous regionalization. As the name suggests, this logic is based on a series of dichotomous (usually yes/no) decisions (see Figure 8.3). Unlike the other two methods, which fill a large area with contiguous regions, dichotomous regionalization may result in only one region within the context of a larger area. The region thus defined is the only sub-area that meets all the conditions specified for the particular type of region; all other sub-areas are unclassified.

Principles of regionalization

Because regionalization is a type of classification, the principles of classification can be applied, within limits, to regionalization. The following list is a tourism-oriented interpretation of a number of classification and taxonomy principles compiled by Grigg (1965). These principles are a set of 'ideals' that should be considered when defining regions. There are, however, some difficulties (as is often the case with ideals) that will prevent a tourism analyst from adhering fully to all of

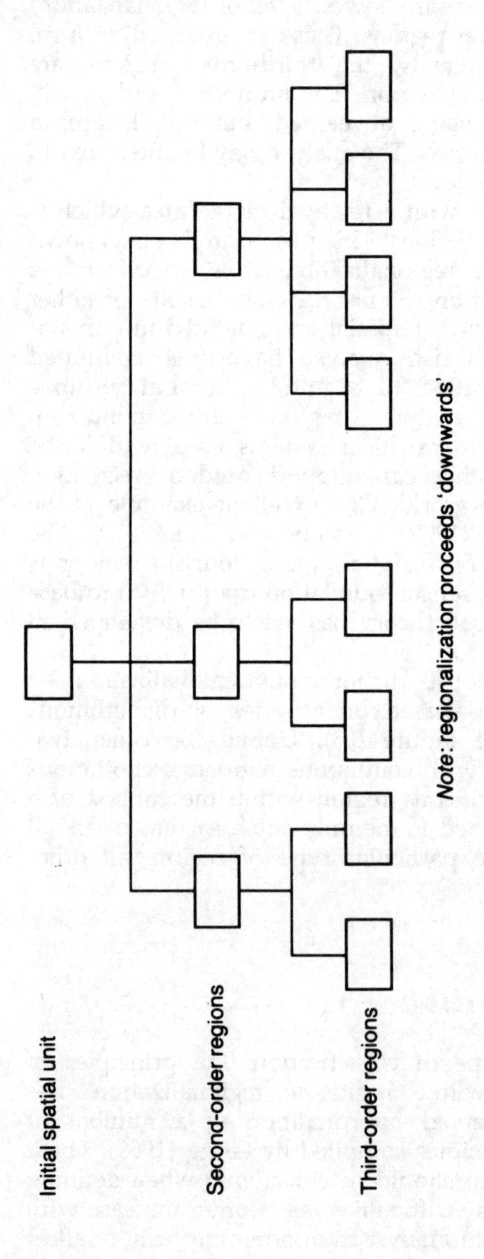

Figure 8.2 Analytical regionalization

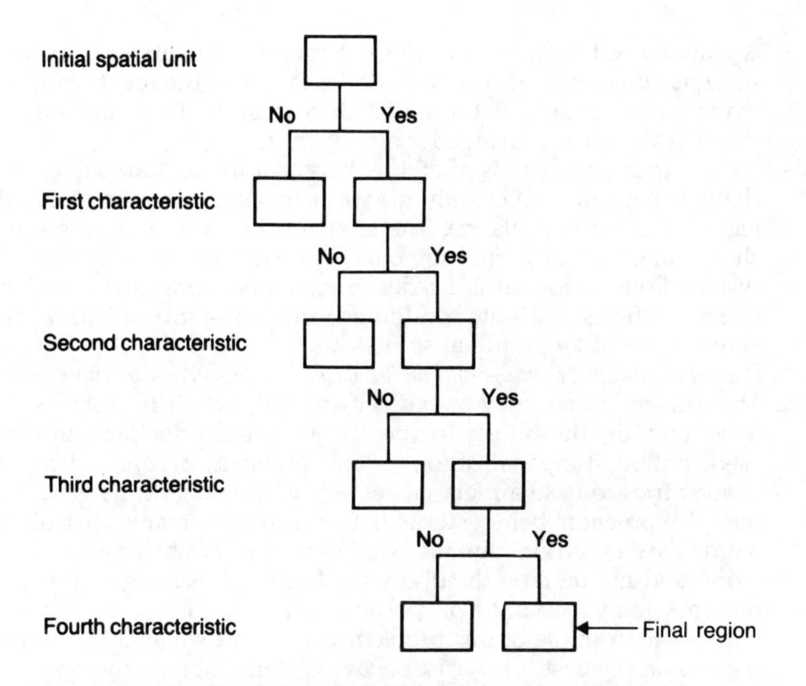

Figure 8.3 Dichotomous regionalization

them. The more important difficulties are discussed briefly along with the relevant principle.

1 *Regional systems should be specific to the problem at hand.* The system developed should be based on a set of criteria and method appropriate to the task before the analyst. A regional system developed for one tourism problem will rarely work well for another problem.
2 *Areas on earth that differ fundamentally in kind should not be subjected to the same regionalization procedure.* Many areas differ from each other in the degree to which they posses certain attributes. They belong to different regions, but they still possess the same general set of characteristics. Other areas, however, will be so different that they cannot be accommodated within the region system: land versus water areas, for example. An oceanic regional system would make no sense applied to land areas. The point is to be sure of the appropriateness of whatever system you develop for all the areas to which you intend to apply it.
3 *Regionalizations should be changed as one gains a better understanding of the areas classified.* This principle may sound like common sense, but there are two difficulties that must be recognized. Although most analysts would like to keep their regional systems up to date, there

is also a need to maintain some degree of stability within those systems. Regionalizations form a basis for subsequent work, so chaos would ensure if the basis for a large body of planning or research work were changed too frequently.

The other problem is that the very nature of tourism regions changes over time. Not only do we learn more about regions, the regions themselves change. The effect of this is to add pressure to alter fundamentally the regional system, not just update the system. Some trade-off is needed between ensuring that a regional system reflects current conditions and constant tinkering that would prevent longitudinal studies.

4 *Regionalizations of places should be based on properties of those places.* This principle, too, may sound self-evident, yet Griff observes that it is one of the most frequently violated principles in areal classification (regionalization). The problem occurs when the analyst tries to base a regional system on a model of the genesis of the phenomenon being studied. For example, many agricultural regions are based on climatic conditions that are believed to affect crops, and not on crop distributions. The result is a regional system that may fail to serve the purpose for which it is intended. Later we will see an example of this problem where tourism analysts defined regions on the basis of tourist behaviour but intended the system to be used as a structure by industry. Even if there is some strong underlying force that influences the characteristics of the places you are studying, it is better to base your regions on the characteristics themselves than on underlying forces.

5 *Regional divisions should be exhaustive.* Although this is a widely held principle in taxonomy, its relevance to tourism can be challenged. The issue here is whether there are places that a tourism analyst would consider non-regions. If the regional characteristics used for classification are a series of climatic variables, the regional system would divide the entire earth's surface into regions, because all areas have climate. On the other hand, if the regions are those places capable of supporting tourism development, many places would conceivably not be classified at all.

6 *Regional divisions should be exclusive.* The same place cannot logically be assigned to two regions. The characteristics used to define regions must be chosen carefully to make sure no inconsistency would violate this principle. In practice, such a violation will usually be obvious and can be corrected by a careful restatement of the relevant criteria.

7 *Regionalizations based on synthetic procedures should use only one differentiating characteristic at each level of division, and each differentiating characteristic must be related logically to the characteristics of the next higher and lower levels.* The first part of this lengthy principle holds that, in hierarchical classification, the divisions made at one level of the hierarchy must be based on the same characteristics across the level. For example, one might begin to define tourism regions in a country on the basis of the predominant

type of attractions: wilderness, big cities, wine-producing regions, and beaches and coastal recreation. One might then observe that each of these divisions implied other divisions. Wilderness areas might be found in interior, mountainous places; big cities on lowlands and near international boundaries; wine-producing regions in a gently rolling fertile agricultural region near lakes; and beaches and coastal recreation in the southern part of the country. There may be a temptation in the next level of the hierarchy to include geomorphology (mountains versus hills versus lowlands), land use (agriculture versus urban development), and geographic position (southern versus northern or interior versus coastal) all in the same level of hierarchy. It should be apparent that a usable and logical regionalization could not be achieved by trying to incorporate all these features at the same level.

The second part of this principle indicates that one must be careful in the choice of criteria used at subsequent levels. To continue our example, if the first level of regionalization were based on predominant type of attraction, the next level should be based on some characteristic that makes logical subdivisions of the four original classifications. The level of existing development (nil, minor, substantial) would be one possible characteristic. In contrast, dividing the four attraction classes into geomorphological units would not be logical.

8 *The characteristics used for regionalization should be relevant to the purposes of regionalization.* Application of this apparently obvious principle ultimately requires assessment of the importance of regional features. This depends, in turn, on one having a sound understanding of the purpose of doing the regionalization. A system of tourism regions based on geomorphological units might be relevant for ski resorts but probably would not be meaningful for meetings and conventions.

9 *Characteristics used to define regions at more general levels should be more important than characteristics used to define regions at more specific levels.* This principle is related to the preceding one. When two or more characteristics are used to define regions, the divisions should be made initially on the most important characteristic. It is not possible to give a fixed ranking of different criteria. This decision has to be made for each exercise. But the basic point is that one should begin the regionalization with the most important features and then refine the divisions with successively less important features.

With regard to the question of selecting and ordering characteristics, be aware how the number of regions grows exponentially with the number of characteristics and the numbers of divisions of each characteristic. If you had a regional system based on five characteristics with five classes each, you would potentially face a system with $5^{(5-1)} = 625$ regions. The need to be conservative in the choice of the number of characteristics and classes should be evident.

10 *The use of more than one differentiating characteristic in either analytic or synthetic regionalization produces a hierarchy.* The existence of hierarchies in regionalization has already been mentioned; what is at issue here is whether or not the hierarchy is valid. Any regional hierarchy is valid if, and only if, the preceding principles have been observed. Failure to do so may result in an ambiguous or meaningless hierarchy. Logical validity, though, is not enough. Hierarchical classifications are representative of reality in the case of biological taxonomy, but one can raise doubts as to whether any regional tourism hierarchy is truly representative of anything. They may be no more than an artefact of a researcher's statistical procedures.

These ten commandments could be supplemented by an eleventh: 'The first ten principles will be subject to pragmatic needs.' As we have suggested in our opening comments, there are some practical difficulties in adhering to all these ideals. When regionalization is only an early step in a much larger planning process, it is unlikely that the analyst will have the inclination to spend much time or money on regionalization alone. At best, these guidelines should alert you to potential dangers in the development of a regional system. With these in mind, you can make acceptable trade-offs between the demands of practical regionalization in tourism and the ideals of regionalization logic.

Regionalization in tourism research

Before moving to a discussion of some specific regionalization procedures, it will be helpful to examine some practical uses of regionalization in tourism to add a degree of 'reality' to the preceding discussion of principles. Analysts define and use regions for many purposes: marketing, administration, promotion, planning and research. One subject in which regionalization plays a significant but often unrecognized role is the evaluation of tourism impacts. Good or ill, the effects of tourism development are limited in space. Further, the nature of those effects varies greatly from place to place as a function of regional conditions, such as the size and complexity of the local economy, the number of tourists, the size of the host population, and the cultural context of development. Redefine or change the scale of a region wherein you are measuring impacts and you may change the conclusions drawn from those impacts.

A more obvious use of the regional perspective in tourism analysis comes from marketing. A number of authors use regions as the starting point in their analysis of variations in how people look at other parts of the world. Examples include Gartner and Hunt's (1987) assessment of the images of four western US states; Embacher and Buttle's (1989) analysis of English perceptions of Austria as a tourist destination; and McCool and Menning's (1993) study of US and Canadian tourists' perceptions of the state of Montana. Such studies are not just of academic interest. By identifying what people look for in a vacation destination and comparing this with what they see in a given destination region, tourism planners can help improve the competitive position of their study region.

A twist on the study of regional images is the a priori recognition that residents in some origin regions will tend to have more favourable images of a destination than other origins, or that the residents of those origins will receive more intense advertisements from competing destinations. In other words, the focus is not on variations in destination images, but on variations in the origins. If these variations are known, tourism marketers can work to reallocate promotional budgets so that those areas with the greatest likelihood of generating future trips receive the greatest share of the promotional budget. One such a study can be found in *Anon* (1980). Holder (1991) examined the likely impact of advertising in the US and Canada by destinations that compete with the Caribbean in her assessment of the future of tourism in that region and identification of strategies to meet that competition. Liu (1986) studied variations in spending patterns among visitors to Hawaii from different origin regions to identify which origins tend to provide the greatest net economic benefits to that state.

Regionalization also provides the basis for identifying spatial variations in the extent of existing development as well as the potential for future development. The study by Smith and Thomas (1983), discussed in the previous chapter, in which the authors identified regions in Ontario capable of supporting golf course development is one example. Fesenmaier and Roehl (1986) used a regional approach to assess the potential for expansion in the campground industry in Texas. Linley (1993) adopted a regional perspective to assess the financial performance of country inns.

The oldest use of regionalization in tourism is the definition of regions to provide a basis for regional planning and to guide tourism development. Bekker (1991) provides one example of this approach. He identifies areas along the British Columbia coast with significant natural and cultural tourism resources that need to be protected to ensure a balanced, long-term approach towards tourism development. Smith (1992) identifies and discusses the role of tourism regions in Luxembourg; and Klaric (1992) explores several different methods for defining tourism regions in Croatia. Notable among the planners who work from a regional perspective is Gunn (1972, 1994). He has been a strong proponent of a regional approach in planning and we will examine some of his work later in this chapter.

Finally, there is the most common and least sophisticated form of regionalization: the delineation of tourism regions in a political area such as a province by a tourism ministry. These regions are usually based on local perceptions or political mandates. They serve a useful purpose by helping to define tourism regions in what would otherwise be a large and amorphous area. These divisions, though, are rarely objective or verifiable – sometimes bordering on the romantic or even silly. This can become a problem because such regions are often used as the basis for forming industry or marketing associations and even serve as a framework for collecting and reporting tourism statistics. Arbitrary and subjective formation of regions needlessly complicates such tasks. Ideally, tourism regions should be defined on the basis of objective research rather than politics. Smith (1987) demonstrated one such method, based on an analysis of tourism resources in Ontario, Canada. The regions he identified reflected the real distribution of tourism attractions and resources in that province and, significantly, showed that the politically determined regions bore no resemblance to resource patterns.

We now turn to an examination of several regionalization methods that can be used to define reliable and meaningful tourism regions.

Cartographic regionalization

Description

Cartographic regionalization is a method of defining regions by drafting and then superimposing a series of maps showing the distribution of important areal characteristics. The procedure may be used to divide a large area into a number of smaller regions or to delimit a single region within a much larger area.

The procedures required for this method are relatively simple, but their successful application requires thought and careful work. The most difficult step of all can be the combination of the series of maps. The problem is not just the physical or computerized overlaying of two or more maps; rather the challenge is to develop a valid system for measuring and combining variables (which may be very different in kind) in a common unit that will allow them to be meaningfully combined into a single map. One possible solution is described in the example below.

Another step that may be valuable in this procedure is to identify and assess any 'critical' features. This step is similar to the identification of 'critical' locational variables in Chapter 7 in the section on the use of the checklist method for site selection. In this case, critical regional features are those characteristics that might cause a place to be excluded from any further consideration for tourism planning. Environmentally sensitive areas that are candidates for protection might be one example. The spatial distribution of any critical features should be mapped before any other features so that any place failing to meet the criteria for inclusion for further study can be eliminated at the outset.

As we shall see, cartographic regionalization in practice has elements of the three forms of regionalization described previously. The task of collecting data for individual places and then generalizing them to larger areas is similar to the logic of synthetic regionalization. Once a combined map has been prepared and the pattern on this map divided into a series of new regions, the analyst is engaged in analytic regionalization. Finally, the use of any critical features to identify places that should be excluded from development is an example of the logic of dichotomous regionalization.

Procedure

1 Define the basic study area. This is usually some political unit such as a county, province or nation. The advantage of using political units is that data are usually collected and reported by political unit.
2 Identify and operationally identify appropriate regional characteristics.
3 Define a measurement scale for each characteristic. The number of classes on the scale is arbitrary; it may range upward from a minimum of two (such as present/absent). Most analysts find that four or five classes is the most practical number. Weights may be used to adjust individual characteristic scores to reflect their relative importance.
4 Collect the data for each characteristic and map separately.
5 Combine the individual maps. This may be done by using transparencies of the individual maps or through a computer using a geographic information systems package.
6 Define regions on the basis of the combined patterns.

Example

Gunn (1979) provides an example of the cartographic approach in some of his work in Texas. A research team was asked to identify tourism regions in central Texas that would have the potential for automobile touring. The team began by identifying nine features important for this type of tourism: (1) water and wildlife; (2) topography, soils and geology; (3) vegetative cover; (4) climate and atmosphere; (5) aesthetics; (6) existing attractions, industries and institutions; (7) history, ethnicity, archaeology, legend and lore; (8) service centres; and (9) transportation and access. Weights reflecting the relative importance of each feature were estimated by dividing a total of 100 points among the nine features. The resulting weights were assumed to represent a ration scale.

The weights of each feature were translated into a five-step scale, ranging from 'very weak' to 'very strong'. Table 8.1 illustrates the results of this exercise. Note, for example, that 'water-wildlife' has a weight of 8; these eight points were divided into five classes to make up the five-step scale.

The team then collected data for the study area and prepared nine maps showing the distribution of values for each feature. Figure 8.4 is a draft of the map for 'water-wildlife'. The nine maps were combined by summing the scores for each part of the study area and a composite map was produced using a computer graphics package to produce a map of regions varying in touring potential (Figure 8.5).

Table 8.1 Index scales for Texas automobile touring

Factor	Weighted **index**	Very **weak**	**Weak**	**Moderate**	**Strong**	Very **strong**
			Scale			
Water, wildlife	8	0	1–2	3–4	5–6	7–8
Topography, soils, geology	10	1–1	2–3	4–6	7–8	9–10
Vegetative cover, pests	7	0	1–2	3–4	5–6	7
Climate, atmosphere	3	0	1	1	2	3
Aesthetics	13	0–1	2–4	5–7	8–10	11–13
Existing attractions, industries, institutions	10	0–1	2–3	4–6	7–8	9–10
History, ethnicity, archaeology, legend, lore	9	0–1	2–3	4–5	6–7	8–9
Service centres	15	0–2	3–5	6–9	10–12	13–15
Transportation, access	25	0–4	5–9	10–15	16–20	21–25
	100					

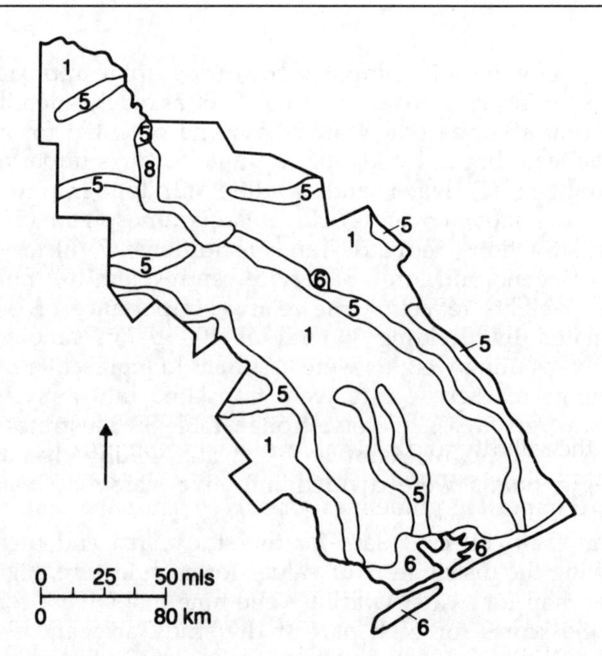

Figure 8.4 Importance of water and wildlife for 'touring–tourism' potential: Texas example (Gunn 1979)

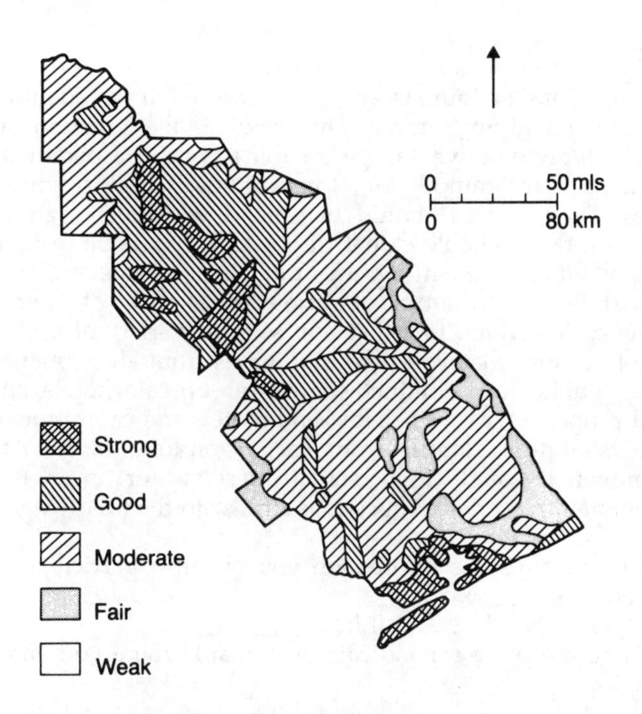

Figure 8.5 Sum of natural and cultural resource factors – destination potential: Texas example (Gunn 1979)

Perceptual regionalization

Description

Perceptual regionalization is basically the mapping of opinion. A survey is conducted to identify how tourists label the area they are visiting as well as the extent of the area covered by the label. A region is then defined by identifying a consensus of opinion on these points. The same process can then be used for business people or public officials involved in tourism.

Because the results of perceptual regionalization are mapped opinions, the validity and reliability of a perceptual region depends on the adequacy of the sampling design and the interview technique. Some of the more important guidelines for survey sampling and interviewing are covered in Chapter 3 of this book.

Perceptual regions are deceptively similar to a priori regions. Both are defined in terms of images or opinions. The difference, and it is an important difference, is that a priori regions are usually the subjective perceptions of one planner, analyst, official or committee. A perceptual region is the objective determination of the 'average' perception of many individuals whose perceptions are relevant to the analyst's project.

Procedure

1 If the opinions of tourists are of interest, begin by identifying an appropriate sampling frame. This involves the selection of places where a representative sample of tourists can be obtained: major attractions, accommodation, restaurants, visitor information centres. These places should be spread widely enough to cover more than the probable extent of the area you hope to identify so that a valid assessment of the full extent of the region can be obtained. Be sure to sample at various times of the day and days of the week. Be certain, too, that you select a variety of each type of venue to ensure that you obtain opinions from all segments of the visiting public (e.g. conduct surveys at upscale hotels and mid-priced properties, as well as low-end motels and campgrounds).

2 Select a random sample from each sampling locale. Ask each respondent (through a personal interview or in a mail-back questionnaire) a series of questions similar to the following:

   (a) Do you have the impression you are in a particular vacation area? Yes ____ No ____
   (b) If 'yes', what do you call it? ________
   (c) Name the place (or indicate on a map) where you entered the region. ________

3 Map the distribution of responses by sampling locale. Map the distribution of responses by entry point into the region.

4 Determine the regional boundaries by inspection and determine if there is a common name.

   If your task is to identify the regional boundaries as perceived by local tourism planners and operators, you can also conduct a survey of them. Because you will typically be working with fewer respondents than if you were interviewing tourists, it may be feasible to ask each respondent to draw a boundary of the tourist region on a base map you provide. An average boundary can then be estimated by inspection of the individual maps.

Example

Gunn and Worms (1973) defined a perceptual region centred on Corpus Christi, Texas, by interviewing local business people. They began by drawing up a list of potential subjects to interview. This list was then shown to key individuals in the Corpus Christi area. All names that received a positive recommendation from at least three independent sources were selected for interview. The final list was 41 local tourism operators. Each was asked the question, 'We are trying to determine how broad a geographic area should be included in the Corpus Christi region from the viewpoint of tourist use. What area would you include in the region?' The respondents were then given a map showing communities, roads and major attractions, and were asked to sketch

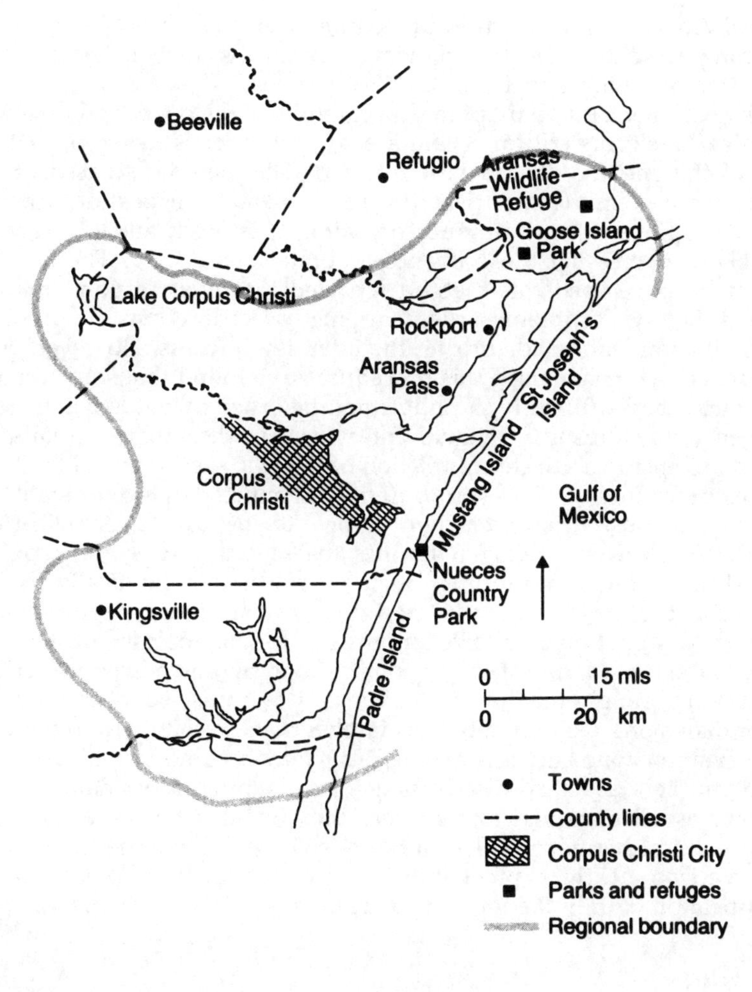

Figure 8.6 The perceptual tourism region around Corpus Christi, Texas (Gunn and Worms 1973)

their boundary on it. The researchers then combined all maps to produce the average perceived regional boundary as shown in Figure 8.6.

Cognitive mapping

Description

Cognitive mapping is a form of perceptual mapping. This particular approach, developed by Fridgen *et al.* (1983) and refined by Fridgen (1987), is a method for identifying and aggregating travellers' images of places. An advantage of this procedure over the form of perceptual mapping described in the previous section is that it permits the definition of multiple regions.

As with perceptual mapping, a carefully developed and executed sampling design is critical. There are several other issues, too, that the user of this method should bear in mind. The method is based on the assumptions that: (1) tourists do have mental images of different tourism regions in any province or nation they visit; and (2) they are capable of expressing those images accurately on a map. The research team who developed this method reported that their respondents had little difficulty completing the mapping procedure on the basis of personal comments made during the interview process. However, other researchers who have used this procedure have found that some tourists do seem to have difficulty in understanding exactly what is being asked of them. As a result, personal interviews (rather than a mail-back survey) to ensure accurate completion of the task seems desirable.

This procedure permits you to divide an entire province or state into tourism regions. There are two issues to be aware of with this application, however. First, it implies that every part of the province should be assigned to a tourism region. This may be politically correct, but in practical terms, not every part of a province has enough potential to merit designation and promotion as a formal tourism destination. Second, the boundaries of the regions defined by this procedure tend to be drawn through the middle of areal data units, such as counties, rather than along their borders. This implies that the county (or other unit) is a transition zone between regions. This may be conceptually accurate, but it can be a practical disadvantage for planning. Most data sources, industry associations and tourism marketing organizations are organized along political boundaries. The authors recognize both issues and, in the 1987 version of their procedure, limited cognitive mapping to the identification of only the top tourist areas in the state of Michigan.

Procedure

1. Choose a sampling frame to produce a representative sample of tourists or potential tourists.
2. Provide each respondent with an outline map of the area to be divided into regions. A typical map might be a provincial or state outline with county boundaries. Major cities and attractions might also be shown. Ask each respondent to circle three to five multi-county regions they perceive to be tourism destination regions. Each respondent should also be asked to place an 'X' in the county they see as the 'heart' of the region.
3. Define a tourism location score for each county:

$$\text{TLS} = (0.4A + 0.4B + 0.2C) \times 1000 \qquad [8.1]$$

where: TLS = tourism location score;
A = number of times a county received an 'X';

B = number of times a county was fully circled;
C = number of times a county was partially circled.

4. Determine county scores by summing the TLSs calculated from each respondent's map. Plot the total scores on a new map.
5. The map produced in step 4 can be interpreted as a topographic map of hilly terrain. Unlike a regular topographic map, where points of equal height are joined by contour lines, the TLS regional map may be produced by drawing lines along the 'valleys' to delimit areas of high TLS scores from each other. Alternatively, you may identify just the leading tourism centres and regions using the following step.
6. Identify the 'tourism centres', i.e. those counties that are seen as a tourism destination and that stand out among surrounding counties. Fridgen suggests the following criteria for a tourism centre: (a) the county's tourism location score is in the top 20 per cent of the TLS distribution; (b) the county has a TLS at least 25 per cent larger than the majority of contiguous counties; and (c) the county is not part of another, higher level region. (A higher level region is a group of counties associated with a regional centre with a larger TLS than the county in question.)
7. Identify the extent of the region surrounding the tourism centre by mapping the *B* factor, i.e. the number of times a county was fully circled by the respondents. Counties to be included in the region are those that are contiguous with the centre (or other county also included in the region) and whose *B* score is at least 80 per cent of the tourism centre's *B* score.

Example

Fridgen (1987) applied this procedure to a study of visitors to Michigan who stopped at state-operated travel information centres. Visitors were handed a map and asked, 'Please circle the counties on the map that you feel go together to form distinct tourism or recreation regions.' County boundaries and names were on the map, but with no other feature. Respondents were encouraged to place at least three to five circles on the map and to place an 'X' in the county that represented the centre of tourism and recreation activities in each region. Visitors were also divided into two groups, based on their self-assessed familiarity with the state. Separate maps were produced for travellers familiar with Michigan and those unfamiliar with the state. Figure 8.7 is the map for respondents familiar with the state. A total of nine regions were identified. Fridgen then went on to explore some of the probable causes for the perceptions of the respondents (noting, for example, that eight of the nine regions were located along a Great Lake coastline). He also explored some marketing and planning implications of the findings.

Fridgen also identified two issues associated with the application of the method. The weights used in the calculation of the TLS are arbitrary, i.e. they do not have any theoretical or empirical foundation.

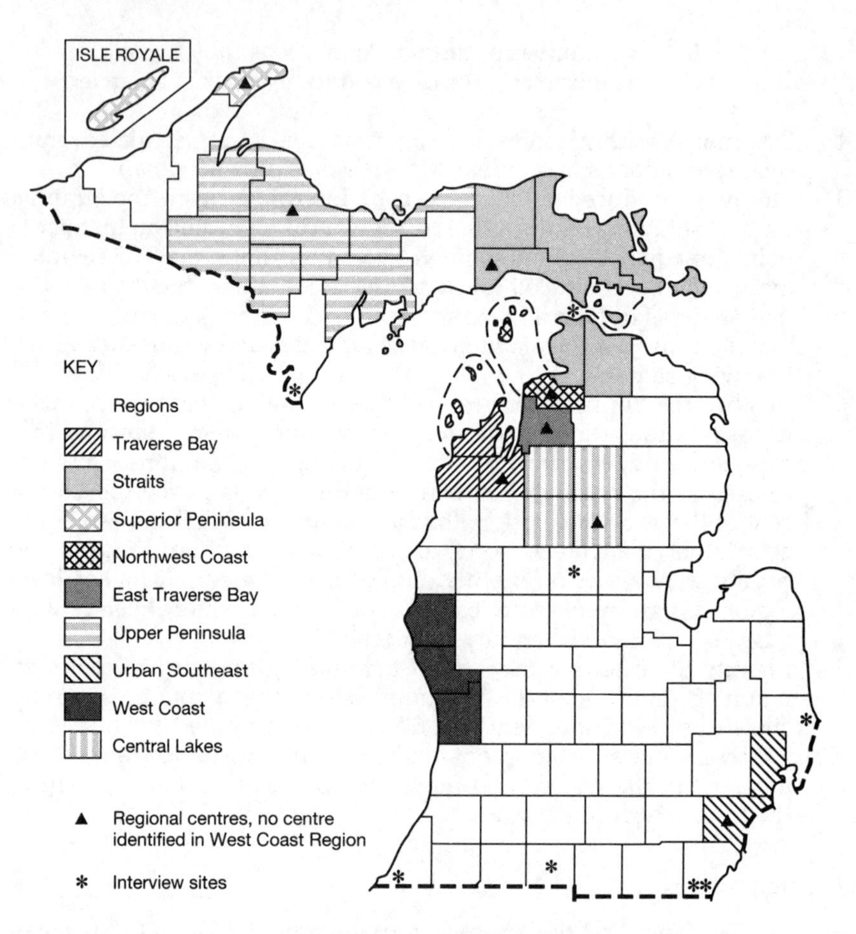

Figure 8.7 Cognitive tourism regions in Michigan (Fridgen 1987)

Experimentation with different weights, however, showed that such differences had relatively little effect on the overall regional patterns that emerged. Further, Michigan displays marked seasons, as do many tourism destinations. His sample was drawn during the summer months, and thus the regions reflect warm weather tourism and recreation activities. Whether significantly different patterns would emerge if a similar study was done in the winter remains a topic for exploration.

Functional regionalization

Description

The concept of functional regionalization is based on the notion that one can identify a set of tourism regions by examining patterns of personal travel. The basis of functional regionalization stems from Philbrick's (1957) exposition of his principle of 'areal functional organization'. At the risk of oversimplifying Philbrick's ideas, one of his points is that there is a close, interdependent relationship between spatial structure and movement patterns within a spatial system. If you can identify the movement of commodities or people for a particular activity within a spatial system (such as an urban area or province) you can deduce some understanding of the structure of that system. Berry (1966) pioneered the application of this notion to national commodity flows, while Goddard (1970) and Stutz (1973a, b, 1974) demonstrated its utility for intra-urban structures. Functional regionalization can be applied to any scale ranging from international systems to neighbourhoods. Tourism analysts are likely to find that functional regionalization is most useful at provincial/state or national levels.

From the perspective of tourism, functional regions may be interpreted as the result of trade-offs travellers make between the desire to have access to a variety of desirable destinations (which tends to pull people away from home and to disperse them across the landscape) and the desire to minimize travel costs (which tends to keep people close to home and to minimize wandering about on the landscape). Unlike cartographic, perceptual and cognitive regions, functional regions are not necessarily characterized by internal homogeneity. In fact, many physical and social characteristics of functional regions will vary as much within regions as between them. The regions are distinguished from each other on the basis of a closely linked set of internally consistent travel patterns. In practical terms, a functional region is defined as a portion of the landscape that contains a set of common origin–destination pairs and the transportation routes between them.

Calculation of functional regions requires a familiarity with factor analysis. However, the application of factor analysis to functional regionalization also reveals a phenomenon that is not likely to be familiar to analysts who may have experience with factor analysis in other contexts. Each factor is associated with a distinct region – except for the final factor. Goddard (1970) explains that factor analysis tends to produce one factor more than the number of meaningful regions. This final factor is a mathematical artefact that summarizes 'residual' travel patterns that are independent of the regions defined. As a result, the pattern implicit in this final factor is usually ignored when mapping the regions. The steps for extracting regions from the other factors are described below.

Procedure

1 Obtain data describing the number of trips made between each of a number of areas, such as provinces or states. These data will form an asymmetrical matrix where the rows represent origins and the columns represent destinations. The matrix is asymmetrical because the number of people leaving origin i to travel to destination j is not necessarily the same as the number of people leaving j to visit i.

In coding the data, the origins are considered to be cases; destinations are considered to be variables.

2 Reduce the data matrix through factor analysis. The conventions of an eigenvalue cut-off of 1.00 and varimax rotation are appropriate. Record the factor loadings for all variables (destinations) on each factor.

3 Obtain factor scores for each origin.

4 Identify regions by one of two methods:

   (a) For each factor list all areas with high loadings (important destinations) and all areas with high scores (important origins). The definition of 'high' is arbitrary; but you might use a threshold of ±0.7 for factor loadings and ±1.0 for factor scores. Locate these origins and destinations on a map and link with straight lines. The pattern of these lines defines a functional region.

   (b) Alternatively, group areas on the basis of their factor scores to define regions of destinations with common origins. This grouping is usually done with a hierarchical clustering algorithm such as that defined in Chapter 4 in the section on factor–cluster segmentation.

Example

Functional regionalization was used by Smith (1983) to define a set of tourism regions for Canada and the US. The number of vacation trips between each of the 50 states, the District of Columbia, and the 10 provinces were recorded in a 61 × 61 matrix. Factor analysis reduced this matrix to a set of 18 factors explaining about 77 per cent of the original variance. Table 8.2 lists the high loadings (greater than ±0.50) and high scores (greater than ±1.00) for the first factor. These were then

Table 8.2 Partial linkages for region 1

State/province	Loading (destination)	Score (origin)
New Brunswick	0.87	23.13
Newfoundland	–	9.65
Nova Scotia	0.71	22.16
Ontario	0.52	9.77
Prince Edward Island	0.82	16.42
Quebec	0.81	12.80
Connecticut	–	2.07
Maine	0.68	7.72
Maryland	0.75	2.25
Massachusetts	0.89	2.21
New Hampshire	–	5.17
Vermont	–	3.40

Source: Smith 1983.

Figure 8.8 Origin–destination linkages for one North American functional tourism region (Smith 1983)

plotted and connected as described in step 4(a) to produce the map in Figure 8.8. A strong regional pattern emerges, but mapping 16 other factors (not counting the final, 'non-regional' factor) produced some overlap among the regions. To identify exclusive regions, step 4(b) was followed. This led to the identification of the seven-region solution shown in Figure 8.9. The names given to each region were somewhat arbitrary, chosen to reflect either traditional regional identities or the presence of a geographical feature that seemed closely associated with the location and orientation of each region.

Destination zone identification

Description

Destination zone identification is a type of regionalization based on an inventory of qualitative characteristics of a region relevant to a given

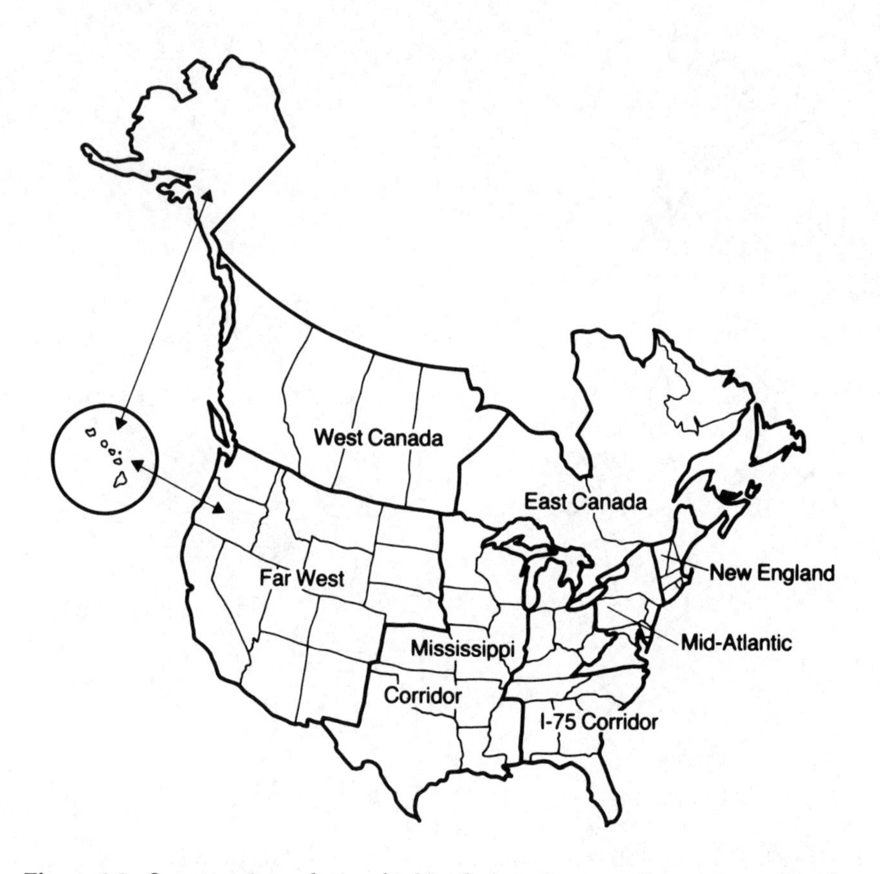

Figure 8.9 Seven-region solution for North American vacation patterns (Smith 1983)

tourism planning problem. Although it does not have the formal objectivity usually associated with statistical regionalization methods, this procedure does permit a planner or analyst to include many different types of variables and regional qualities in the regionalization system, which quantitative approaches cannot handle.

The definition of a destination zone involves the type of logic used for dichotomous regionalization. You begin with a large area and systematically identify those subsections that have certain features and those that do not. Working through a list of criteria, you slowly narrow the study area into one or more regions that meet all the criteria established for a destination zone.

The concept of a destination zone is useful, but imprecise. A number of misperceptions are sometimes associated with their use. We will look at these misperceptions after the presentation of the procedure and an example of the application of the method in Saskatchewan, Canada.

Procedure

An analyst must rely more on intuition and judgement than on quantitative skills to define destination zones. The definition of a destination zone is essentially a matter of specifying those characteristics the region should have and then identifying those areas that meet the criteria. Each planner will select criteria to the particular planning situation, but the following list, drawn from Gunn (1979), is typical.

1 The region should have a set of cultural, physical and social characteristics that create a sense of regional identity.
2 The region should contain an adequate tourism infrastructure to support tourism development. Infrastructure includes utilities, roads, business services, and other social services necessary to support tourism businesses and to cater to tourists' needs.
3 The region should be larger than just one community or one attraction.
4 The region should contain existing attractions or have the potential to support the development of sufficient attractions to draw tourists.
5 The region should be capable of supporting a tourism planning agency and marketing initiatives to guide and encourage future development.
6 The region should be accessible to a large population base. Accessibility may be by road, scheduled air passenger service, or cruise ships.

Example

An example of the establishment of destination zones may be found in the tourism plan for the Canadian province of Saskatchewan (Balmer, Crapo and Associates, undated). The province provided the consultants with the following criteria for tourism zones:

1 Regions should have common features upon which to develop an identity, such as important historical, cultural or physical resources. Regions may attract pleasure, business or personal business travellers, or some combination of these.
2 The region must be large enough to have access to the financial and human resources necessary to develop and promote tourism on a regional basis, but must not be so large that it is impractical to plan or manage.
3 Each part of the zone must have the strength of the whole zone. That is, no one city or town should dominate the whole zone.
4 Zones should follow present municipal or county boundaries.
5 Each zone should possess a sufficient quantity of tourism plant facilities.

The intent of each of these criteria was discussed by the consultant

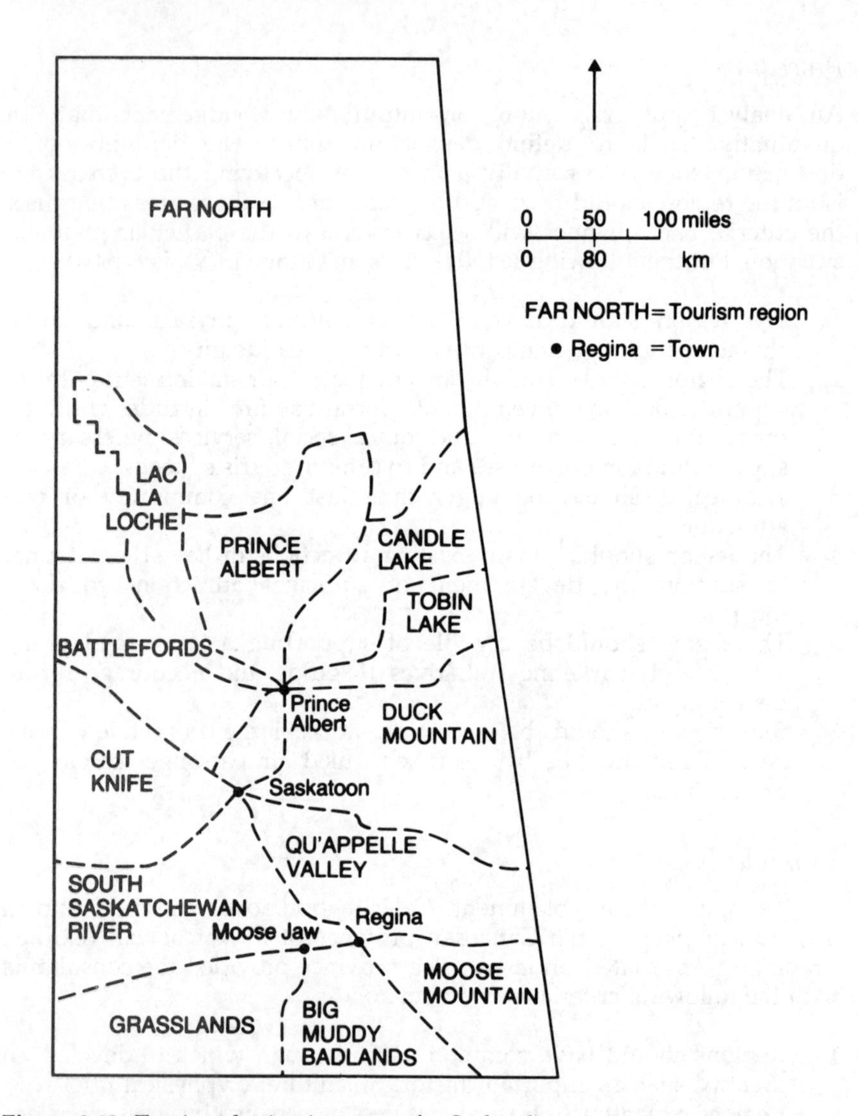

Figure 8.10 Tourism destination zones for Saskatchewan (Redrawn from Balmer, Crapo, and Associates)

with the Ministry of Tourism and Small Business. The firm then collected information relating to each criterion from existing data sources, maps and regional development reports as well as from a series of discussions with public and private agencies.

The consultants came to the conclusion that the regions should not be centred on towns or cities because travellers might tend to visit only those towns and ignore the rest of the province. This would violate the intention of the third criterion. Destination zones and the accompanying promotional material should be designed to encourage side-trips off main highways. This would lead, in the view of the consultants, to longer tourist visits and greater expenditure. Their approach called for making towns the 'hubs' of several zones simultaneously. Each zone could then draw visitors and residents alike from the 'hub' for day, weekend or week-long trips.

The zones defined this way had highways as their borders (a violation of criterion 4 that was agreed to by the province). A couple of exceptions to this new rule were made when the areas immediately adjacent to the proposed highway borders were too similar on either side of the highway to warrant separation into separate zones or when side-road development was too sparse to permit side-trips. In general, though, development zones were delimited by highways, with towns at their corners. Figure 8.10 is a map of the resulting pattern.

The destination zones in Saskatchewan were originally intended to serve both as promotional regions and for a series of industry associations. It was discovered after defining the zones that, while they did form plausible regions for promotion, the level of tourism development in most was insufficient to form a workable industry association. The problem of dividing up the businesses in hub towns into two or more industry associations or asking them to join two or more associations was neither anticipated nor resolved by the consultants.

The concept of destination zones is useful for many purposes: marketing, promotion, industry organization, planning, and policy formation. Gunn (1982) has attempted to clarify some of the confusion and ambiguity surrounding the phrase 'destination zone' as it is used in these diverse contexts. This list of 'corrected misconceptions' is based on his work for Tourism Canada.

- *Misconception 1: Destination zones have precise boundaries.* In fact, the boundaries of destination zones are wide transitional belts. There may be practical cartographic or administrative reasons for drawing narrow and precisely located lines delimiting adjacent zones on a map, but it is naïve to really believe that two plots of land separated – to use the Saskatchewan consultants as an example – by 6 m of macadam pavement belong to distinct regions.
- *Misconception 2: Destination zones exist forever.* Zones not only have fuzzy boundaries, they have fuzzy life expectancies. They do not exist in perpetuity; they may be created through human intervention and they can be dissolved through the same process. Natural landscape evolution may also change boundaries or even terminate their existence.
- *Misconception 3: Developers and local communities many simply define themselves to be a destination zone.* Local initiative can go a long way towards creating the conditions needed for tourism growth, but enthusiasm and advertising are not enough. These must be coupled with physical development and success in drawing tourists. Promotion has its place, but there has to be more to draw tourists to a destination for an extended stay of repeat visits than just a destination marketing organization agency with high ambition.

- *Misconception 4: Destination zones are exclusively tourist areas.* Destination zones are created to aid tourism development, but they require more than just tourist attractions. They must also contain utilities, hospitals, fire and police protection, tour operators, transportation services, and a variety of other business and personal services. To be sure, there are isolated resorts that do not depend on local community structures for utilities and services, but these are the exception.
- *Misconception 5: Identification of destination zones is the same as doing a feasibility study.* The identification of destination zones can be a useful part of a feasibility study, but zone identification is not always necessary and it is never a substitute for a feasibility study. Locating a tourism business in an area designated as a tourism region is no guarantee of success. Planning, design, financing, promotion and management are all more important in the success or failure of a business than just being associated with a designated region.

Summary

Tourism analysts often find it useful to define regions for research or planning. There are a number of methods available. Each has its own strengths and best applications. Destination zones, for example, are useful for a province-wide tourism development strategy; functional regions are a valuable empirical tool for analysing the linkages in mass tourism flows; perceptual and cognitive regions provide insights into tourists' images.

These methods have been developed through application and evaluation. Each is designed to assist you in the completion of larger tasks. Neither tourism analysts nor the industry have paid sufficient attention to the principles of logic required for effective regionalization. We have examined both the logic of regionalization and its application to the real world in this chapter. As a result, you should be better able to balance scientific rigour and logic with the need to produce a workable regional system. You are also in a better position to evaluate the reliability of regional systems proposed by others.

Further reading

Gunn C A and **Larsen T R** 1988 *Tourism potential – aided by computer cartography*. Centre des Hautes Etudes Touristiques, Aix-en-Provence, France.

Lovingood P E and **Mitchell L S** 1989 A regional analysis of South Carolina tourism. *Annals of Tourism Research* **16**: 301–17.

Mitchell L S and **Murphy P E** 1991 Geography and tourism. *Annals of Tourism Research* **18**: 57–70.

Opperman M 1992 Intranational tourist flows in Malaysia. *Annals of Tourism Research* **19**: 482–500.

Pearce D 1989 *Tourist development*, 2nd edn. Longman, Harlow, Essex, UK.
Teye V 1988 Geographic factors affecting tourism in Zambia. *Annals of Tourism Research* **15**: 487–503.
Walmsley D J and **Jenkins J M** 1992 Tourism cognitive mapping of unfamiliar environments. *Annals of Tourism Research* **19**: 268–86.

CHAPTER 9

Describing tourism regions

Introduction

Good research depends on good description. However, description is so much a part of everyday life that it is easy to overlook its importance to the development of new knowledge. Progress in the scientific understanding of tourism begins with descriptive procedures such as naming, classifying, measuring, comparing and summarizing. Without accurate description, we miss much of what the world has to teach us.

Consider a stroll through a forest. Almost everyone will have some sort of emotional response to the solitude, grandeur, fragrances, mystery or loneliness of the deep woods. Now, add a talkative companion who is also a professional who understands some aspect of the forest. In addition to the sensory impact of the forest, your appreciation will be different depending on your companion's expertise: botany, ecology, economics, entomology, ornithology, pedology, silviculture. Certain aspects of the forest will acquire vivid new meaning while others fade into the background. Your companion's description of the environment has changed your perception of it. The actual change depends on the methods you use to observe and the things you (or the one you are with) choose to describe. Ideally, you will not have lost your ability to respond emotionally as your view becomes more scientific, but rather you will have gained a new and keener appreciation of some part of reality.

The history of science over the last 400 years can be viewed as a history of the improvement in description. In fact, the genius of the Newtonian synthesis, which marks the birth of modern physical science, was the recognition that the central challenge to scientists was to **describe** the physical forces of the universe accurately and quantitatively and not to attempt to answer unanswerable questions about why these forces – or the universe – exist.

The challenge for a tourism analyst is similar: to provide the most accurate description possible. It is common to hear some research criticized as 'me[...] blem is often not that th[...] [...]ption is not informative or c[...] [...] to be useful. Description works best when it is tied to another problem. Description for its own sake is rarely sufficient in any social science. It must be undertaken with the spirit of trying to uncover some new, hidden relationships or patterns that will teach us new, useful things about the world. Thus the use of any method discussed in this chapter will not tell us much about tourism unless it is used to answer a question that extends beyond the mere calculation of the descriptive statistic. These methods are only means to an end. With that in mind, we examine some descriptive tools available to tourism analysts.

Mean centre

Description

A basic problem in tourism analysis is how to summarize the spatial distribution of facilities, resources or tourists. One of the simplest methods to do this is the mean centre – the spatial equivalent of the arithmetic mean. The mean centre is located by a pair of spatial coordinates that provide the point most typical of a set of points representing some tourism phenomenon.

The mean centre may be determined for one set of data or several mean centres may be determined for several sets and then compared to each other. For example, determining the mean centre of the distribution of tourists at an attraction for each hour the facility is open may reveal valuable insights into the behaviour or movement of those visitors. The mean centre might also be mapped for different sets of facilities – perhaps hotels, motels, restaurants, travel agencies – to compare the relative location of each. From this summary information you may be able to achieve a better understanding of some of the forces affecting the location and operations of each type of business.

It should be noted that any measure of central tendency, including the mean centre, necessarily loses much information. One cannot work backwards from the mean centre to reproduce the original pattern of points. Different patterns can yield identical mean centres. A major implication of this weakness is that it is often wise to present the mean centre with additional information such as a map of the actual array of points and the standard distance or the standard deviational ellipse (described later in this chapter).

The mean centre is based on the abstraction of actual facilities or other observations to dimensionless points. Many times this presents no problem, but if the units mapped occupy a significant portion of the base map or are long, linear features, the use of points, and thus the use of the mean centre, may be inappropriate.

Another problem with the mean centre is its sensitivity to extreme values. Because the value of the mean centre is based on the sum of

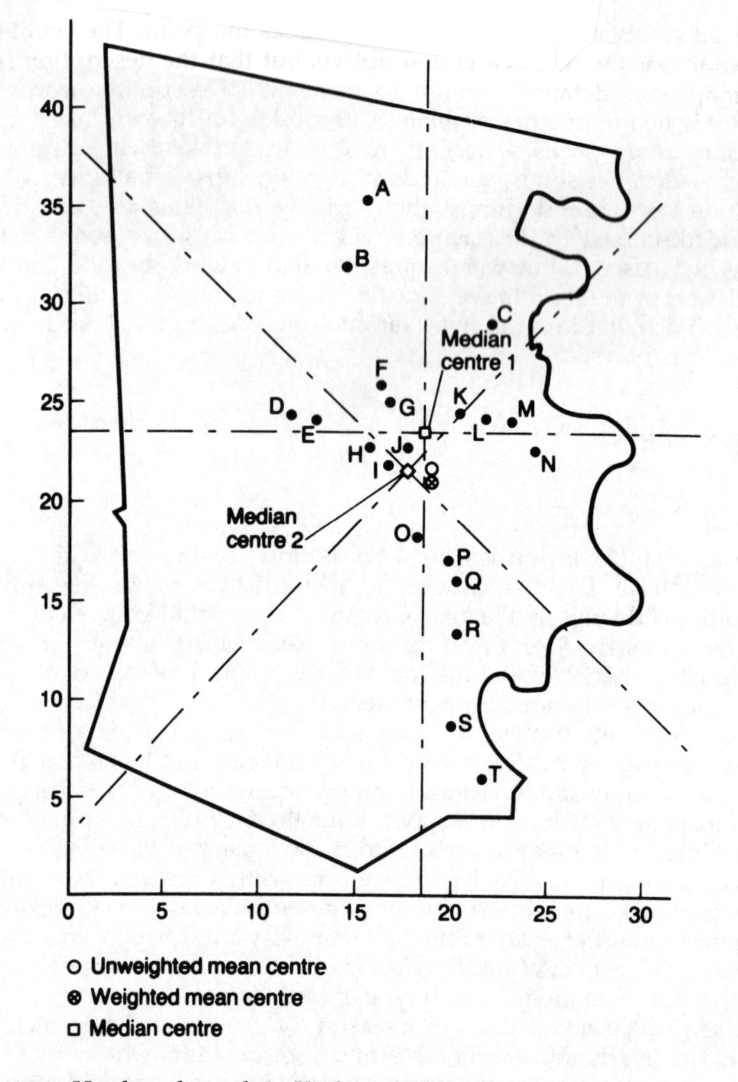

Figure 9.1 Hotels and motels in Kitchener–Waterloo, Ontario: two mean centres and two median centres

coordinates, the addition of a coordinate with a very large value will have much more effect on the location of the mean centre than the addition of several points with small coordinates. The existence of one or more extreme values can usually be detected by examining a plot of the actual data points or by the use of the standard distance statistic. If such points exist, you may wish to consider excluding them from your analysis. At a minimum, acknowledge their biasing effect in your discussion of the mean centre.

It may have occurred to you that if you can calculate a mean centre, it should be possible to calculate a median centre. Such a measure does exist. It is defined as the intersection of two perpendicular lines, each of which divides the point pattern into halves (Cole and King 1968). An example may be seen in Figure 9.1. (It should be noted that some geographers define the median centre as the point of minimum aggregate travel, such as Neft (1966) and King (1969). This is not the definition used here.) A problem with the median centre is that different median centres are obtained depending on the orientation of the perpendicular lines. An example of the lack of stability in the median centre's location may also be seen in Figure 9.1. In this case, a second median centre is determined by rotating the perpendicular lines by 45°. The lack of reliability in the median centre limits its usefulness in tourism analysis. A procedure developed by Seymour (1968) can be used to reduce the lack of reliability; his method is based on a converging algorithm that iteratively estimates the location of the point that most closely matched the statistician's notion of a median. A brief and clear discussion of the median centre problem can also be found in Taylor (1977: 30–2).

Procedure

1 Prepare a map of the study area and plot the locations of the features as points.
2 Superimpose a square grid as a coordinate system. The grid may be oriented as desired.
3 Determine the coordinates for each point. List these on a work sheet similar to that in Table 9.1. For some types of problems it may be appropriate to weight each point. Accommodation firms such as hotels, for instance, might be weighted by the number of rooms.
4 Sum the *X*-coordinates and divide by the number of points to obtain the arithmetic mean:

$$\bar{X} = \frac{\Sigma X_i}{n} \qquad [9.1]$$

where: $\bar{X}$ = *X*-coordinate of mean centre;
X_i = horizontal coordinate of point *i*;
n = number of points.

Repeat for the *Y*-coordinates.

If weighted points are used, multiply each point by its weight before calculating the mean coordinate:

$$\bar{X} = \frac{\Sigma(W_i X_i)}{\Sigma W_i} \qquad [9.2]$$

where: W_i = weight of point *i*;
other variables are as defined previously.

Table 9.1 Calculation of unweighted and weighted mean centre

Point	Weight	X	Y	Weighted X	Weighted Y
A	81	15.2	34.8	1 231.2	2 818.8
B	30	14.2	31.5	426.0	945.0
C	29	21.6	28.6	626.4	829.4
D	17	11.1	24.1	188.7	409.7
E	22	12.8	23.8	281.6	523.6
F	10	16.0	25.5	160.0	255.0
G	22	16.5	24.8	363.0	545.6
H	130	15.5	22.5	2 015.0	2 925.0
I	44	16.5	21.6	726.0	950.4
J	47	17.5	22.5	822.5	1 057.5
K	36	20.1	24.1	723.6	867.6
L	40	21.4	23.9	856.0	956.0
M	21	22.6	23.8	474.6	499.8
N	44	24.0	22.2	1 056.0	976.8
O	45	18.0	18.0	810.0	810.0
P	102	19.6	16.8	1 999.2	1 713.6
Q	25	20.0	15.8	500.0	395.0
R	122	20.0	13.8	2 440.0	1 683.6
S	40	19.9	8.4	796.0	336.0
T	20	21.5	5.8	430.0	116.0
Total	927	356.0	432.3	16 925.8	19 614.4

$\bar{X}: \frac{364.0}{20} = 18.2$ $\quad Wtd\bar{X}: \frac{16\,925.8}{927} = 18.3$

$\bar{Y}: \frac{432.3}{20} = 21.6$ $\quad Wtd\bar{Y}: \frac{19\,614.4}{927} = 21.1$

Coordinates of unweighted mean centre = 18.2, 21.6

Coordinates of weighted mean centre = 18.3, 21.1

Example

Table 9.1 and Figure 9.1 provide an illustration of this calculation – in this case the pattern in the location of hotels and motels in Kitchener–Waterloo, Canada. Each establishment has been mapped on an outline map of the urban region. An arbitrary coordinate system was overlaid and the coordinates of the points, designated by the letters A to T, noted in the table. Weights equal to the number of rooms are also given. Locations of both the unweighted and weighted mean centres were then calculated. This may be compared to the locations of two median centres, which are also shown on the map.

Standard distance

Description

As noted previously, the mean centre loses information about the distribution of facilities or tourists in a region. A useful supplement to that statistic is the standard distance, the spatial equivalent of standard deviation. This is a measure of the variation in facility or tourist locations around their mean centre. The more widely spread the locations, the greater the standard distance.

Standard distance is also a function of the size of a region. If the distribution of hotels or other accommodation in Greater London, UK, were compared to those in Kitchener–Waterloo, you would find a much greater standard distance for London. The difference in values does not necessarily indicate anything directly about the relative dispersal of accommodation in a large British city and a medium-sized Canadian urban area. Instead, the difference would probably reflect more directly the differences in the areas of the two cities. Hesitate, therefore, before comparing standard distances between regions. If the areal extent is the same, a direct comparison might be made. If their sizes are significantly different, you must convert the standard distances to a common scale. One way to do this is to calculate the standard distance for the population of the city as well as that for the phenomenon you are studying. A simple method for estimating the standard distance of the population of the city as an array of points located at the geographic centres of enumeration areas or other census divisions. These points are then weighted by the population they represent to provide a weighted standard distance for the population. The standard distance of the point pattern being studied is then divided by the weighted standard distance for the population:

$$RD = \frac{SD_x}{SD_{pop}} \qquad [9.3]$$

where: SD_X = standard distance of the point pattern as defined below in equation [9.4];
SD_{pop} = standard distance of the population;
RD = relative dispersal.

The relative dispersal of the two regions might then be compared. This approach requires, of course, that you have sufficient information to calculate the standard distance of the population. Lacking this information, you might interpret the standard distance as the radius of a standard circle. The standard circle contains all the points within one standard distance of the mean centre. Additional concentric circles can be drawn at intervals of one standard distance until the entire region is covered. You then compare the percentage of points actually found within each ring with the frequencies found in other regions or with some hypothetical pattern.

Table 9.2 Calculation of standard distance

Point	Distance to mean centre	Distance squared
A	5.9	34.81
B	4.7	22.09
C	3.4	11.56
D	3.3	10.89
E	2.7	7.29
F	2.1	4.41
G	1.7	2.89
H	1.4	1.96
I	0.9	0.81
J	0.7	0.49
K	1.3	1.69
L	1.6	2.56
M	2.0	4.00
N	2.3	5.29
O	1.5	2.25
P	2.0	4.00
Q	2.5	6.25
R	3.4	11.56
S	5.5	30.25
T	6.7	44.89
		S = 209.94

Procedure

1. Prepare a map of the distribution of facilities or tourists as described in the procedures for the mean centre. Calculate the mean centre.
2. Set up a work sheet similar to that in Table 9.2.
3. Measure the distance between each point and the mean centre. Use either 'map distance' (the distance you obtain by directly measuring on the map) or real-world distance.
4. Calculate the standard distance:

$$SD = \sqrt{(\Sigma D_i^{\,2}/n)} \qquad [9.4]$$

where: SD = standard distance;
D_i = distance between each point and the mean centre;
n = number of points, not including the mean centre.

Example

Figure 9.2 and Table 9.2 illustrate the calculation of standard distance. As in the previous example, the data came from a study of the distribution of commercial accommodation firms in Kitchener–Waterloo. The above procedure was followed, using an arbitrary scale for measuring the distance between each point and the mean centre. The

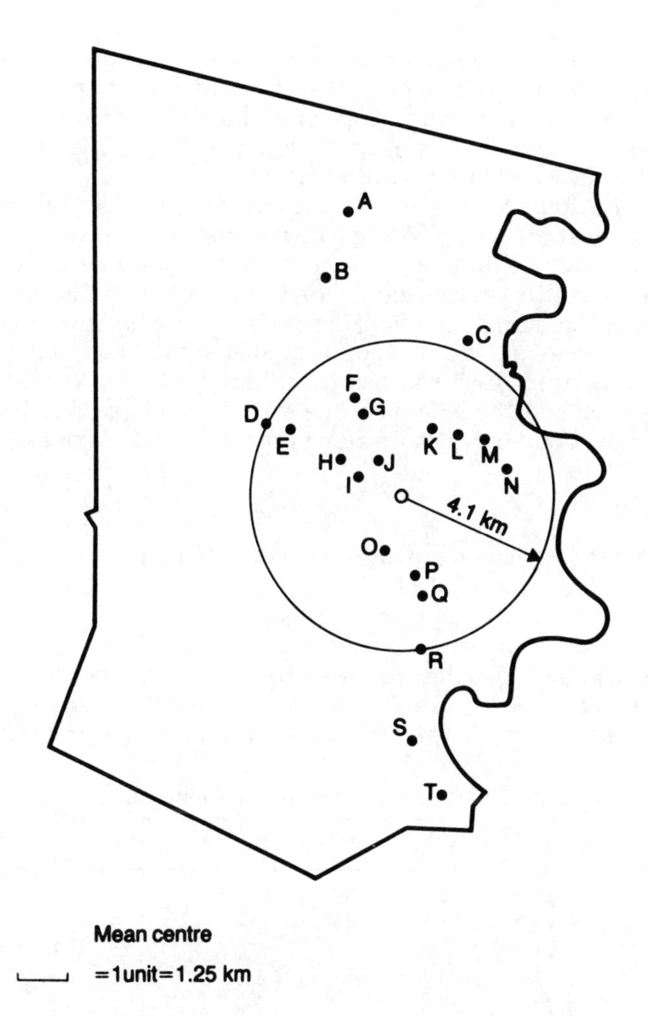

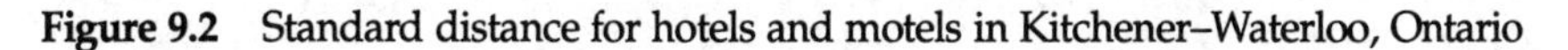

Figure 9.2 Standard distance for hotels and motels in Kitchener–Waterloo, Ontario

result was a standard distance of 3.24 units. To convert this to real-world distance, 3.24 units was multiplied by the representative fraction, 1.25 km per unit, to obtain a standard distance of 4.05 km.

Standard deviational ellipse

Description

As you probably noticed in the map of accommodation in Kitchener–Waterloo, the dispersion of points is often greater in one direction than another. In the current example, accommodation is spread further north–south than east–west. The standard distance is unable to reflect this quality because it is a single value. A more precise measure of point dispersion is the standard deviational ellipse. This measure reflects not only dispersion but also orientation. It was developed in 1926 by Lefever; Ebdon (1977) and Kellerman (1981) refined the measure into its current form.

There are four components to the standard deviational ellipse that reflect key features of the point pattern. First, the ellipse is centred on the mean centre of the pattern. The lengths of the long axis and short axis, the next two components, reflect the degree of dispersion in the maximal and minimal directions, respectively. Finally, the orientation of the ellipse represents the angle of directional bias in the point pattern.

The procedures described below indicate how to calculate the angle of orientation and the lengths of the axes. Each ellipse axis is twice the length of the standard distance along the minor and major axes.

Procedure

1 Prepare a map showing the distribution of points of the phenomenon being studied.
2 Prepare a work sheet similar to that in Table 9.3. Seven columns are required. The first two list the X- and Y-coordinates for each point; the next two list the distances between each coordinate and the mean of X or Y ($\bar{X}$ and $\bar{Y}$). These differences ($X - \bar{X}; Y - \bar{Y}$) are labelled X' and Y'. The fifth and sixth columns list the values of X'^2

Table 9.3 Data for calculating the standard deviational ellipse

Point	X	Y	X'	Y'	X'^2	Y'^2	$X'Y'$
A	15.2	34.8	–3.0	13.2	9.0	174.2	–39.6
B	14.2	31.5	–4.0	9.9	16.0	98.0	–39.6
C	21.6	28.6	3.4	7.0	11.6	49.0	23.8
D	11.1	24.1	–7.1	2.5	50.4	6.3	–17.8
E	12.8	23.8	–5.4	2.2	29.2	4.8	–11.9
F	16.0	25.5	–2.2	3.9	4.8	15.2	–8.6
G	16.5	24.8	–1.7	3.2	2.9	10.2	–5.4
H	15.5	22.5	–2.7	0.9	7.3	0.8	–2.4
I	16.5	21.6	–1.7	0.0	2.9	0.0	0.0
J	17.5	22.5	–0.7	0.9	0.5	0.8	–0.6
K	20.1	24.1	1.9	2.5	3.6	6.3	4.8
L	21.4	23.9	3.2	2.3	10.2	5.3	7.4
M	22.6	23.8	4.4	2.2	19.4	4.8	9.7
N	24.0	22.2	5.8	0.6	33.6	0.4	3.5
O	18.0	18.0	–0.2	–3.6	0.4	13.0	0.7
P	19.6	16.8	1.4	–4.8	2.0	23.0	–6.7
Q	20.0	15.8	1.8	–5.8	3.2	33.6	–10.4
R	20.0	13.8	1.8	–7.8	3.2	60.8	–14.0
S	19.9	8.4	1.7	–13.2	2.9	174.2	–22.4
T	21.5	5.8	3.3	–15.8	10.9	249.6	–52.1
MEANS:	18.2	21.6		*TOTALS:*	192.0	930.3	–181.6

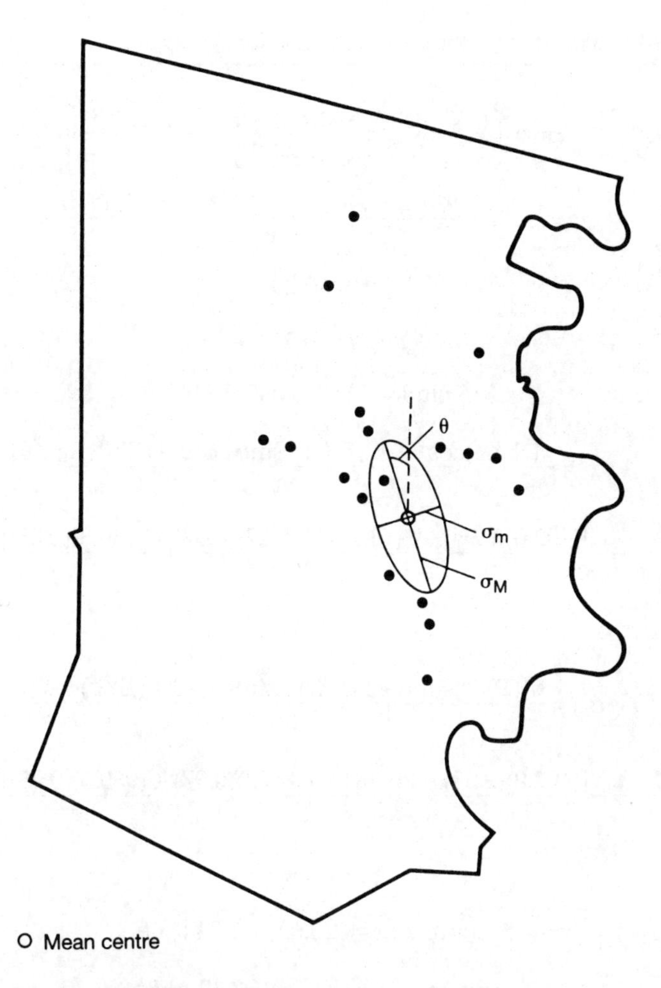

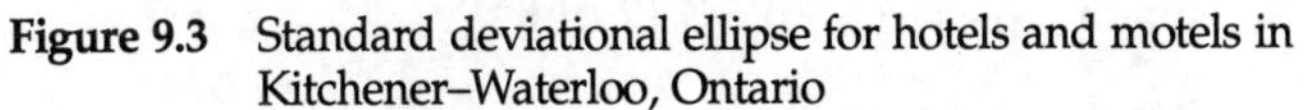

Figure 9.3 Standard deviational ellipse for hotels and motels in Kitchener–Waterloo, Ontario

and Y'^2. The last column contains the values of $X'Y'$. Totals are needed for X, Y, X'^2, Y'^2 and $X'Y'$.

3 Calculate the angle of rotation, θ:

$$\tan\theta = \frac{(\Sigma X'^2 - \Sigma Y'^2) + \sqrt{[(\Sigma X'^2 - \Sigma Y'^2)^2 + 4(\Sigma X'Y')^2]}}{2(\Sigma X'Y')} \qquad [9.5]$$

where: $\tan\theta$ = tangent of the angle of rotation, θ; all other variables are as defined previously.

Table 9.4 Calculation of the standard deviational ellipse

$$\tan\theta = \frac{(\Sigma X'^2 - \Sigma Y'^2) + \sqrt{[(\Sigma X'^2 - \Sigma Y'^2)^2 + 4(\Sigma X'Y')^2]}}{2(\Sigma X'Y')}$$

$$= \frac{(192.0 - 930.3) + \sqrt{[(192.0 - 930.3)^2 + 4(-181.6)^2]}}{2(-181.6)}$$

$$= -0.23$$

$$\approx -13^\circ$$

$$\sin\theta = -0.22 \quad \cos\theta = 0.97$$

$$\sigma_M = \frac{\sqrt{[\Sigma(X'^2)\cos^2\theta - 2(\Sigma X'Y')\sin\theta\cos\theta + (\Sigma Y'^2)\sin^2\theta]}}{n}$$

$$= \frac{\sqrt{[(192.0)(0.95) - 2(-181.6)(-0.22)(0.97) + (930.3)(0.05)]}}{20}$$

$$= 0.62$$

$$\sigma_M = \frac{\sqrt{[(\Sigma X'^2)\sin^2\theta + 2(\Sigma X'Y')\sin\theta\cos\theta + (\Sigma Y'^2)\cos^2\theta}}{n}$$

$$= \frac{\sqrt{[(192.0)(0.05) + 2(-181.6)(-0.22)(0.97) + (930.3)(0.95)]}}{20}$$

$$= 1.56$$

minor axis = 2(0.62) = 1.24 units

major axis = 2(1.56) = 3.12 units

4. Calculate the standard deviation along the minor axis of the ellipse:

$$\sigma_m + \frac{\sqrt{[\Sigma(X'^2)\cos^2\theta - 2(\Sigma X'Y')\sin\theta\cos\theta + (\Sigma Y'^2)\sin^2\theta]}}{n} \qquad [9.6]$$

where: σ_m = standard deviation along the minor axis; all other variables are as defined previously.

5. Calculate the standard deviation along the major axis of the ellipse:

$$\sigma_M = \frac{\sqrt{[(\Sigma X'^2)\sin^2\theta + 2(\Sigma X'Y')\sin\theta\cos\theta + (\Sigma Y'^2)\cos^2\theta]}}{n} \quad [9.7]$$

where: all variables are defined as previously.

6. Plot the standard deviational ellipse by centring the ellipse on the mean centre. The lengths of each axis are obtained by doubling the standard distances, σ_m and σ_M. Orient the ellipse according to the angle measured clockwise from 'north' (the Y-axis passing through the mean centre). If it is negative, measure the angle anticlockwise. It is often possible to verify the orientation by visual examination of the ellipse and the scatter of points.

Example

Tables 9.3 and 9.4 and Figure 9.3 illustrate these calculations using our Kitchener–Waterloo data. Equation [9.5] yields a tangent value for θ of –0.23, which is associated with angles of approximately 13° and 347°. Because the tangent is negative, the ellipse is rotated anticlockwise 13° from 'north' (or clockwise 347°).

Equations [9.6] and [9.7] provided values for σ_m and σ_M that, when doubled, give the lengths of the greater and lesser axes of the ellipse. These values, combined with the angle of rotation and mean centre, produced the ellipse shown in Figure 9.3. As previously, the units of measurement are arbitrary; they can be selected to conform to any scale appropriate to the problem at hand.

Defert's **Tf**

Description

Defert's *Tf* ('tourist function') (Defert 1967) is a simple measure of the importance of tourism within a regional economy. Specifically, *Tf* is the ratio between the number of tourist beds and the resident population. Because it is only a surrogate for the actual importance of tourism in a region, it must be used with caution. If the following warnings are observed, however, *Tf* can provide a useful measure of tourism development in a region.

The first warning is that comparison of *Tf* among cities of greatly different sizes can lead to some logical but misleading conclusions. Very large cities such as Paris, Tokyo or Mexico City will have small *Tfs*, especially in comparison to *Tfs* of resort towns. This suggests that tourism is only a small sector of the economy of these large cities. This is true, but it should not be taken to mean (as it easily might) that tourism in large cities is unimportant for the region or nation in which the city is located. Although the *Tf* for Paris, for example, will not be the highest in France, Paris is responsible for more tourist-nights than any other location in France. Care should be taken when making inferences from the size of a *Tf* about the importance of a city as a tourism generator in the context of a region or nation.

Also, *Tf* is vulnerable to a seasonal bias. Regions that have a large percentage of their tourist capacity as hotels and motels will usually have *Tf*s closely correlated with the actual size of the tourism industry in that region. On the other hand, regions that have a substantial portion of their capacity in campsite or seasonal resorts may have *Tf*s that are much higher than the actual importance of tourism would dictate.

In some locales the regional capacity of accommodation may yield a *Tf* that is much smaller than you would expect to find. This occurs in regions where a high percentage of accommodation comprises privately owned cottages or condominiums. Tourism – in the form of second home use – can be a very important source of local economic activity, but the *Tf* ratio will usually miss this because private second homes are typically excluded from the calculation of *Tf*.

The definition and collection of accommodation data also poses problems. The availability of accurate data frequently puts a limit on the accuracy of analysis. Careful thought is called for in the definition of certain types of accommodation capacity. Campsites, for example, rarely have any prescribed limit on the number of people that can use them as long as they are members of the same party. This is true, too, for cottages. The most practical solution for estimating the capacity of these types of accommodation is to determine from surveys the average number of occupants per unit and then to interpret that as the basis for measuring capacity. This was done in the following example from New Brunswick. If any accommodation is open for only part of the year, you might weight its capacity by the fraction of the year for which the business operates.

Procedure

1. Identify the range of tourist accommodation in the study region: hotels, motels, cottages, resorts, campgrounds and so on. Obtain reliable counts of the nightly capacity (in terms of 'beds' or 'person-nights') for each type of accommodation. Sum these and designate the total as *N*.
2. Obtain a reliable estimate of the location population, *P*.
3. Calculate *Tf*:

$$Tf = \frac{100(N)}{P} \qquad [9.8]$$

where the variables are as defined above.

Table 9.5 Calculation of *Tf* for selected counties in New Brunswick, Canada

County	Hotels and motels (beds)	Cottages (beds)	Campgrounds (sites)	Total capacity	Population	*Tf*
Albert	567	1 952	2 709	5 228	21 946	23.8
King's	651	8 128	3 651	12 430	43 137	28.8
Westmorland	5 574	10 200	6 321	22 095	102 617	21.5

Source: Keogh (1982).

Example

Table 9.5 is an example of the calculation of *Tf*. These data come from Keogh's (1982) study of the New Brunswick (Canada) tourism industry. The number of hotel and motel rooms and campsites was collected for each county in the province (only three counties are shown here). The number of commercial cottages was also obtained from government sources. The number of rooms and campsites was multiplied by three, the average number of guests per room or at a campsite; the number of cottages was multiplied by four, the average number of guests staying at a cottage. The number of beds in hotel and motel rooms was not weighted. These estimates were then totalled and transformed using equation [9.8] to obtain county values of *Tf*.

Compactness index

Description

Tourism research is often concerned with describing the characteristics of destination regions: size, climate, attractiveness, level of development, etc. One quality that is often overlooked but that can be of value is the shape of the region. Shape is a simple and valid measure of the overall internal accessibility of the region. The more compact the region, the easier it will be to ship commodities or to move tourists around the region, *ceteris paribus*.

Many shape definitions, if given at all, are qualitative: Brazil is triangular; Italy looks like a boot. Such qualitative descriptions clearly are not useful to planners or analysts, yet quantitative measures have been slow in developing. There are two major reasons for this slowness.

First, there are formidable technical problems to developing a usable measure of shape. Some of the indices that have been proposed involve integral calculus, a level of mathematics usually missing from most tourism curricula. Others are based on a lengthy list of every change of direction in the boundary line. This latter procedure is commonly used in land deeds, but is not practical as a shape index.

The other reason for the lack of interest in the shape of tourism regions is that shape is determined by the border of the region – and the border is the edge of the irrelevant. Once a planner gets to the border, he is usually not interested in looking any further. The lack of interest in what lies beyond the border may unintentionally be translated into a lack of interest in the border itself – and thus in the shape of the region.

A number of shape indices have been developed by geographers. An overview of these is found in Coffey (1981: 101–4). One of the measures described by Coffey that shows some promise for tourism analysis is the compactness index, C. The use of C is appropriate because of the emphasis placed by planners on providing services to tourists in a region. The difficulty of service provision tends to vary inversely with a region's compactness. The measure is also useful as an indicator of the relative degree of physical contact with surrounding regions (potential origins or destinations). The less compact the shape, the greater the relative boundary length and the greater the degree of contact with adjacent regions.

The compactness index is a ratio scale. This makes it available for use in many other applications. Like other measures of shape, though, it is incomplete. It indicates nothing about size, smoothness of the border, the presences of 'holes' in the region, or of the separation of the region into two or more subregions. The impossibility of developing a single measure of shape that contains all this information means that additional information, including maps, should be used to complement the compactness index.

Procedure

1 Obtain a base map of the region. Calculate the area of the region. A planimeter is a useful device for doing this if you have one, or you can use the tedious but simple procedure of overlaying a fine grid and counting squares within the border to estimate area. Designate the area as A.
2 Measure the greatest diagonal, D', of the region. This diagonal is the longest straight line that can be drawn between any two points of the boundary of the region.
3 Calculate the diameter, D, of a circle with the same area as the region:

$$D = 2\sqrt{(A/\pi)} \quad [9.9]$$

4 Calculate the compactness index, C:

$$C = \frac{D}{D'} \quad [9.10]$$

The extreme values of C are 0.00 if the region were a line, and 1.00 if the region were a circle. The higher the value of C, the more compact the region.

Example

Equations [9.9] and [9.10] have been applied to two countries that illustrate the extremes of shape: a compact country, Zimbabwe, and an elongated country, Chile. Table 9.6 and Figure 9.4 summarize these calculations. Chile's index of 0.21 is much smaller than Zimbabwe's index of 0.92, reflecting the less compact shape of Chile.

Table 9.6 Calculation of index of compactness

CHILE
Area measured from map: 56 square units

$$D = 2\sqrt{[A/\pi]} = 2\sqrt{[56/3.1416]} = 8.44$$

$$C = \frac{D}{D'} = \frac{8.44}{40} = 0.21$$

ZIMBABWE
Area measured from map: 234 square units
D′ measured from map: 21 units

$$D = 2\sqrt{[A/\pi]} = 2\sqrt{[234/3.1416]} = 17.26$$

$$C = \frac{D}{D'} = \frac{17.26}{21} = 0.82$$

Connectivity index

Description

As previously noted, the significance of the compactness index is that researchers sometimes want to examine the relative compactness and internal accessibility of a region. More compact regions have a greater degree of internal accessibility, *ceteris paribus*. Other things, however, rarely are equal. A direct measurement of accessibility can be a useful complement to the compactness index and may be useful in its own right.

One measure of accessibility with relevance to tourism is based on the fact that travel generally follows established routes. These routes, links connecting origin and destination nodes, form a transportation network. The connectivity index describes the overall accessibility of a region in terms of the level of interconnectivity among nodes in the network. In general, the greater the connectivity, the better it is for tourism.

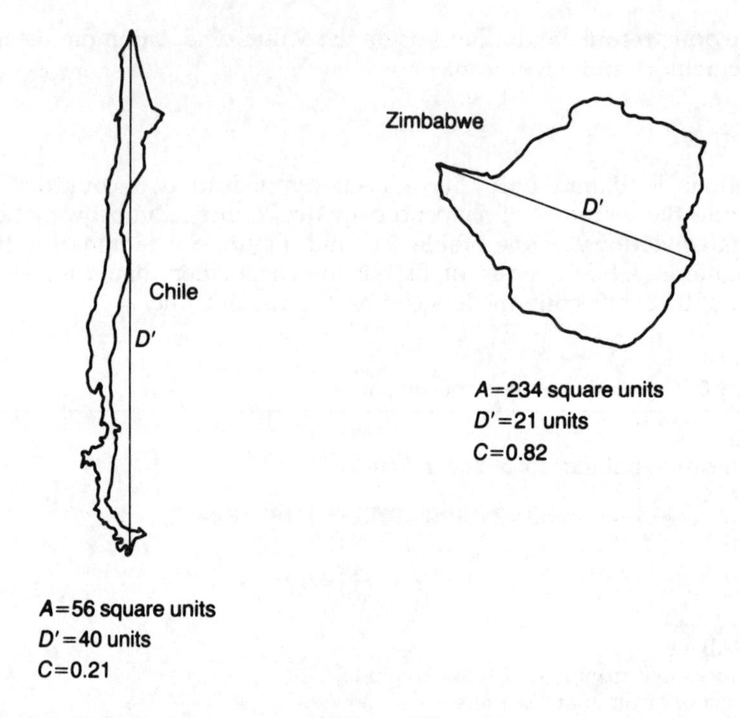

Figure 9.4 Indices of compactness: Chile and Zimbabwe

The connectivity index comes from a branch of mathematics known as graph theory. Graph theory is concerned with the properties of networks. Taylor (1977: 58–65) discusses six statistics used by graph theorists. Of these six, the gamma index, γ, also called the relative connectivity index, is the most useful for tourism. Gamma is independent of the absolute size of the network (a desirable quality when you want to compare different levels of network development) and is relatively simple to calculate.

Gamma is based on the ratio of actual to possible linkages; it indicates nothing about the ease of travel, travel time, or the lengths of individual line segments. The higher the degree of connectivity, the higher the value of γ.

This simple statistic can produce deceptive results if you use it to draw conclusions about networks of greatly different physical size. Two networks might both have indices of 0.75, indicating a relatively high degree on connectivity, but if one region has links averaging 10 km and the other has links averaging 100 km, the conclusion of equal accessibility must be properly interpreted. In this type of comparison, equal connectivity is not the same as equal travel time or cost.

Procedure

1 Obtain a map showing the major transportation corridors between

towns, resorts, and other significant nodes. The definition of 'major corridor' and 'significant nodes' is a matter of judgement. Because the index is based only on the topological properties of the network, the map need not be drawn to scale. In fact, a rough sketch or a stylized cartogram showing routes and nodes is usable.

2 Count the number of direct links between pairs of nodes. Label this L.
3 Count the number of points, P.
4 Calculate γ:

$$\gamma = \frac{L}{3(P-2)} \qquad [9.11]$$

Equation [9.11] is basically the ratio between the actual number of links in a network and the total possible number given the existing points. Extreme values are 0.00 for a system of points totally unconnected with each other, and 1.00 for a system where every possible connection is made. The index is meaningful for any network with three or more points.

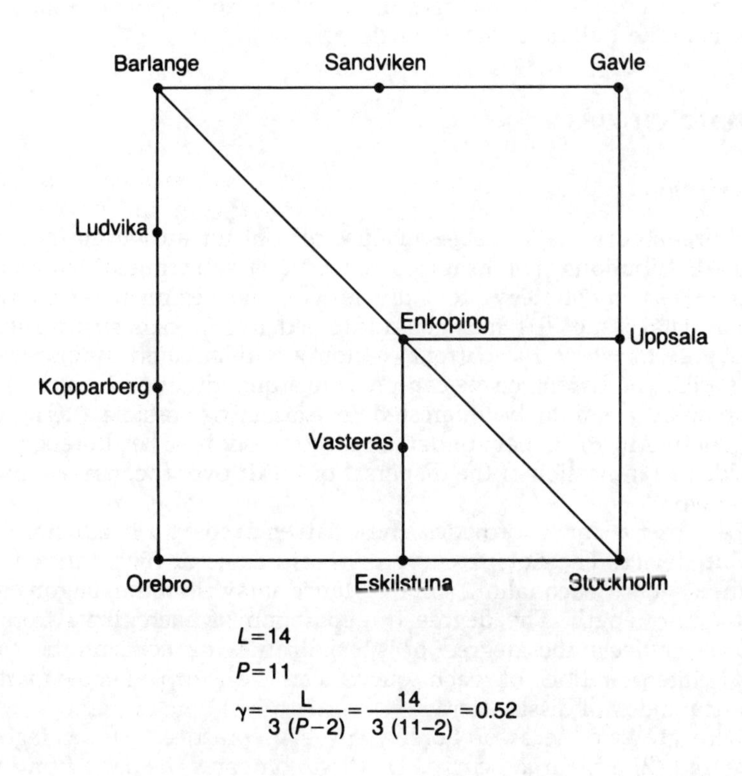

Figure 9.5 Index of linkage connectivity: a Swedish road network

Example

Figure 9.5 is an example of the calculation of γ for a portion of the highway network in Sweden. The value of 0.52 for γ indicates a system with only a moderate degree of connectivity. This can easily be seen by noting the relatively large number of towns that are linked to only two other towns.

You should note that equation [9.11] is appropriate only for networks that can be represented on a flat sheet of paper – planar networks. This is normally not a problem if we are studying road or rail linkages for which every intersection is meaningful. Many times, though, tourism networks have linkages that cross but do not form meaningful nodes. The flight paths of airlines are an example. These networks are non-planar and require a modification of equation [9.11]:

$$\gamma = \frac{L}{(0.5)P(P-1)} \quad [9.12]$$

If we apply equation [9.12] to our example, $\gamma = 0.25$. This is much lower than the previous value of γ. The reason for the drop is that there are many more possible linkages if we are free to 'hop' over established lines to connect all points on the network.

Lorenz curves

Description

The Lorenz curve is an especially useful tool for answering questions about distributions. For example, tourism is sometimes identified by planners as a strategy for diversifying a region's economy. A fundamental issue that must be addressed in this case is determining the degree to which the current economy is dominated by one or two industries. The Lorenz curve can provide a quantitative measure of this. Other analysts might be interested in examining traffic patterns on a road network to better understand congestion. The Lorenz curve provides an indication of the dispersal of traffic over the various links in the network.

The Lorenz curve provides these diverse comparisons through a graphic device. The Lorenz curve consists of one or more curved lines within a square, each falling some distance away from the diagonal over most of its length. The degree of separation of each curve from the diagonal reflects the degree of specialization or concentration. The visual interpretation of each curve can be supplemented with a statistical index of dissimilarity.

There are two types of Lorenz curve in practice: (1) a categorical curve and (2) a bivariate curve. The first compares the concentration of some phenomenon over a range of categories to a uniform distribution in those categories. The second compares the concentration of some

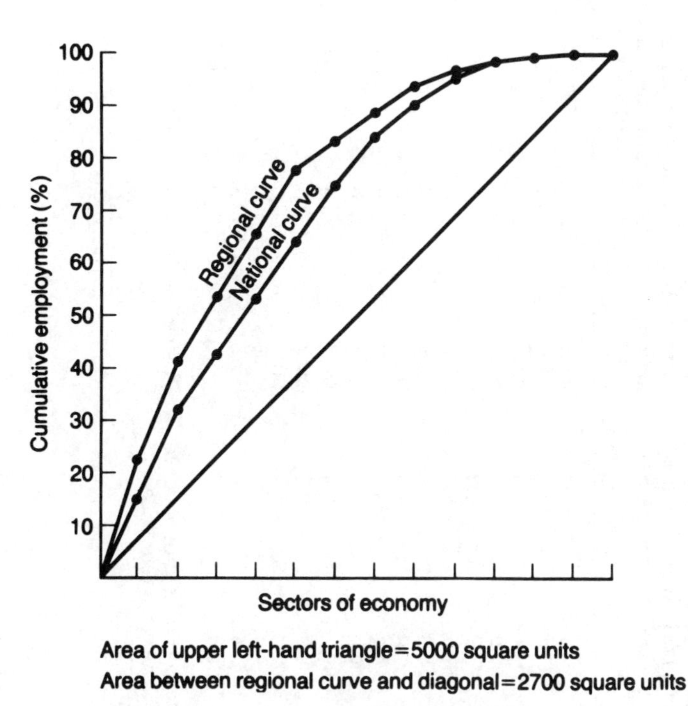

Figure 9.6 A categorical Lorenz curve for employment

phenomenon to an independent phenomenon. This distinction will become clearer when we look at the examples below. The procedures for the two different curves are slightly different, so we will examine them separately.

Procedure for the categorical Lorenz curve

1 Data used for a Lorenz curve must meet certain requirements. The data must: (1) be expressible as percentages; (2) make sense when the percentages are added; (3) consist of positive values only; and (4) be capable of being assigned to discrete categories. Types of data that *cannot* be used include annual changes in the numbers of tourists, population or tourist density, attitudinal measures, and measures of regional attractiveness. In contrast, actual numbers of tourists, tourists' expenditures, business revenues, and numbers of trips can be described with a Lorenz curve. Once you have determined that your data meet these requirements, arrange the data in a table similar to that in Table 9.6.

2 Set up a square graph similar to that in Figure 9.6. The variable you wish to describe, Y, is plotted along the vertical axis; the categories are placed along the horizontal axis. Divide the horizontal axis into equal intervals from 0 to 100 per cent. Divide the horizontal axis

Table 9.7 Hypothetical employment data for a 13-sector economy

Sector	Raw percentages		Ranked percentages				
	% in region	% in nation	% in region	Cumulative region	% in nation	Cumulative nation	Cumulative percentages for diagonal (even distribution)
Fisheries	0.1	0.2	23.0	23.0	17.2	17.2	7.7
Forestry	0.2	0.5	18.5	41.5	14.6	31.8	15.4
Mining	0.7	0.1	13.4	54.9	11.3	43.1	23.1
Agriculture	2.8	3.5	12.0	66.9	11.1	54.2	30.8
Manufacturing	18.5	14.6	11.0	77.9	11.0	65.2	38.5
Construction	3.3	5.4	5.7	83.6	9.9	75.1	46.1
Transportation, communication	5.7	11.3	4.7	88.3	8.7	83.8	53.8
Trades	13.4	11.1	4.6	92.9	6.5	90.3	61.5
Finance, insurance, real estate	4.7	6.5	3.3	96.2	5.4	95.7	69.2
Business and personal services	23.0	17.2	2.8	99.0	3.5	99.2	76.9
Commercial services	12.0	9.9	0.7	99.7	0.5	99.7	84.6
Other services	11.0	8.7	0.2	99.9	0.2	99.9	92.3
Public administration, defence	4.6	11.0	0.1	100.0	0.1	100.0	100.0
				$A = 1023.8$		$R = 955.2$	$R = 699.9$
				$M = 13 \times 100 = 1300.0$			$(100/13 \cong 7.7)$

Graphic method for regional curve:

$$\text{Index of dissimilarity} = \frac{\text{area between curve and diagonal}}{\text{area of upper left–hand triangle}}$$

$$= \frac{2700 \text{ square units}}{5000 \text{ square units}}$$

$$= 0.54$$

Arithmetic method for regional curve:
Using national curve as reference:

$$I = \frac{A - R}{M - R}$$

$$= \frac{1023.8 - 955.2}{1300.0 - 955.2}$$

$$= 0.20$$

Using the diagonal as a reference:

$$I = \frac{1023.8 - 699.9}{1300.0 - 699.9}$$

$$= 0.54$$

into an appropriate number of equal-sized units (the number being determined by the number of categories).

3 Sum the values of *Y*. Convert the individual values into percentages of this total. Order the percentages from high to low.

4 Plot the cumulative percentages of *Y* over all categories on the graph, beginning in the lower left-hand corner. The curve will first rise steeply but become less steep as it approaches the upper right-hand corner.

Example

Table 9.7 and Figure 9.6 illustrate the construction of a categorical Lorenz curve. In this case the data describe levels of employment in various sectors of a national and regional economy, represented by two separate curves. If employment in either the nation or the region were evenly distributed over all sectors, their curve would fall on the diagonal. On the other hand, if all employment were concentrated in one sector, the curve would form a right-angle triangle covering the upper left-hand half of the square. In most cases the curve will fall between these extremes.

Note that in Figure 9.6 the regional curve is further from the diagonal than the national curve. This indicates a higher degree of economic specialization in the region. A more precise measure of the degrees of specialization may be obtained in either of two ways. The simplest conceptually, but often more tedious in practice, is to compare areas on the graph:

1 Calculate the area of the triangle formed by the diagonal, the left-hand border, and the top border.

2 Calculate the area between the curve and the diagonal.

3 Divide the area from step 2 by the area from step 1. The quotient is an index of dissimilarity that ranges from 0.00 (no concentration) to 1.00 (complete concentration in one category).

An alternative method requires a bit more arithmetic but allows for more flexibility and avoids the task of having to calculate the areas:

1 Create a new column in the original data table showing the cumulative percentages. Sum the values of the cumulative percentages and designate the total as *A*.

2 Determine the maximum possible value of *A*, i.e. the value if the pattern were one of complete specialization. This would be represented by a value of 100 per cent in the first category; all subsequent categories would thus also show a value of 100 per cent. With 13 categories, the total would be $13 \times 100 = 1300$. Designate this value as *M*.

3 Compute the cumulative percentage total for a reference curve, usually (but not necessarily) the diagonal. Alternatives to the diagonal could include using a national curve as a reference for a

regional curve. Designate the total cumulative percentage for the reference curve as R. With 13 categories, each category would have 100/13 = 7.7 per cent. The cumulative percentages in our example are shown in Table 9.7.

4 Calculate the index of dissimilarity, I, with the following equation:

$$I = \frac{A - R}{M - R} \quad [9.13]$$

Note that the value of I depends on which curve is used as the reference curve. In any case, the lower the value of I, the greater the similarity between two curves. A value of 0.00 indicates a perfect match; a value of 1.00 indicates no similarity between the two curves, i.e. complete specialization.

The index of dissimilarity has been calculated for our example with both methods. Indices using both the diagonal and the national curve as the references are determined for the regional curve as well. Notice that the index of dissimilarity should have the same value for the graphic and the arithmetic method when the diagonal is used as the reference curve.

This example illustrates a limitation in the Lorenz curve we have not yet mentioned. If we were to combine forestry, mining and fisheries into a single category called 'primary industries', both the region and the nation would show a more balanced economy. The reason for this is not, of course, due to any structural change in the economy but only to statistical reasons: the new, aggregated sector has a larger percentage of the total employment than any of the three original sectors. Care must be taken, therefore, to make sure the scope of the categories used to report the data are the most appropriate for the problem at hand.

Procedure for the bivariate Lorenz curve

The major difference between the categorical and the bivariate curves is that the bivariate curve uses a second variable, X, in place of the categories. Both variables must meet the same requirements described previously.

1 Begin with the construction of the Lorenz curve as described above, except mark both axes in percentages.

2 Before ranking the percentage values of Y in decreasing order, divide the value of Y for each observation by its corresponding X value. This step is necessary to ensure that the Lorenz curve will have the desired shape.

3 Plot the ranked percentages of Y against the corresponding percentages of X. Remember to use the ratio between Y and X to rank Y, but plot the cumulative percentages of X and Y, not the ratios.

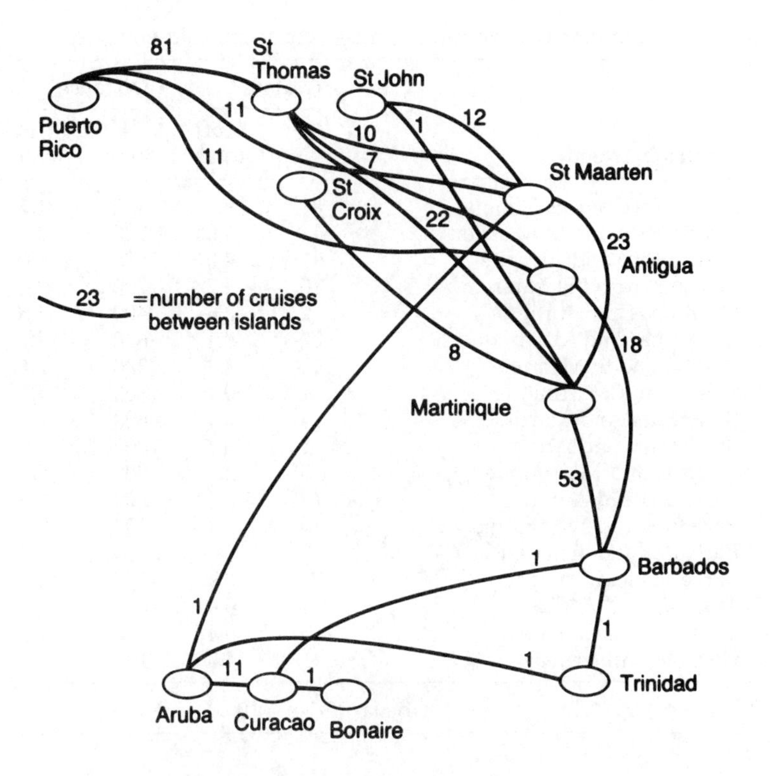

Figure 9.7 Cruise ship routes and voyages in the Lesser Antilles

4 If desired, calculate an index of dissimilarity using either method described previously.

Example

As an example of the calculation of the bivariate curve, consider the distribution of cruise ships and voyages in the Lesser Antilles. A quick glance at Figure 9.7, a sketch-map of the relative locations of islands and connecting ship routes reveals that there is a very uneven distribution in the number of voyages.

Table 9.8 contains a summary of the number and distribution of cruises. Distances between islands and the percentage distribution of distances is also shown. The ratios between the number of cruises and distances was then determined (Table 9.9) and used to rank the percentages of the number of cruises. Finally, a Lorenz curve was plotted (Figure 9.8) and the index of dissimilarity calculated.

Table 9.8 Distribution of a sample of cruise ship routes and voyages

		Cruises		Distance	
Link	**Routes between**	**No.**	**% of total**	**km**	**% of total**
1	Puerto Rico and St Thomas	81	29.7	100	1.2
2	Puerto Rico and St Maarten	11	4.0	300	3.6
3	Puerto Rico and Antigua	11	4.0	500	6.0
4	St Thomas and St Maarten	10	3.7	200	2.4
5	St Thomas and Antigua	7	2.6	400	4.8
6	St Thomas and Martinique	22	8.1	600	7.1
7	St John and St Maarten	12	4.4	200	2.4
8	St John and Martinique	1	0.4	600	7.1
9	St Maarten and Martinique	23	8.4	600	7.1
10	St Maarten and Aruba	1	0.4	1000	11.9
11	St Croix and Martinique	8	2.9	500	6.0
12	Barbados and Antigua	18	6.6	500	6.0
13	Barbados and Martinique	53	19.4	200	2.4
14	Barbados and Trinidad	1	0.4	400	4.8
15	Barbados and Curaçao	1	0.4	1100	13.1
16	Trinidad and Aruba	1	0.4	1000	11.8
17	Curaçao and Bonaire	1	0.4	100	1.2
18	Curaçao and Aruba	11	4.0	100	1.2
	Totals	273	100.0	8400	100.0

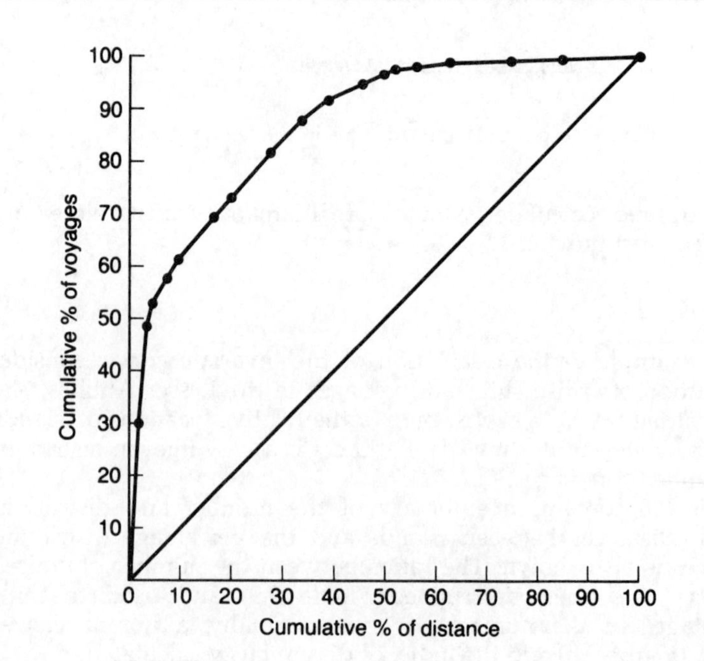

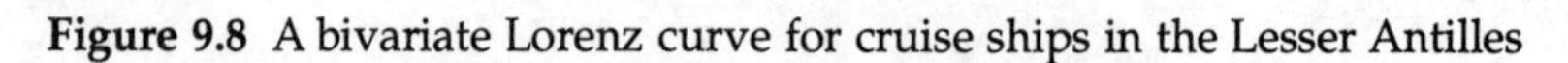

Figure 9.8 A bivariate Lorenz curve for cruise ships in the Lesser Antilles

Table 9.9 Rankings of percentages of cruise ship voyages and route lengths

Route	Ratio[a]
1	81.0
2	3.7
3	2.2
4	5.0
5	1.8
6	3.7
7	6.0
8	0.2
9	3.8
10	0.1
11	1.6
12	3.6
13	26.5
14	0.3
15	0.1
16	0.1
17	1.0
18	11.0

[a]Ratio = number of voyages/distance in 100 km units.

	Cruises		**Distances**	
	% Voyages	**Cumulative %**	**% Distance**	**Cumulative %**
1	29.7	29.7	1.2	1.2
13	19.4	49.1	2.4	3.6
18	4.0	53.1	1.2	4.8
7	4.4	57.5	2.4	7.2
4	3.7	61.2	2.4	9.6
9	8.3	69.5	7.1	16.7
2	4.0	73.5	3.6	20.3
6	8.1	81.6	7.1	27.4
12	6.6	88.2	6.0	33.4
3	4.0	92.2	6.0	39.4
5	2.6	94.8	4.8	44.2
11	2.9	97.7	6.0	50.2
17	0.4	98.1	1.2	51.4
14	0.4	98.5	4.8	56.2
8	0.4	98.9	7.1	63.3
16	0.4	99.3	11.8	75.1
10	0.4	99.7	11.8	86.9
15	0.4	100.0	13.1	100.0
		$A = 1441.9$		

$M = 18 \times 100 = 1800$

$R = 950.7$ (5.56% per route accumulated over 18 routes)

Table 9.9 *Continued*

$$I = \frac{A - R}{M - R}$$

$$= \frac{1441.9 - 950.7}{1800.0 - 950.7}$$

$$= 0.58$$

Nearest-neighbour analysis

Description

A researcher examining a map of the location of tourism facilities may sometimes wonder whether the pattern shows any semblance of order. Such a question is more than just idle interest. If you know that a pattern is clustered, for example, you may have a clue that will eventually lead to a better understanding of the forces that have shaped the pattern. Occasionally a simple visual inspection of a map is sufficient to indicate whether a pattern is clustered, uniform or random. More often, though, visual inspection is not reliable because all three patterns may be present to some extent. A method to determine more precisely and objectively the nature of a point pattern is nearest-neighbour analysis. This procedure, originally developed by ecologists (Grieg-Smith 1952; Clark and Evans 1955) to study the distribution of vegetation, compares the average observed distance between each point and its nearest neighbour to a theoretical average distance based on the assumption of a random distribution. The decision of whether a pattern is clustered, random or uniform is based on the value of that ratio. If the ratio between the observed distance and the theoretical distance is less than 1.00, implying that the points are closer together than expected, one concludes that the pattern is tending towards clustering. A value of greater than 1.00 indicates a tendency towards uniform spacing (dispersal). A value equal to 1.00 indicates a random pattern.

The concept of nearest-neighbour analysis is relatively simple, but there are several issues that need to be emphasized regarding its application. First, the ideal shape for delimiting an area to be studied is a square. It is not uncommon, though, to find that a square will include regions that may be devoid of interest to an analyst. We will look at an example of this problem when we examine the distribution of spas in Europe. The square we place over Europe includes portions of the Atlantic Ocean and the Mediterranean Sea. The nearest-neighbour statistic applies to the entire area of the square, not just the land portion. Under such circumstances, geography can bias the results towards a conclusion of clustering that may not reflect reality.

An alternative to the use of the square is to use the actual boundaries of the study area. While this is, in fact, a common practice, there is a

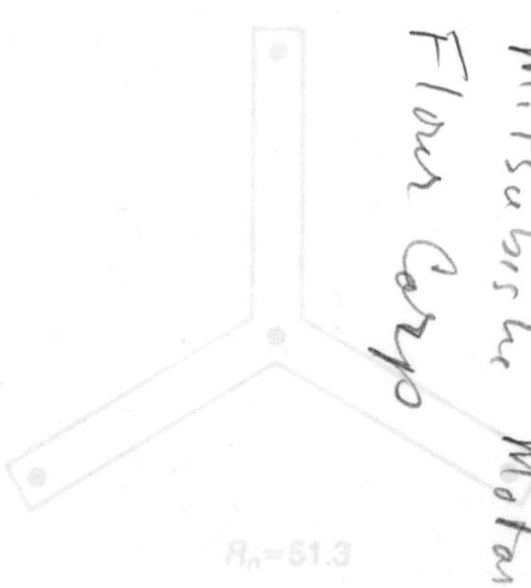

$R_n = 51.3$

Figure 9.9 A figure capable of producing an extremely high R_n value

potential danger because certain shapes can bias the results of the analysis. An extreme example is given in Figure 9.9. The theoretical maximum for the nearest-neighbour ratio is about 2.15; in this particular example the ratio is 51.30.

Another difficulty that may be encountered in interpreting the value of the nearest-neighbour ratio is based on the fact that the ratio is calculated from averaged distances. Figure 9.10 illustrates three different patterns that have low ratios, indicating clustering. None of these, however, represent what most people would consider to be a clustered pattern.

One of the longest standing problems with nearest-neighbour analysis is known as the boundary effect. This effect arises when a boundary is drawn delineating, as a sample, a subregion causing the true nearest neighbours of some points to fall outside the boundary. If these points are ignored (which they should be) the nearest-neighbour ratio will be biased high. The problem becomes greater with smaller numbers of points and smaller squares. The boundary problem had long been a serious obstacle to effective use of the method. A variety of procedures to correct it were ineffective, unreliable or too complicated. The problem was ultimately resolved by Pinder (1978) who observed

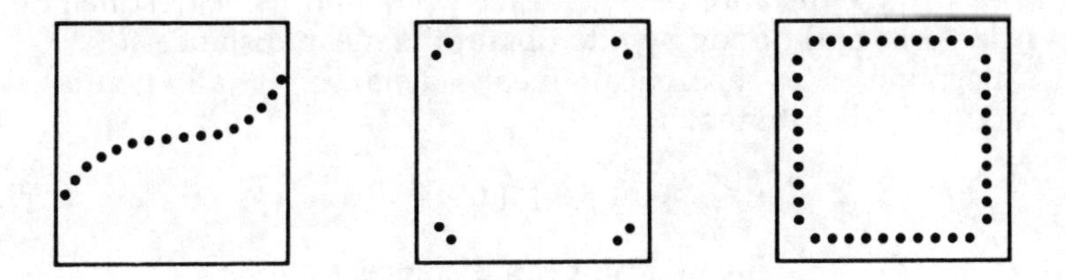

Figure 9.10 Patterns capable of producing small R_n values

that the boundary effect is due to the procedure for estimating the theoretical distance. He developed a different method for estimating the theoretical distance by accounting for variable numbers of points and variable square sizes. His procedure has been incorporated into the method given below.

Finally, distances between nearest neighbours are measured as straight lines on a map. In reality, this ignores the effects of perceptions, borders, physical barriers and transportation networks. For example, a spa on the northern coast of France may have its nearest neighbour in England, across the English Channel. The effects of a political boundary, language differences, and the physical barrier of open water, may mean the two spas are functionally more distant than two spas physically further apart but still in France. Little work has been done to develop the use of perceptual distance or accessibility as substitute, so this remains an area in need of research.

Procedure

1 Obtain a map of the region being studied. Plot the location of the features you are studying as points.
2 Draw a square to encompass the distribution of points. The size of the square is important. The larger the square, the more likely you are to produce a low nearest-neighbour ratio, and thus conclude the pattern is clustered. There is no rule to determine the size of the square to be used, but some guidelines are possible. If you are studying a sample space within a larger region, draw the square of such a size that it fits well within the total region and adequately samples the point pattern in the entire region. If, on the other hand, you are examining the entire population of points, choose between two strategies on the basis of project characteristics. A square may be drawn that is just large enough to encompass all points, ignoring natural boundaries, and oriented so that the size of the square is minimized. This is appropriate if there is little chance the pattern will ever grow beyond its current range. Should you expect the pattern to grow, however, then it is more appropriate to use a square just large enough to encompass the entire study region. Again, the square should be oriented to minimize its size.
3 Calculate the area, a, of the square.
4 Count the number of points, n, within the square.
5 Measure the distance between each point and its nearest neighbour. Add these and divide by n to obtain the mean distance, d_o.
6 Determine the theoretical mean distance if all points were randomly distributed:

$$d_r = C\,[\sqrt{(a/n)}] \qquad [9.14]$$

where: d_r = theoretical mean distance;
C = $0.497 + 0.127[\sqrt{(a/n)}]$

7 Calculate the nearest-neighbour ratio:

$$R_n = \frac{d_o}{d_r} \qquad [9.15]$$

8 Interpret the value of R_n. Interpretation is based on the null hypothesis of a random distribution indicated by $R_n = 1.00$. Values significantly larger than 1.00 indicate a tendency towards regularity or uniform spacing. A perfectly uniform hexagonal pattern (the pattern with the maximum degree of dispersal) has an R_n of 2.15. Values significantly less than 1.00 indicate a tendency towards clustering. Complete clustering (when all points occupy the same location) produces a value of 0.00. A significance test for R_n has been developed by Pinder (1978) and may be found in the Appendix at the end of this book.

Example

An illustration of nearest-neighbour analysis can be seen in Figure 9.11 and Table 9.10. Twenty-one major spas in Europe have been plotted and the entire region delineated by a square. For the purpose of this example, the square includes much of Western Europe and some of

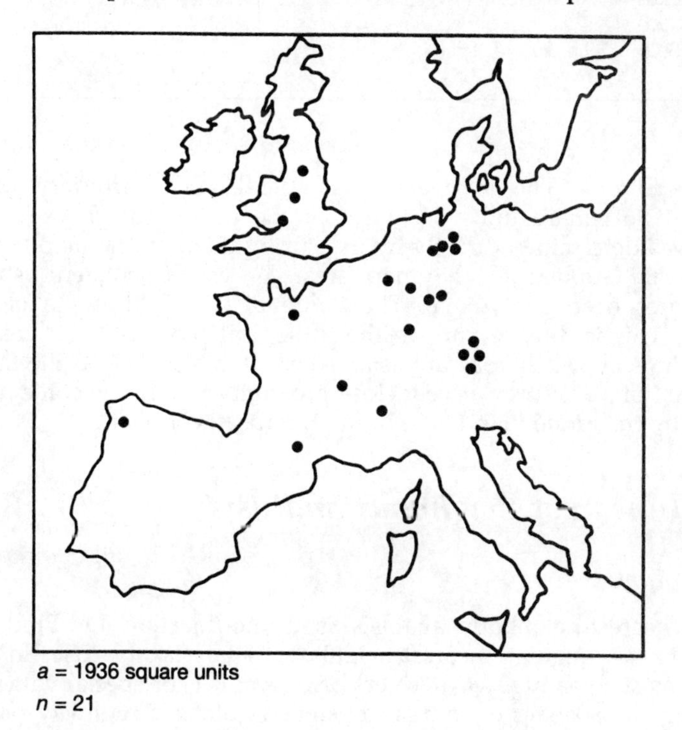

Figure 9.11 Major spas in Europe

Table 9.10 Nearest-neighbour analysis for the pattern in Figure 9.11

Nearest-neighbour distances (arbitrary units)	
5.5	$C = 0.497 + 0127\ \sqrt{(a/n)}$
3.0	$= 0.497 + 0.127\ \sqrt{(1936/21)}$
3.0	$= 1.716$
1.0	
1.0	$d_r = C\sqrt{(a/n)}$
1.0	$= 1.716\sqrt{(1936/21)}$
1.0	$= 16.48$
2.5	
1.0	$R_n = \frac{d_o}{d_r}$
1.0	
1.5	$= 1.9/16.48$
1.5	$= 0.12$
0.5	
0.5	
0.5	
0.5	
0.5	
6.5	
2.0	
2.0	
2.0	
2.0	
$\Sigma = 39.5$	$39.5/21 = 1.9 = d_o$

Eastern Europe. The map area was calculated in arbitrary units; it covers 1936 square units. Distance to the nearest neighbour of each point was determined directly by measuring them on the map using the same scale applied to the map area. R_n was calculated using the procedures described above. The value of 0.12 indicates a clustered pattern. This is due, in part, to the geological processes that create the hot springs at which spas are established. It is also due to the fact that spas are 'forced' into relatively close proximity within the context of the square by the limited land area of the European continent.

Linear nearest-neighbour analysis

Description

Linear nearest-neighbour analysis is a modification by Pinder and Witherick of regular nearest-neighbour analysis for use in linear situations such as highways, rivers or coasts. It is of special value when studying the spacing of tourism businesses along a highway or urban street.

The distances between tourism businesses or other landscape features are usually measured in physical units. Other measures, however, are possible. Travel time, travel cost, and even the number of intervening features might be adapted for use if they are pertinent to your problem and can be defined on an objective, ratio scale.

The same precautions about the use and interpretation of the nearest-neighbour ratio generally apply for the linear version, with the exception of the question of the shape of the study region. Because the linear version is applied to a line, no study area – in a two-dimensional sense – is defined.

Procedure

1 Begin with a dot map showing the distribution of facilities along a linear feature. Measure the length of the line, L. This line may be defined by reference to either natural end-points such as where a highway crosses a political boundary, or by reference to arbitrary end-points. The most common arbitrary end-points are the two extreme observations in the distribution. If these are used as the termini for the line, they must not be used in the subsequent analysis as observation points. To do so would bias the ratio value high, leading you to conclude a greater tendency towards regularity than is correct.
2 Count the number of observation points, n.
3 Measure the distance between each point and its nearest neighbour along the line. Determine the mean distance, d_o.
4 Estimate the theoretical distance between points under the condition of a random distribution:

$$d_r = 0.5[L/(n-1)] \qquad [9.16]$$

5 Calculate the linear nearest-neighbour ratio, LR_n:

$$LR_n = \frac{d_o}{d_r} \qquad [9.17]$$

6 Interpret the results. A value of 1.00 indicates a random distribution, while a value of 0.00 indicates complete clustering. A perfectly uniform spacing produces a ratio of 2.00. A graph for determining the significance level of LR_n may be found in the Appendix.

Example

Figure 9.12 is a map of Interstate Highway 65 between Montgomery and Mobile, Alabama (USA), showing the locations of service stations at various exists. A linear nearest-neighbour analysis was conducted on this stretch of highway to see if the spacing of service stations was

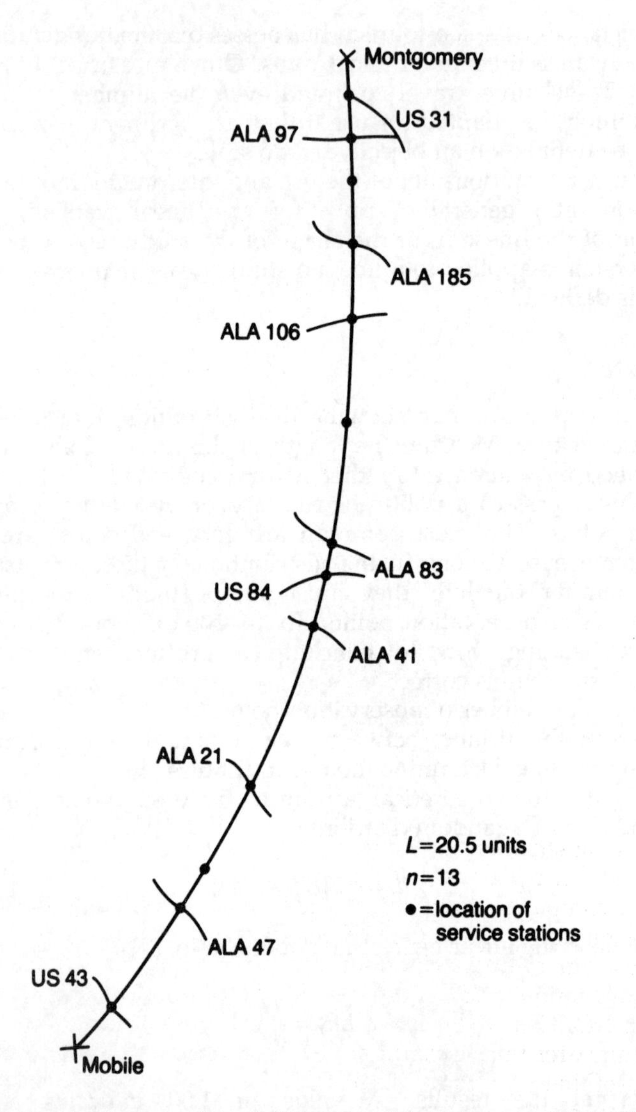

Figure 9.12 Service stations along I-65 between Montgomery and Mobile, Alabama

random, clustered or uniform. Table 9.11 contains the calculations involved for the 13 observations. The ratio between the observed nearest-neighbour distance and that expected was 1.3. Referring to the significance graph in the Appendix, a value of 1.3 for 13 observations falls just within the area that indicates randomness in spacing.

Table 9.11 Linear nearest-neighbour analysis for the pattern in Figure 9.12

Nearest-neighbour distances (arbitrary units)		
2.3		$n = 13$
0.9		$L = 20.5$
0.9		
1.8		
1.0		$d_r = 0.5[L/(n - 1)]$
0.6		$= 0.5(20.5/12)$
0.6		$= 0.9$
0.6		
2.0		
1.4		$LR_n = d_o/d_r$
1.4		$= 1.2/0.9$
0.7		$= 1.3$
0.7		
0.7		
0.7		
$\Sigma = 15.0$	$15.0/3 = 1.2 = d_o$	

Spatial association index

Description

The spatial association index, developed by Lee (1979), is a method for comparing point pattern distributions. Like nearest-neighbour analysis, this index compares an expected pattern to an observed pattern to reach a conclusion about the degree of clustering, dispersion or randomness. It differs from nearest-neighbour analysis in one important way. The spatial association index compares the distribution of one set of points to the distribution of a second set; regular nearest-neighbour analysis merely compares the distribution of a single set to a theoretical pattern.

The usefulness of the spatial association index might be seen in a simple example. Assume you have the task of analysing the distribution of fast-food franchises in a large urban area. You suspect that the outlets of any one chain, say, 'Bubba's Burgers', tend to avoid locating near other outlets of the same chain, but seek locations close to competitors from other chains. Regular nearest-neighbour analysis can tell you whether Bubba's Burgers tend to disperse with respect to each other, but it cannot tell you anything about the pattern of Bubba's in comparison to other fast-food outlets. This is when you would use the spatial association index.

The values of the spatial association index range from 0.00 upwards. Its interpretation is similar to that of nearest-neighbour analysis: a value of 1.00 indicates a random distribution; values significantly less than

1.00 indicate clustering of one set of points with reference to the other set; values significantly greater than 1.00 indicate spatial avoidance. There is no upper limit to the value of the index. The reason for this is that the maximum value is a function of two separate point patterns and their respective densities.

Procedure

1 Follow steps 1 to 4 of the proceedure for nearest-neighbour analysis. Be sure, however, to use separate symbols for the two sets of points you plot. Also be sure to obtain separate counts for the number of points in each set. Label these counts n_1 and n_2. Designate their total $(n_1 + n_2)$ as N.

2 Calculate the relative proportions of the two sets of points:

$$m_1 = \frac{n_1}{N} \qquad [9.18]$$

$$m_2 = \frac{n_2}{N} \qquad [9.19]$$

note that m_1 and $m_2 = 1.00$.

3 Measure the distance between each point in the set you are studying and its nearest neighbour in the reference set. Total these distances and divide by the number of points in the study set to obtain the mean distance, d_o.

4 Determine the theoretical average distance if all points were randomly distributed:

$$d_r = \frac{m_1}{2[\sqrt{(n_2/a)}]} + \frac{m_2}{2[\sqrt{(n_1/a)}]} \qquad [9.20]$$

5 Calculate the spatial association index:

$$R^* = \frac{d_o}{d_r} \qquad [9.21]$$

6 R^* *may be interpreted the same way as* R_n, described previously. A significance test for R^* *can be found in the Appendix.*

Example

A comparison of the distribution of two sets of restaurants will illustrate the application of the spatial association index. Figure 9.13 is a map of the locations of a hamburger franchise (Bubba's Burgers) and other fast-food restaurants in a portion of an urban area. The procedure described above was followed, and the results are summarized in Table 9.12. The finding was a spatial association index of 0.40, indicating that

Bubba's tends to locate its outlets close to other fast-food establishments.

Table 9.12 Calculation of spatial association index for Figure 9.13

Area = 59.29 square units
5 Bubba's Burgers = n_1
13 Other restaurants = n_2
18 = N

$$m_1 = \frac{5}{18} = 0.28$$

$$m_2 = \frac{13}{18} = 0.72$$

Nearest-neighbour distances for Bubba's

0.4 units
0.3
1.2
0.5
0.7
3.1 ÷ 5 = 0.62 = d_o

$$d_r = \frac{0.28}{2\sqrt{(13/59.29)}} + \frac{0.72}{2\sqrt{(5/59.29)}}$$

$$= 0.30 + 1.24$$

$$= 1.54$$

$$R* = \frac{d_o}{d_r} = \frac{0.62}{1.54} = 0.40$$

Peaking index

Description

A characteristic of many tourism businesses is that the number of customers varies dramatically over time. Restaurants boom on weekends and holidays, but are quiet in the early part of the week. Resorts are booked solid for a few weeks or months and then operate at a much reduced level for the rest of the year – or may even close. It can be helpful for a business planner or recreation programmer to be able to measure quantitatively the tendency of people to use a facility or visit a destination in one time period as opposed to other time periods. Tabulations of frequencies for different time periods can be of use, but these do not provide a succinct index. A measure that summarizes a substantial amount of data on temporal use patterns into a single value is the peaking index (Stynes 1978). This measure is an open-ended scale with a minimum value of 0.00. The greater the degree of concentration (peak use), the greater the value of the index. The index is derived from a graph called an exceedance curve – so-called because the curve illustrates the number of times a particular use level is reached or exceeded.

By itself, P_n is not especially informative. The value depends not only on the degree of peaking but also on the total volume of business and

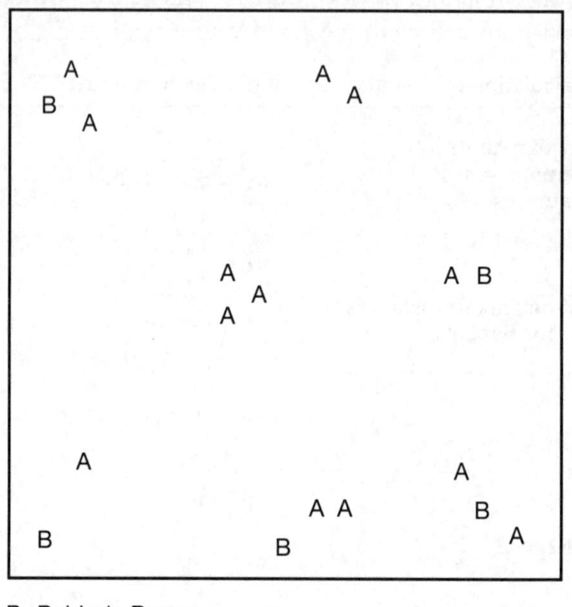

Figure 9.13 Hypothetical pattern of fast-food restaurants

on the choice of time periods used for analysis. The primary use of the index, therefore, is for comparison between businesses or for examination of trends in peaking over time in one facility.

P_n varies with the size of the time period chosen for analysis. For example, if you are working with hotel occupancy data, you will probably work with monthly averages. If the busiest month had an occupancy rate of 90 per cent and the slowest month had a rate of 40 per cent, some days in the busiest month will have exceeded 90 per cent while some days in the slowest month will have fallen short of 40 per cent. The use of precise daily data would probably lead to a different index value than that reached with monthly averages. There is insufficient information and experience to suggest the degree to which aggregation of time periods into larger units changes the value of P_n, but in general the value of the index declines – indicating less peaking – as longer time periods are used for analysis.

In the absence of objective guidelines for selecting the time period for analysis, the choice is made on the basis of data availability, the purposes of the study, and common sense. It would not be wise, for example, to collect hotel occupancy data on an hourly basis if you are interested in annual patterns. Not only would the volume of data be unwieldy, rooms are let (in most reputable establishments) on a nightly basis, not hourly. On the other hand, an analysis of use levels in an amusement park on a typical weekend might well be based on hourly data.

Procedure

1 Obtain use data for the business or facility in question. Group these by day, week, or other appropriate time period. The data must cover a reasonably long time period, often an entire season or year. Either actual numbers of users or occupancy rates may be used.
2 List use levels and their associated time periods in decreasing order of use.
3 Plot the data on a graph similar to that in Figure 9.14. The graph itself provides a quick visual indication of the relative peaking of use at the business; the sharper the drop-off in the curve, the greater the concentration of visitors in a short period of time.
4 A more precise measure of peaking is obtained through the equation:

$$P_n = \frac{V_1 - V_n}{(n-1)V_1} \times 100 \qquad [9.22]$$

where: P_n = peaking index;
V_1 = number of visitors during the busiest period;
V_n = number of visitors during the nth period;
n = reference period for comparison to the busiest period (1 = busiest period).

P_n equals 0.00 when the number of visitors is the same in all periods. Its value increases as use levels concentrate in certain periods. As explained previously, the upper limit of P_n depends on the total level of use and the choice of n. The value of n, the time period used for comparison with the busiest period, is largely a matter of personal choice. You might choose the mid-point in the data or else a value of n reflecting some natural division in the facility's schedule. If a resort finds its year divided into four seasons of about three months each, with different visitor patterns, activities, and marketing programmes in each season, you might select an n of 3 (the length of one season) as a useful basis for analysis.

Example

Table 9.13 and Figure 9.14 provide an example of the calculation of P_n. In this case, the occupancy rates are for hotels in Vancouver, British Columbia (Canada). The busiest month in Vancouver is traditionally August, with an average occupancy rate of 96 per cent. This drops off until a low is reached in December. A value of 6 for n, reflecting the mid-point in the number of months over a year, was chosen. Application of equation [9.22] produced a value of 2.90 for P_n. Again, recall that this index is more meaningful when compared with peaking indices from other cities or other years.

Table 9.13 Hotel occupancy rates for Vancouver, Canada

Month	Occupancy rate (%)
August	96
July	91
September	91
June	89
October	85
May	82
April	74
March	72
November	70
February	60
January	53
December	49

$$P = \frac{V_1 - V_6}{(n-1)V_1} \times 100 = \frac{96 - 82}{(5)96} \times 100 = 2.9$$

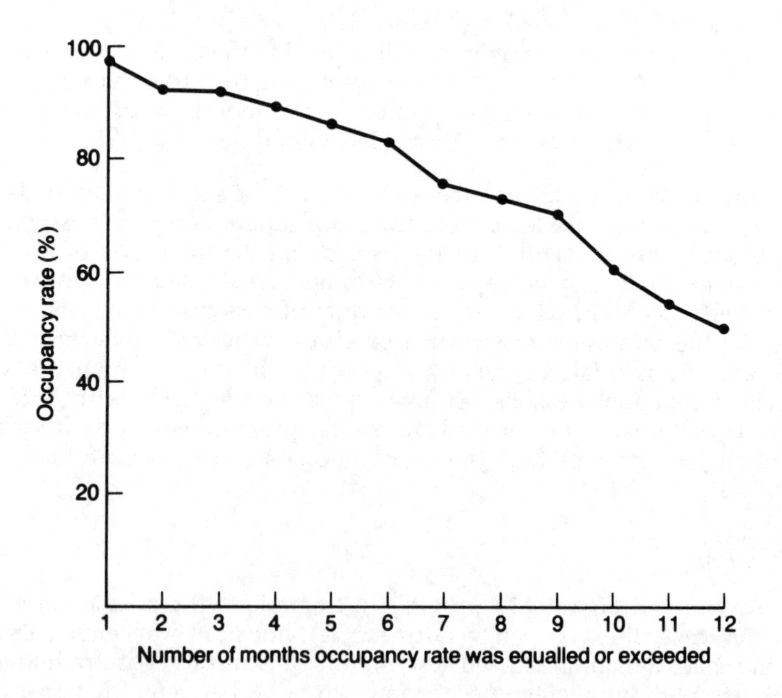

Figure 9.14 Exceedance curve for tourist accommodation in Vancouver, British Columbia

Directional bias index

Description

Vacation travel, whether that of individuals, groups, or entire populations, often shows a predilection for one particular direction. The percentage of Canadians travelling south in the winter, for example, is much higher than the number heading north. The directional bias index, developed by Wolfe (1966), is a simple measure of this tendency. It is an origin-specific index because it summarizes the travel patterns of an origin with respect to each of its destinations. The index can be used as one component of a large systematic description of the travel patterns of a region or population, or it can be used as an independent variable in modelling travel flows and in testing hypotheses about the forces affecting travel patterns.

The index is only a measure of the proportion or relative distribution of trips from one origin to each destination. It does not reflect the net balance of travel between two regions. For those situations in which you might wish to study the reverse flow, the index may be easily generalized to cover this need. All regions may be considered in turn as an origin and then as a destination. The index, though, still does not indicate anything about net travel between two regions. If this is of interest, it must be calculated separately. It should be noted that net flow by itself does not indicate a directional bias. Both statistics, net flow and directional bias, can be used together to obtain a fuller picture of travel flows than either would provide by itself.

The directional bias also ignores the actual route of a trip, and it does not indicate anything about travel time or costs. The index is not useful for vacations that do not have a single destination, such as touring vacations.

On the positive side, the index has ratio scale properties; it can be subjected to a wide range of statistical analyses. Although its applications are limited, it is a useful and reliable index of one particular tourism phenomenon.

Procedure

1. Construct an origin–destination matrix, similar to that in Table 9.11. Each origin is represented by a row; each destination by a column. The individual cells contain the number of trips (person-trips, person-nights, or other unit) for an origin–destination pair. Designate each cell as T_{ij}, where i is the origin and j is the destination.
2. Calculate the totals for each row and column. Designate these as ΣT_i and ΣT_j respectively.
3. Obtain the directional bias index, D, for each origin as:

$$D = \frac{10^5(T_{ij})}{\Sigma T_i \Sigma T_j} \qquad [9.23]$$

where: 10^5 is a scaling factor; any factor of appropriate size may be substituted as desired.

Example

Table 9.14 and Figure 9.15 offer an illustration of the calculation of the directional bias index. The data here represent flows among four hypothetical origins and four destinations. We can see in Table 9.14 that the number of trips from origin *a* to destination *B* is 30. Origin *a* generates a total of 115 trips distributed over all destinations, while destination *B* receives 135 trips from all origins. Using equation [8.23], we obtain index values for trips from *a* to *A* of 300, *a* to *B* of 193, and so forth.

Table 9.14 Data for calculation of a directional bias index

Origins *i*	Destinations *j* A	B	C	D	$\Sigma T_i = i$
a	50	30	10	25	115
b	60	90	15	5	170
c	10	10	80	15	115
d	25	5	30	75	135
$\Sigma T_j =$	145	135	135	120	

Tourism attractiveness index

Description

An important problem in regional tourism planning is the assessment of the potential of candidate regions for attracting tourists. A region with few historic sites, little natural beauty, poor climate, no beach, limited recreational or shopping opportunities, and little potential to develop

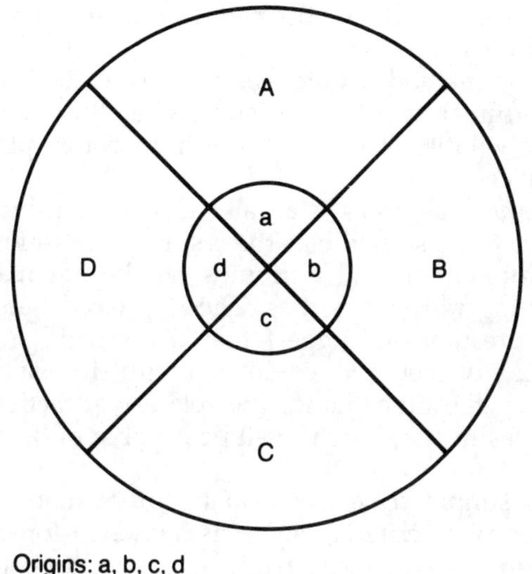

Origins: a, b, c, d
Destination: A, B, C, D

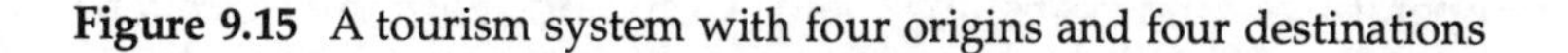

Figure 9.15 A tourism system with four origins and four destinations

these will be a poor choice for investment. A planner familiar with a number of regions being considered for development may have an intuitive feel for the relative attractiveness of different regions, but personal impressions are not always reliable for business or policy decisions. A better method would be to use the evaluations of a number of informed individuals, objective data, and a systematic evaluation procedure. In the best of all possible worlds, such a method might also use the expressed preference of actual and potential visitors to obtain insights into visitors' decision-making criteria. An adequate sample of the general travelling public, however, can be costly. Surveys of visitors and potential visitors would also require the assumption that there is a strong correlation between expressed views and travel behaviour.

A more practical solution is to use a relatively small panel of experts familiar with aggregate tourist demand and tastes, and with the regions being considered for development. Such a panel, properly administered, can provide valuable information about the tourism attractiveness of different regions. The following method is an outline for conducting this type of expert survey; it is based on a procedure developed by Gearing *et al.* (1974).

Their procedure involves asking a panel of experts to assign weights to a series of attributes to reflect their overall importance for tourism development. Once these weights are established, the researcher then asks the panel to evaluate each of a number of tourism regions on these attributes. This evaluation is done with a numerical scale so that regional scores can be multiplied by attribute weights. Relative attractiveness scores are then obtained by adding up the individual attribute scores for each region.

The user of this method should be aware of certain assumptions on which the procedure rests. Anyone employing this index should take care to verify the validity of these assumptions before placing too much faith in the results.

First, the method assumes that all relevant attributes have been identified. Failure to do so will bias the results in an unknown direction. Further, it is assumed that all attributes can be adequately evaluated with a scale having ratio properties. The key word here is 'adequate'. No one would presume to suggest that the artistic and architectural heritage of Rome, for example, could be completely summarized on a 10-point scale. On the other hand, the relative attractiveness of Rome *vis-à-vis* other cities in Italy might well be capable of being summarized on such a scale.

The use of a simple scale can also be problematic when you are summarizing complex criteria, such as climate, for regions whose characteristics run the entire spectrum of possibilities. For example, if you were evaluating the states of the USA on the basis of climate, what type of meaningful comparison could be made between Hawaii with warm temperatures and sunshine, and Colorado with an excellent winter sports climate? When this type of comparison causes problems, the solution is to make a more precise designation of the criteria and of the type of development to be used as a reference for evaluating the regions.

A user also assumes the scales are reliable and valid measures of the criteria. One test that can be used to check on the reliability of the experts' judgements is to examine the overall similarity of standardized criteria weights across all judges. Kendall's coefficient of concordance (described in many introductory statistical texts) is often an appropriate measure to use for this. The question of validity of the experts' judgements is more difficult. Indeed, it is virtually untestable. Your best assurance here is to use the most reputable, able and sincere experts available.

Finally, this entire procedure rests on the assumption that the total attractiveness of a region may be estimated by the addition of weighted attribute scores. This assumption means that there is no significant interaction effect between the various regional characteristics that affect attractiveness. It further means that the failure of a region to provide any level of some quality (indicated by a score of 0.00) does not eliminate that region from consideration as a candidate for development.

With respect to interaction effects, you could examine each attribute in relationship to every other attribute to identify any obvious problems such as two qualities that measure the same thing, one attribute that would negate the possibility of another or two attributes whose combination produces a result greater than the simple sum those two variables would otherwise imply. If any such pairing is found, you might identify new attributes that avoid the problem.

Any user of this procedure needs to maintain a balance between the blind and uncritical belief in the potential for quantifying the attractiveness of a tourism region and a defeatist scepticism about the value of any quantitative tool for providing usable attractiveness information. If this sense of balance is achieved, and if the research design is followed carefully with attention paid to the need to identify and evaluate carefully all relevant regional characteristics, the tourism attractiveness index can be a valuable planning tool.

Procedure

1 Establish a list of attributes for judging the attractiveness of tourism regions. These attributes will be specific to the regions being considered. The list should be as comprehensive as possible. The attributes should be stated as precisely as practical and should be independent of each other, not implying the existence of each other or not contradicting the existence of each other. The list can be developed by you if you have adequate expertise or it can be developed with the aid of experts.

2 Group the individual criteria into a small number of major categories. Gearing *et al.* (1974) used the following: (1) **natural factors** – (a) natural beauty and (b) climate; (2) **social factors** – (a) architecture, (b) festivals, and (c) other folk cultural attractions; (3) **historical factors** – (a) ancient ruins, (b) religious shrines and practices, and (c) historical importance; (4) **shopping and recreational resources** – (a) sport opportunities, (b) museums, zoos, aquaria and gardens, (c) health and relaxation opportunities, and (d) stores and shops; (5) **tourism infrastructure** – (a) adequate roads, utilities and health services, and (b) adequate food and lodging facilities.

3 Select a panel of experts to assign weights to the criteria. The number and make-up of the panel is somewhat arbitrary; aim for competence, diversity of backgrounds, and an ability to work in a group. Gearing *et al.* chose a panel of 26 experts from travel agencies, government tourism policy offices, airlines, hotels and universities.

4 Present the panel with the criteria and instructions for assigning weights. Numerous weighting procedures are possible; a method developed by Churchman *et al.* (1957) has received widespread acceptance and is appropriate for this type of problem. Details of their procedure are provided in the Appendix. The method involves a systematic comparison of weights between pairs of attributes to produce a series of weights, W_i, ranging from 0.00 to 1.00.

5 Once the panel has assigned weights to each criterion, have them evaluate the degree to which each region meets the criterion. Each region should be rated on a 0.00 to 1.00 scale. Obtain the mean

ranking across all panel members for each criterion in each region. Designate the mean score for each criterion in each region as S_{ij}, where i refers to the criterion and j refers to the region. Multiply each score, S_{ij}, by its weight, W_j, to obtain a measure of the attractiveness of each region in terms of any criterion.

6 Finally, sum each region's A_{ij} values to derive a single measure of the relative tourism attractiveness of that region, A_j.

Example

Gearing *et al.* (1974) conducted their study on 65 regions in Turkey. The procedure outlined above was followed for 17 criteria, resulting in the weights shown in Table 9.15. Each of the regions was then evaluated by a panel of experts to produce relative attractiveness scores. Scores of one region are also shown in Table 9.15, along with the weighted attractiveness score calculation.

Table 9.15 Calculation of a touristic attraction index

Criteria	Weight ×	Regional evaluation =	Weighted score
Natural beauty	0.132	0.80	0.106
Infrastructure	0.131	0.80	0.105
Food and lodging	0.125	0.70	0.087
Climate	0.099	0.50	0.050
History	0.065	0.60	0.039
Archaeology	0.057	0.90	0.051
Local attitudes	0.054	0.70	0.038
Religious significance	0.053	0.70	0.037
Art and architecture	0.051	0.50	0.026
Sport facilities	0.046	0.30	0.014
Nightlife	0.045	0.30	0.013
Shopping	0.036	0.30	0.011
Peace and quiet	0.032	0.80	0.026
Festivals	0.029	0.80	0.023
Local features	0.026	0.70	0.018
Educational facilities	0.015	0.50	0.007
Fairs and exhibits	0.011	0.60	0.007
		Touristic attractiveness score =	0.658

Summary

Tourism analysts have a wide choice of descriptive procedures. The possibilities include simple tabulation and percentage summaries as well as the better-known univariate statistics such as mean and variance. Tourism analysts, however, are not limited to these methods; more specialized and more informative descriptive tools are available. Many of the more important ones have been discussed in this chapter. A few of these – Defert's *Tf*, the peaking index, the directional bias index, and the tourism attractiveness index – were developed specifically for tourism and recreation. Still others have been borrowed from various social sciences, especially geography. The importance of the geographic perspective in tourism may be seen in the fact that 10 of the 15 measures are spatial or regional statistics.

This does not mean that only geographic problems may be studied with these tools or that the geographic perspective is the only relevant one in tourism description. These tools have the potential for application in a wide variety of tourism topics. One of the challenges to analysts is not just to use the tools presented here, but to combine them, elaborate them, and build on them to provide more powerful and useful methods for solving the complex problems of tourism description.

Further reading

Cooper C P 1981 Spatial and temporal patterns of tourist behaviour. *Regional Studies* **15**: 359–71.

Duffield B S 1984 The study of tourism in Britain – a geographical perspective. *GeoJournal* **9**: 27–35.

Gunn C A 1993 *Tourism planning*, 3rd edn. Taylor and Francis, Bristol, Pennsylvania.

Keogh B 1984 The measurement of spatial variations in tourist activity. *Annals of Tourism Research* **11**: 267–82.

Mitchell L S 1984 Tourism research in the United States: A geographical perspective. *GeoJournal* **9**: 5–15.

Murphy P E and **Brett A C** 1982 Regional tourism patterns in British Columbia: a discriminatory analysis. In Singh T V, Kaur J, and Singh D P (eds) *Studies in tourism – wildlife – parks – conservation*. Metropolitan Book Company, New Delhi.

Pearce D 1987 *Tourism today*. Longman, Harlow, Essex, UK.

Pearce D 1989 *Tourist development*, 2nd edn. Longman, Harlow, Essex, UK.

Smith S L J 1983 *Recreation geography*. Longman, Harlow, Essex, UK.

Smith S L J and **Brown B A** 1981 Directional bias in vacation travel. *Annals of Tourism Research* **8**: 257–70.

CHAPTER 10

Shadow prices and non-market valuation

Introduction

A challenge facing many policy-makers and managers who are responsible for allocating public resources is that there are usually more uses for a resource than the supply can support. A given parcel of open space may be desired by park planners, shopping centre developers, and residential subdivision developers. An historic building might be the object of competing demands from an architectural preservation group, a retail developer interested in an attractive facility for boutiques, and a municipal government that is being pressured to provide additional parking. The decision about which use a particular resource should ultimately serve can be complicated and contentious, involving arguments from economists, special interest groups and the general public, as well as restrictions imposed by legislation, zoning, and urban or regional plans. A policy-maker or manager needs information on the value of a resource in competing applications to help resolve the debates.

The question of the value of a resource can be difficult. First of all, 'value' has different meanings for different people. If you were to ask, what is the value of the trip that I took to a conference last week, you would get a variety of answers depending on whether you asked me, my students, my department head, the organization that held the conference, or the manager of the hotel where I stayed. I would describe my answer in terms of the benefits of exchanging information with peers and the opportunities for networking. My students would describe the insights I brought back into my courses. The department head would emphasize the increase in the profile of the department and the potential to lead to research contracts. The organization would speak of the objectives of the conference itself. The hotel manager would talk about the economic returns of the conference from room rentals, accommodation, food and beverage services, and audio-visual equipment rentals.

The meaning of value is sometimes synonymous with 'benefit'. Many decisions about whether to initiate public projects are based on the results of a benefit–cost analysis (or cost–benefit analysis). In this context, the benefits of the project are quantitative estimates of the contributions to gross domestic product and certain other quantifiable benefits (described in Chapter 11) that would be created by the proposed project.

In other circumstances, however, 'value' and 'benefit' have different connotations. In the case of my trip to a conference, I tend to think of the benefits of the conference in subjective, intangible terms such as knowledge gained and social contacts; the hotel will focus on the value of revenues received. This distinction is sometimes expressed in terms of 'use value' versus 'exchange value'.

The use value of a resource is the subjective, non-quantitative benefit an individual receives from the use of a resource, whereas exchange value is the price that resource can command in the market-place. For example, the use value of diamonds as ornaments is relatively small compared to the use value of water for drinking, and yet diamonds are expensive while water is inexpensive. The explanation is that diamonds are scarce, their availability on the market is also regulated by an international cartel, and they are prized for their beauty whereas water is abundant in many localities and sometimes can be obtained free of charge by collecting rain-water or digging a well. If a region were to suffer a drought or be deprived of access to water because of political hostilities or war, the exchange value of water can suddenly soar, reflecting its substantial use value. Exchange value thus reflects not only the demand for a resource, but its availability. And because exchange value represents how much some would be willing to pay – or exchange – for a resource, it is a quantitative measure of the value of that resource. Use value is not easily translated into an objective, measurable number, and thus is a less useful indicator of 'value' for the purposes of resource allocation.

Exchange value is used in the following way as a tool for allocating resources. If you owned a plot of land and you were trying to decide whether to put that land into housing, tourism or agriculture, you would want to know the marginal utility of that land in each use, i.e. how much money you would make from each hectare of land in each use. You would rationally put the land to whichever use provided you with the greatest benefit – the greatest economic return. If all land were optimally allocated among society's needs, the return to you would be the same regardless of which use you selected. If the return were not the same – let us assume that the return would be greatest through some form of tourism development – you would conclude that there was a local shortage of land for tourism development and that this shortage had driven up the potential returns to be derived from land used for that purpose. Conversely, if the returns from putting that land into agriculture were low, you could conclude that there was a surplus of agricultural land – reflected in the low value of economic returns associated with agricultural development.

As a private landowner, you could rely on the market-place to help you make your decision. Although the intricacies of governmental regulations, financing, zoning, strategic plans, and many other issues will make your decision more difficult and distort the functioning of the market, the market will give you a reasonably good indication of the value of your land in alternative uses and thus direct you to the best decision.

When resources are publicly owned, the same principle applies, except that the returns are assessed in terms of overall social benefits, not individual profit. The public agency controlling resources should allocate those resources to maximize public benefit derived from those resources. A basic problem, of course, is that public resources are rarely allocated through the market-place, so traditional economic measures of value in competing uses are not available. A public planner should not ignore the principle of economic efficiency (i.e. allocating resources to maximize their returns) so he must turn to non-market mechanisms to estimate the values of public resources. Many of these mechanisms are based on the concept of 'willingness to pay'. Even though there is no actual transfer of funds, one attempts to obtain an estimate of what typical users would be willing to pay for specified resources in specified applications. These willingness-to-pay measures are sometimes known as 'shadow prices'. Two basic approaches are used to derive shadow prices for public goods: (1) the direct method (also known as the contingent valuation method), which involves some form of directly asking users about the value of a resource; and (2) the indirect method, which involves inferring willingness to pay for a resource by observing the expenditures of money on activities associated with the private use of public resources.

We examine these two approaches in more detail in this chapter. Before looking at them, though, we must first review some basic concepts.

Basic concepts

The phrase, 'willingness to pay' is self-descriptive, although the methods used to estimate consumers' willingness to pay for a publicly provided resource are not so obvious. Willingness to pay, as a concept, includes any price actually paid as well as the extra the consumer would have been willing to pay if necessary. For example, the admission to a national park for a day's recreation might be $10.00, but a visitor might have been willing to pay an additional $5.00, if required: a total willingness to pay of $15.00. The extra amount a consumer would be willing to pay is an important measure of the total benefits received from a resource, and is known as 'consumer surplus'. Measures of this type are most appropriately applied when evaluating the real or potential benefits derived from improvements or additions to some tourism development, compared with the next best alternative (which might be no change in the existing supply or quality of the resource).

Direct valuation methods require great attention to survey method-

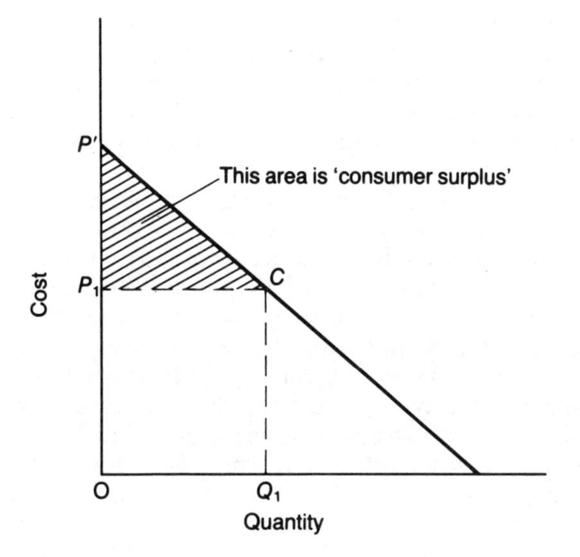

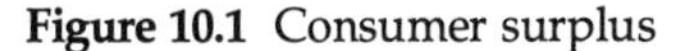
Figure 10.1 Consumer surplus

ology and questionnaire design. Indirect methods are often less sensitive to data-collection difficulties and thus are often more reliable. These latter methods are based on the concept of consumer surplus, mentioned above. To develop a better understanding of consumer surplus, let us begin by referring to Figure 10.1. This figure is a traditional demand curve. Given a cost of P_1, we predict that consumers would purchase Q_1 units of a commodity. Their total expenditure will therefore be OP_1CQ_1. From the demand curve we can observe that consumers would actually be willing to pay more than P_1, although the total quantity consumed would decrease. The additional amount they would be willing to pay is the area above the price line and below the demand curve: $P_1P'C$. This area is the consumer surplus.

If the commodity had no market price, it is still possible to estimate a demand curve and the consumer surplus of the commodity. This can be difficult, but one approach, known as the 'travel cost model', is useful in some tourism contexts. Hotelling first suggested this approach, but it is most often associated with Clawson and Knetsch (1966). The basic strategy in this approach is to interpret the travel costs associated with a visit to a site as an estimate of the willingness of visitors to pay for that site. Given this interpretation, the price axis of the demand curve is replaced by travel costs, and the quantity consumed is the number of trips made at different cost levels. The actual details of how to estimate the consumer surplus of a site using the travel cost approach are described below.

Incidentally, in the following sections, we will use terms such as 'facility', 'site', 'resource', and 'commodity'. Valuation methods generally can be used equally well for any of these. Remember, therefore, if we use one term such as 'resource', the method will normally be applicable to other types of tourism and recreation opportunities.

One final comment is necessary before we turn to the details of the valuation measurement techniques. One of the most important tools developed by resource managers for taking decisions about the feasibility of proposed projects is benefit–cost analysis. Benefit–cost analysis is essentially a paper-and-pencil experiment designed to estimate whether society would be better or worse off if a proposed project were to be implemented. The decision is based on a comparison of the incremental benefits that would be produced by a project and the incremental costs imposed by the same project. The particular decision may be a choice between approving the project or rejecting it, a choice among several alternative projects, or the optimal scale of a proposed project. The procedures we examine in this chapter are among those used by analysts in estimating economic benefits of tourism projects. The methods and conventions in conducting a benefit–cost analysis are described in Chapter 11.

Contingent valuation method

Description

The contingent valuation method (CVM) is a research strategy in which individuals are asked directly about their willingness to pay for some improvement in a resource or for additional use of a resource. The origins of CVM can be traced back to at least Ciriacy-Wantrup (1947) who suggested the use of interviews to assess the economic value of various soil conservation initiatives. However, the method is most closely associated with Davis, who explored the potential of a survey design to simulate a market that would put 'the interviewer in the position of a seller who elicits the highest possible bid from the user for the services being offered' (Davis 1963: 245). Davis' work generated interest among resource economists who had long struggled with the task of assigning shadow prices to non-market commodities, especially changes in the quality or quantity of public resources. Among other early pioneers in the development of CVM were Cicchetti and Smith (1973) who examined the value of reducing crowding on wilderness trails, Darling (1973) who applied CVM to urban parks, Hammack and Brown (1974) who explored the value improvements in waterfowl hunting, and Hanemann (1978) in his work on beach water quality. Since these early studies, literally hundreds of applications of CVM have been published, not only in the context of recreation, but also pollution control and the provision of other public goods.

The reliability of the results of this method depends on the ability of the analyst to both execute a carefully designed survey and simulate a market that reflects the essential elements of a market. In other words, the survey must be based on a representative sample and must present questions that are clear and meaningful to the respondent, and do not bias the responses of the respondents. Further, the questions should provide a clear understanding of the nature of the commodity or resource in question, the proposed changes, the mechanism for charging fees, and other relevant aspects of the hypothetical market.

In the context of tourism, contingent valuation is most useful in the following situations:

1. Developing of value estimates for facilities that are frequently visited as part of a multiple-destination trip. In this case the travel cost approach (the primary alternative to CVM) is difficult to apply because of the problem of allocating trip costs among various destinations.
2. Estimating the increase in value resulting from management actions to reduce congestion, perhaps by opening substitutable sites or overflow facilities. A reduction in congestion at a given facility typically increases the value of that facility to visitors, but the travel cost method typically interprets any reduction in the total number of visitors as indicative of decreasing value.
3. Evaluating management plans to improve user satisfaction in a number of sites, or evaluating planned improvements for activities not associated with any particular site. A common example of this latter situation is game-management programmes to increase the number of migratory birds and fish.

The survey method rests on two key assumptions: (1) the ability of tourists to set an exchange value on changes in the quality or quantity of specific resources and (2) that this value can be accurately solicited by direct questioning. With respect to the first assumption, some policy analysts and planners doubt whether consumers are always able to provide realistic estimates of the maximum they would be willing to pay for a resource. As Scott (1965), an early critic of CVM, put it 'ask a hypothetical question and you will get a hypothetical answer'.

Mitchell and Carson (1989: 295) review the experience of many researchers with this method as well as the theoretical foundations for 'strategic responses', i.e. whether individuals are likely to give false answers to CVM questions in order to gain some small advantage for themselves. Their conclusion, on both empirical and theoretical grounds, is that respondents are likely to give valid and meaningful answers if the CVM questionnaire is properly designed and administered. Their conclusion supports that of Arrow (1986: 183) who said, 'Neither the empirical evidence nor the theoretical arguments convinced me that strategic bias is liable to be significant.'

A more pertinent concern is that of the validity of the second assumption. The amount a tourist is willing to pay for some change in the supply of a resource can be greatly influenced by what he perceives to be the alternatives and by the pricing vehicle, i.e. how the money is to be collected. As a result, users of this method must make certain that the respondents clearly understand what the alternatives will be under different levels of funding.

If you can ask a question about what someone is willing to pay for an increase in the quality or quantity of a resource to obtain an estimate of its value, could you obtain the same information by asking what someone would have to be paid to accept a decline in the quality or quantity of a resource (willingness to accept or WTA, instead of willingness to pay or WTP)? Mitchell and Carson (1989) have examined this issue as well, noting that although several theoretical studies argue that WTA approaches are conceptually valid but that empirical studies have sometimes yielded WTA estimates as much as four times higher than WTP estimates. The explanations for such discrepancies are still a matter of debate, but the explanation appears to be linked to how citizens perceive their rights to access or use of public resources, and the fact that access to public resources is non-transferable. Whatever the reason, current guidelines are that willingness to accept (or willingness to sell) questions are best avoided.

Based on their extensive review of the theoretical foundations of CVM, Mitchell and Carson (1989: 50–52) suggest the following be taken into account when designing a CVM scenario for a survey.

1. The respondent should be reminded to consider his or her current level of disposable income (income after taxes and other long-term obligations) before responding. If the unit of analysis is the household, household income should be considered, not just the respondent's income.
2. In the case of publicly provided goods, the respondent should be reminded that he or she is currently paying for a given level and quality of supply through taxes or other mechanisms. A quantitative estimate of that amount can be offered if it can be calculated from public budget data.
3. The exact nature of the improvement should be made explicit. If improvement in water quality for swimming is the amenity being studied, the respondent might make some inference that the improvement in water quality might also improve fishing. The surveyor should explicitly note that the quality improvements refer to swimming only, and no other implicit environmental change.
4. If a proposed change in an amenity level would cause a significant change in the prices of other commodities, this should be explained to the respondent. For example, improvements in water quality at a beach might require the construction of a new water-treatment plant whose costs might eventually translate into higher prices at local restaurants.
5. The scenario should explain whether the improvement is permanent or temporary (and if temporary, how long the improvement will last). The mechanism by which the improvement will be made available, if not obvious, should be made explicit. The payment vehicle (how the charges are to be collected), frequency of payments, and who will have access to the good should all be made explicit.
6. The wording of the question should ensure that respondents are

expressing their consumer surplus for the good, and not some other value such as 'existence value' or 'fair price'.

Procedure

1 Determine as precisely as possible the context of the problem you wish to study. Do you want to estimate the value of increasing the capacity of a specific site, of increasing the population of a particular species of game bird, of enhancing interpretive services at a museum, or what? The definition of the problem requires identification of the group from which you intend to draw your sample and to which you intend to generalize your findings.
2 Using the guidelines provided above, develop carefully worded questions aimed at eliciting answers about the willingness to pay. Originally, the design of willingness-to-pay questions was in bidding format. In this design you ask the respondent whether he would be willing to pay some specified price (usually a relatively low price) for some change in the quality or quantity of a resource. If the individual indicates he is willing, the price is increased systematically until the respondent indicates he is no longer willing to pay. Current practice, however, is to ask the respondent for the maximum amount he would be willing to pay.
3 If your study population is easily definable and measurable, such as boat owners or licensed hunters, you can draw a sample directly and administer the survey. The total willingness of the user population to pay can be directly extrapolated from the results. If, on the other hand, your study group is the general population or some large, amorphous group, you may find it necessary to use a two-stage survey. The first stage includes not only the willingness-to-pay question, but also questions about the individual characteristics of respondents that are believed to influence willingness to pay. These characteristics typically include socio-economic variables as well as travel patterns and recreation equipment ownership. A multiple regression equation is then calibrated relating these individual characteristics to the expressed willingness to pay.

   The second stage of the survey involves a larger sample of your study population in order to identify the number of people with characteristics identified in the first stage of the survey. This information is then used in the multiple regression equation to estimate the willingness of individuals in the general population to pay.

Example

Donnelly and Nelson (1983) conducted a study of the net economic (exchange) value of deer hunting in the American state of Idaho using CVM. Although the economic value of hunting is generally recognized, there is a need for greater precision in estimating the actual monetary-equivalence of that activity. Such values, in specific jurisdiction, would help managers and elected officials make better decisions about the allocation of funds and other resources among competing uses. For example, if it can be shown that certain changes in land-use management would improve deer hunting – which would translate into greater value – and that the cost of those changes was less than the increased value, economic efficiency would be improved.

As Donnelly and Nelson (1983: 3) note, the total value of a resource used in producing a commodity consists of several components: consumer and producer expenditures, consumer and producer surpluses, as well bequest, existence and option values. Their study focused on only one of these – consumer surplus. Consumer surplus, or net willingness-to-pay is the standard measure of value in benefit–cost analyses and in most other resource policy contexts.

The sample of this study was drawn from the population of resident and non-resident hunters having an Idaho deer licence in a given year. A total of 1445 hunters were randomly selected and then interviewed through a telephone survey regarding their most recent hunting trip in Idaho. After some preliminary questions on recent hunting activity, the respondents were asked how much they spent on various aspects of the trip. Then they were asked if the trip was worth more to them than they actually spent. Nearly 90 per cent (88.8 per cent, to be precise) indicated, 'yes'. Table 10.1 presents the willingness-to-pay portion of the questionnaire.

Note that if the respondent indicated that the trip was worth more than he paid, he was asked if – assuming costs increased because of transportation costs or some other reason – he would pay up to 20 per cent more. If so, the bid was raised to 50 per cent more than costs actually paid, then 100 per cent, and then increasingly higher by 100 per cent increments until a ceiling was hit. Then the surveyor worked between the last acceptable 'bid': and the first unacceptable bid to find the maximum.

Table 10.1 Sample willingness-to-pay questions

Please estimate the amount spent on transportation on this trip.

$ ______

Now, estimate the amount spent on accommodation on this trip.

$ ______

Was this trip to ______ worth more than you actually spent?

No: *Stop here*

Yes : Next, I would like to ask some hypothetical questions about this trip to ____________. Assume that the trip became more expensive, perhaps due to increased travel costs or something, but the general deer or elk hunting conditions were unchanged. You indicated that $ ______ were spent on this trip for your individual use.

Table 10.1 *Continued*

Would you pay $ __________ *(insert estimate 20 per cent higher than amount paid)* rather than not be able to hunt deer or elk at this area?

Protests – will not answer

Record why: 1. *It's my right*
2. *My taxes already pay for it*
3. *No extra value*
4. *Like to, but not able*
5. *Refuse to put a dollar value*

No: *Work between 0 and 20 per cent to find highest acceptable value. Split the difference in half until you reach nearest $1 (less than $10) or nearest $5 (greater than $10).*

Yes: *Continue to next question.*

Would you pay $ ______ *(insert estimate 50 per cent greater than amount paid)* rather than not be able to hunt deer or elk at this area?

No: *Work between 20 and 50 per cent to find highest acceptable value. Split the difference in half until you reach nearest $1 (less than $10) or nearest $5 (greater than $10).*

Yes: *continue to next question.*

Would you pay $ ______ *(insert estimate 100 per cent greater than amount paid)* rather than not be able to hunt deer or elk at this area?

No: *Work between 50 and 100 per cent to find highest acceptable value. Split the difference in half until you reach nearest $1 (less than $10) or nearest $5 (greater than $10).*

Yes: *Keep going in 100 per cent increments until you receive a negative answer. Work between last bid and unacceptable offer to find highest unacceptable limit.*

Is this amount, $ _____, what you personally would pay – not an amount for all members of your party?

No: *Repeat process for personal bids.*

Yes: Now, suppose that instead of ______ deer (*number they said they say, from earlier in the interview*), you could have seen ______ (*double the number*). How much, if any, would you increase your value of $ _______? $ ______

That is all the questions I have for you. Thank you for taking the time to answer these questions. Your responses will be very valuable to us.

Goodbye.

Respondents were also provided with the opportunity to indicate that the trip was worth more than they paid but they would not pay more for various reasons, such as, 'It's my right' and 'My taxes already pay for it.' The surveyor was also asked to ensure that the bids estimated represented *personal* willigness to pay, not travel party. The estimates of actual trip costs and estimates of willingness to pay were examined by the research team to remove outliers (i.e. values that were so extreme that they would unduly affect the results and probably reflect spurious answers). In practice, this meant that any willingness-to-pay estimate exceeding $1000 was examined as a possible outlier. The distance travelled, number of days spent hunting, the number of hours per day spent hunting, and other trip characteristics were examined to determine the credibility of the answer. The results of the study indicated a mean bid by deer hunters of $40.09, for an average trip lasting 2.09 days.

The results were compared to those derived using a travel cost model (described in the next section). The authors noted that the advantage of the CVM approach over the travel cost method was the ease of data analysis. The basic analysis of the results took only three days to complete. On the other hand, the authors noted that the CVM method relies heavily on the ability of surveyors to elicit willingness-to-pay answers. In particular, the bidding process used required effort to avoid both respondent fatigue as well as unrealistic infinite bids (e.g. 'I'd pay whatever I had to'). Further, the CVM approach requires demographic and other descriptive information collected across respondents and then incorporated into some form of regression analysis in order to generalize to the broader population.

Travel cost method

Description

The travel cost method (TCM) is an indirect method for determining the value of a tourism site. It is based on the development of a model for predicting site use from observed consumer behaviour. The method may be used to derive values for the site that provided the original data or for sites similar to the original study site.

TCM usually involves a regression model to relate levels of use to travel costs. Drawing data from visitors facing different travel costs allows the analyst to infer values of a site by observing the relationship between travel costs and the level of use of the site. This is done by estimating a demand curve for which travel costs are a surrogate for market prices.

Variables other than distance and travel costs might also be included in a TCM analysis. Measures of competition, site quality, and various traveller characteristics might be incorporated to broaden the scope of the analysis. For the sake of clarity, however, we will concentrate only on variables directly related to travel cost.

Before we discuss the procedures of the travel cost method, we need to consider how travel costs are to be defined. Analysts generally agree that actual fuel costs such as gasoline for a private automobile should be included. Agreement about the inclusion of other costs is more debated. Possible costs include: (1) the value of travel time; (2) costs of food and accommodation above that normally spent (i.e. incremental costs beyond that spent while at home); (3) equipment costs, such as rentals or a pro-rated (or depreciated) cost of equipment ownership.

Arguments about these costs involve debates as to the accuracy of estimation procedures as well as whether they represent costs associated with a specific trip, as opposed to general 'lifestyle expenses' that should not be attributed to actual trips. Although admission or entrance fees are real costs to the travellers, there is some debate as to the inclusion of charges for admission to public lands, such as provincial or national parks.

The issue of the value of time is particularly difficult. To be sure, time has value. However, time spent on vacation is not clearly a cost – it is part of the experience. In other words, a day spent riding through the Canadian Rockies on the Rocky Mountaineer or several hours relaxing in the first-class section of a Cathay Pacific plane *en route* to Hawaii seems to be more of a desirable leisure experience than a cost. On the other hand, driving a 10-year-old car with two tired, cranky children through a rainstorm for two hours, late at night, while looking for a motel with vacancy, is definitely a cost. A full review of the issues of what to include in travel costs is beyond the scope of this section. The interested reader may find a useful discussion in Walsh (1977). In brief, though, it seems reasonable to include fuel, tolls, food and lodging in excess of what you would spend at home, the costs of consumable supplies such as film or bait, and at least some admission charges.

TCM is based on several assumptions. First, it is assumed that the response of average individuals to a user fee of a given magnitude is the same as their response to a travel cost of the same magnitude. In other words, a traveller is assumed to value a Pound spent on petrol the same as he would value that Pound spent on accommodation or admissions.

It is further assumed that the relationship between travel costs and the number of trips taken is linear. We also assume that visitation rates are not constrained by capacity limits. This is especially critical. If use levels reflect capacity constraints rather than the attractiveness or utility of a site, we cannot use TCM to derive valid estimates of value. The result is a downward bias in our estimate of value because fewer people are visiting that would otherwise be the case.

Use of the TCM is also best limited to trips with single purposes or destinations. If a traveller takes a trip for multiple purposes or visits multiple destinations, TCM will usually over-estimate the benefits associated with the implicit sole purpose of the trip.

The particular procedure described here assumes no competition from other sites. This is not a critical assumption; we will relax this particular assumption in the next section.

Procedure

1 Define the specific site for which you are developing the valuation. Define, too, the origins of users, usually cities or counties. Sometimes it may be appropriate to use more precise origins (such as counties) close to the study site and then more aggregate origins (such as provinces) for more distant origins.
2 Conduct an on-site survey to determine how many users come from each origin.
3 Determine the total population of each origin.
4 Estimate the average travel costs for visitors from each origin. This estimate will be a function of distance between each origin and destination as well as the specific components of travel costs.
5 Determine the visits per capita for each origin by dividing the total number of trips made by residents from each origin by the population of that origin.
6 Plot visits per capita against cost. Fit, by visual inspection or (preferably) with least-squares estimation, a line through the array of points to develop a function for predicting average visitation rates for any particular level of travel cost.
7 Develop an aggregate site demand curve. This curve is based on the assumption that individuals respond to user fees and travel costs in the same way. For example, a person who faces $3.00 in travel costs and a $3.00 admission charge will participate at the same rate as someone facing $6.00 in travel costs and no admission fee.

 We begin developing the aggregate site demand curve by assuming a zero user fee. If there is a fee, add the fee to the total travel costs and continue to work with the assumption of a zero user fee. Recall that in step 6 we developed a function to predict visitation rates for various levels of travel costs. Systematically add costs to the initial level of travel costs and predict, for each increment, the expected use level. Continue this process until all users are priced out of the site, according to your model. The estimated use levels are plotted against the higher costs to generate the demand curve. The area under the demand curve is the consumer surplus.

We will look at two examples of the application of the travel cost method: one that is hypothetical to illustrate the details of the calculations, and one that is drawn from a real world study.

Example 1

Table 10.2 illustrates the type of data needed to estimate the distance decay curve (the individual travel demand curve) as well as the aggregate site demand curve: origin population, number of visits from each origin, per capita visits, and travel costs. A regression analysis of trip costs and per capita visits produced the function:

Table 10.2 Hypothetical data for deriving a trip demand curve for an isolated site

Origin	Population	Number of visits	Visits per capita	Trip costs ($)
A	100 000	50 000	0.50	11.50
B	200 000	200 000	1.00	5.00
C	300 000	75 000	0.25	15.00

$$V_{ij} = 1.373 - 0.0752(C_{ij}) \quad [10.1]$$

where: V_{ij} = visits per capita from origin i to destination j;
C_{ij} = trip costs.

This function has been plotted as Figure 10.2. Note that when travel costs in this example equal $18.25, visits are reduced to zero. Equation [10.1] is then used to calculate the expected number of trips for each origin for a range of travel costs in excess of the initial cost. These calculations are summarized in Table 10.3. The first row of data shows the expected user rates under the condition of no net increased in travel costs. This is, of course, the same as the per capita rates in Table 10.2. Travel costs are then increased $1 at a time until a maximum of $18.25 is reached (the last increment used to impose total travels costs of $18.25 to the visitors from each origin is not necessarily $1.00). Per capita trip rates, visits per origin, and total visits are calculated. The total visits are plotted against the incremental costs to generate the aggregate site demand curve (Figure 10.3). The area under the curve is the consumer

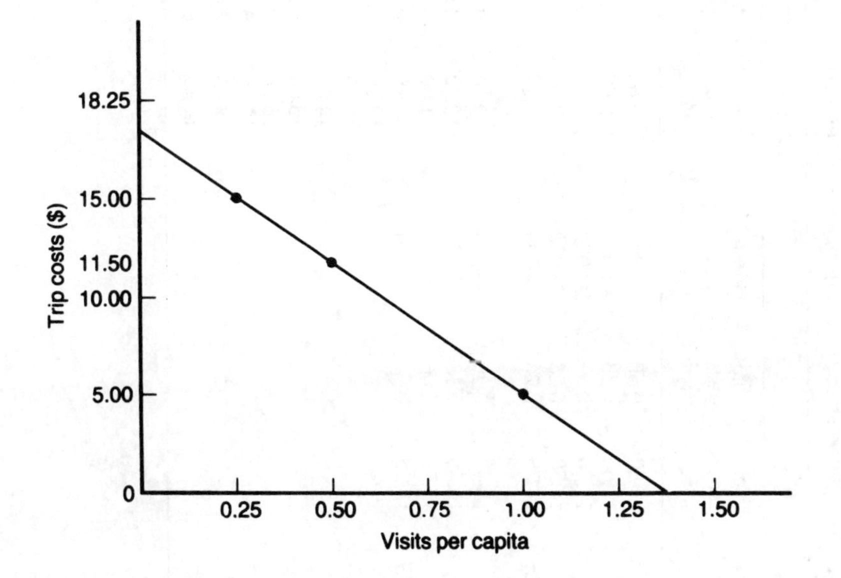

Figure 10.2 Trip demand curve for an isolated site

Table 10.3 Calculation of an aggregate site demand curve for an isolated site

	A			B			C			
User fees ($)	Total costs ($)	Visits per capita	Visits	Total costs ($)	Visits per capita	Visits	Total costs ($)	Visits per capita	Visits	Total visits
0	11.50	0.50	50 000	5.00	1.00	200 000	15.00	0.25	75 000	325 000
1	12.50	0.43	43 000	6.00	0.92	184 000	16.00	0.16	48 000	275 000
2	13.50	0.35	35 000	7.00	0.85	170 000	17.00	0.09	27 000	232 000
3	14.50	0.28	28 000	8.00	0.77	154 000	18.00	0.02	6 000	188 000
3.25	14.75	0.26	26 000	8.25	0.75	150 000	18.25	0	0	176 000
4	15.50	0.20	20 000	9.00	0.70	140 000				160 000
5	16.50	0.13	13 000	10.00	0.62	124 000				137 000
6	17.50	0.06	6 000	11.00	0.55	110 000				116 000
6.75	18.25	0	0	11.75	0.49	98 000				98 000
7				12.00	0.47	94 000				94 000
8				13.00	0.40	80 000				80 000
9				14.00	0.32	64 000				64 000
10				15.00	0.25	50 000				50 000
11				16.00	0.16	32 000				32 000
12				17.00	0.09	18 000				18 000
13				18.00	0.02	4 000				4 000
13.25				18.25	0	0				0

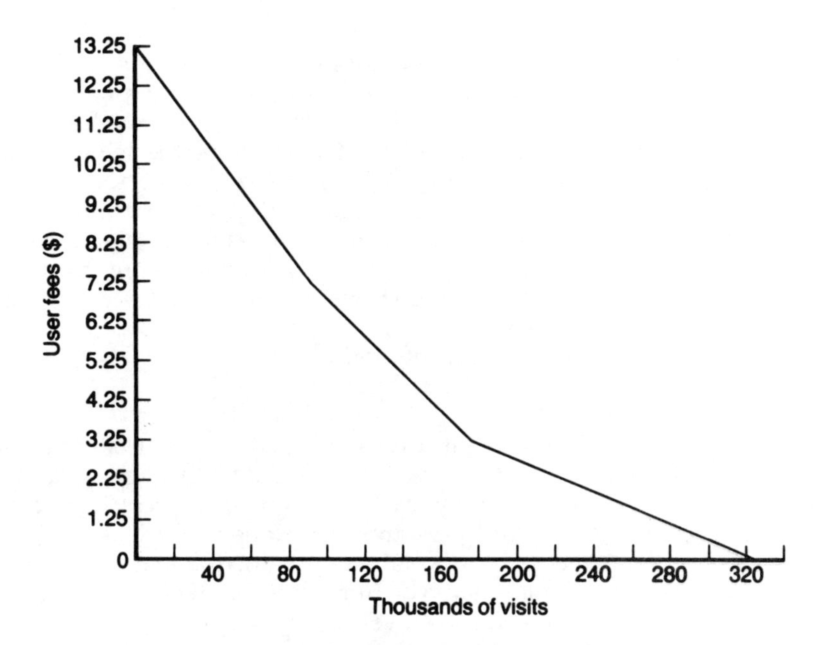

Figure 10.3 Aggregate site demand curve for an isolated site

surplus. This may be obtained graphically or by calculating the consumer surplus for each individual and then summing them. The total consumer surplus in this example is approximately $1.8 million.

Example 2

Donnelly and Nelson (1986), in the same study cited previously, also applied TCM to their analysis of the value of deer hunting in Idaho. In a telephone survey of 1445 individuals who were licensed deer hunters in Idaho (regardless of their residence), the authors collected information on whether hunting was the primary purpose of the trip, the round-trip distance to each site visited, transportation and accommodation expenses, and household income. A basic model relating travel propensity and selected causal variables was defined as:

$$\frac{T_{ij}}{P_i} = b_0 - b_1 D_{ij} + b_2 Q_j - b_3 S_j + b_4 I_i \qquad [10.2]$$

where: T_{ij} = trips from origin i to destination j;
P_i = population of i;
D_{ij} = round-trip distance from i to j;
Q_j = the hunting quality of site j, operationally defined as the 1981 harvest from that area (the number of deer harvested);
S_j = a measure of the cost and quality of substitute hunt

areas (with respect to the area under study); two measures were tested: (1) total number of animals seen and total harvest divided by distance travelled from origin, as calculated for the site with the highest ratio for hunters from a given origin, except for the site under study, and (2) the sum of quality per mile based on all sites visited from a particular origin than the particular site in question. Neither measure produced statistically significant results, so both were dropped from the analysis;

I_i = per capita income of origin i;

$b_0 - b_4$ = statistically estimated coefficients.

This equation defined the per capita demand curve for hunting in Idaho. This equation was then used to calculated the aggregate site demand curve, by systematically increasing D_{ij}, as a proxy for travel costs. Conversion of round-trip distance was done in three steps. First, distance was converted to transportation costs for the average vehicle: estimated costs per mile, derived from the hunters' estimates of transportation costs. The value of travel time was also added, using the figure of one-third of the average wage rate, as estimated by the US Department of Labour (which, at the time of the study, was US $8.00/hour).

Licence and entry fees to public lands are a special case. Although these are out-of-pocket costs for many recreationists, the US Water Resources Council (1983: 78) interprets them as a transfer of consumer surplus from the recreationist to the government. Thus, to more accurately measure economic efficiency, such transfers should not be counted as travel costs but, rather, added to the total willingness to pay. These fees would also be considered to be benefits in a benefit–cost analysis. The actual value of licences and fees is prorated over the number of trips taken to compute benefits per trip.

Once the aggregate site demand curve was calculated from the regression analysis, the value of hunting in each study region was obtained by calculating the area under the curve. Dividing this by the number of sample trips yielded the consumer surplus per trip: $50.23.

The primary strength of TCM, compared to CVM, is that it is based on behaviour rather than on opinion. TCM also allows the analyst to experiment with the effects of changes in other variables, such as hunting quality, on the overall benefits of a site. However, it also involves some difficult data collection tasks and potentially unrealistic assumptions to estimate travel costs. The model also involves substantial analytical work. Donnelly and Nelson (1986: 20) noted that their TCM required two months of work, while their CVM analysis took between two and three days.

In conclusion, Donnelly and Nelson concluded that neither the TCM nor the CVM is superior in all cases. Each has different strengths and weaknesses, and each may be used with reasonable confidence in different situations.

Travel cost method allowing for site substitution

Description

This version of the travel cost method is a generalization of the basic travel cost method to allow for the effects of one or more substitute sites in the same market area as the study site. The procedure allows us to estimate the consumer surplus attributable to both sites, although the method can lead to an apparent under-estimation of the value of both sites because of lowered use rates. The process involves the calculation of a trip demand function and derivation of an aggregate site demand curve as before, except that this procedure is applied to two or more competing sites.

One assumption of this method is that tourists always go to the destination with the lowest overall travel costs. In fact, of course, this is not true. Tourists do not seek simply to minimize travel costs. Whether they are exploring new territory, seeking variety, or simply working with imperfect information, tourists are often observed going to more distant destinations than closer ones. Further, destinations are rarely perfectly substitutable. One often finds significant variations in quality or service between destinations offering ostensibly the same features or attractions.

What all this means is that you might attempt to develop some estimate of the relative market share of each destination in each origin market to obtain an indication of what percentage of tourists from any origin are likely to be diverted by the creation of a new site.

The creation of a new destination theoretically reduces the consumer surplus of an existing site. This loss of value should not be interpreted as the lowering of overall quality or intrinsic merit of a site. Rather, it reflects the loss of monopolistic value: with one or more competing sites, an existing site no longer commands the same value it did when it was the only source of recreation or tourism possibilities.

The creation of an alternative site not only tends to lower the value of an existing site as measured through the TCM, it can also raise its value by reducing congestion. This increase in value, however, is better measured through the CVM, described previously. Alternative sites can also result in direct savings to tourists by reducing the total cost necessary to reach the facilities offered at existing sites. We will examine later a method for estimating the travel savings associated with the creation of a substitute site.

Procedure

1 Follow steps 1–6 outlined in the procedure for the basic travel cost method.
2 Using the travel demand curve derived in step 6, develop aggregate site demand curves for both sites. This may be done using the procedure described in step 7 above, but assume that

Table 10.4 Calculation of an aggregate site demand curve for competing site II

	A			B			C			
User fees ($)	Total costs ($)	Visits per capita	Visits	Total costs ($)	Visits per capita	Visits	Total costs ($)	Visits per capita	Visits	Total visits
0	16.00	–	–[a]	4.00	1.04	208 000	12.50	0.43	129 000	337 000
1				5.00	1.00	200 000	13.50	0.36	108 000	308 000
2				6.00	–	–[b]	14.50	0.28	84 000	84 000
2.50							15.00	0.25	75 000	75 000
3							15.50	–	–[c]	–

[a] Travel costs from origin A to site II exceed travel costs to site I. No trips will be made to site II.
[b] Once travel costs exceed $5.00, residents of origin B stop visiting site II and switch to site I.
[c] Once travel costs exceed $15.00, residents or origin C stop visiting site II and switch to site I.

Table 10.5 Calculation of the aggregate site demand curve for site I

	A			B			C			
User fees ($)	Total costs ($)	Visits per capita	Visits	Total costs ($)	Visits per capita	Visits	Total costs ($)	Visits per capita	Visits	Total visits
0	11.50	0.50	50 000	5.00	–	–[a]	15.00	–	–[a]	50 000
1	12.50	0.43	43 000							43 000
2	13.50	0.35	35 000							35 000
3	14.50	0.28	28 000							28 000
4	15.50	0.20	20 000							20 000
4.50	16.00	0.17	17 000							17 000
5	16.50	–	–[b]							0

[a] Residents of origins B and C do not visit site I because site II is closer.
[b] Residents of origin A switch to site II once travel costs to site I exceed $16.00.

travellers will go only to the site with the lowest travel costs. The area under each curve is a measure of the value of each site.

Example

Assume another site has been constructed near the site used in our first example. The features of the second site are identical to those of the first. Its location is slightly closer to origins B and C but slightly further away from A than the first site. See Figure 10.4 for the basic data.

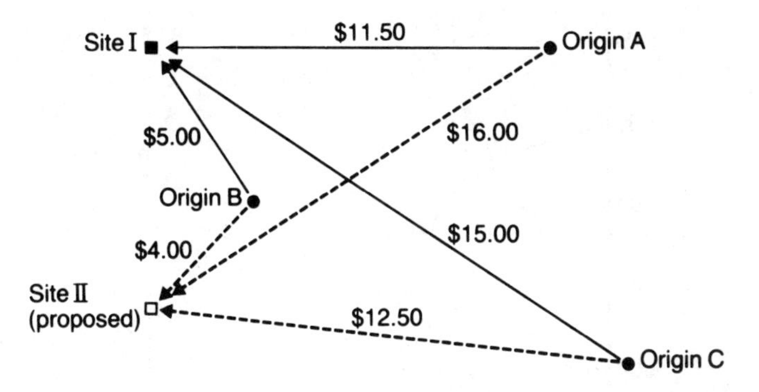

Figure 10.4 A two-site, three-origin travel system

Using the demand function calibrated for the first site, we can predict the number of visits to site II under a variety of user fees. Note that there is no positive user fee that will permit residents of A to visit site II. Even at zero user fees, travel costs to site II exceed those to site I. In the case of origins B and C, users will pay a net maximum of $1.00 and $2.50, respectively, before switching to site I. These calculations are summarized in Table 10.4.

The aggregate site demand curve for site II is calculated using data from Table 10.4, and is shown in Figure 10.5. The area under the curve is approximately $592 000. Note that the consumer surplus for site II is much smaller than for site I without the existence of site II. The reason for this difference is the fact that the consumer surplus for site I without competition reflects the greater value one normally assigns to a resource for which there are no substitutes. The value of site II is lower, not because it is an inferior site, but because consumers now have two sites to choose from, so they value each site less than they would if either were the sole site.

We can also calculate an aggregate site demand curve for site I to see what effect site II has on the value of site I. These calculations are shown in Table 10.5. Here site I draws only from origin A when the user fee is less than $4.50. The new site demand curve for site I is plotted in Figure 10.6. Consumer surplus for site I is reduced to $140 000.

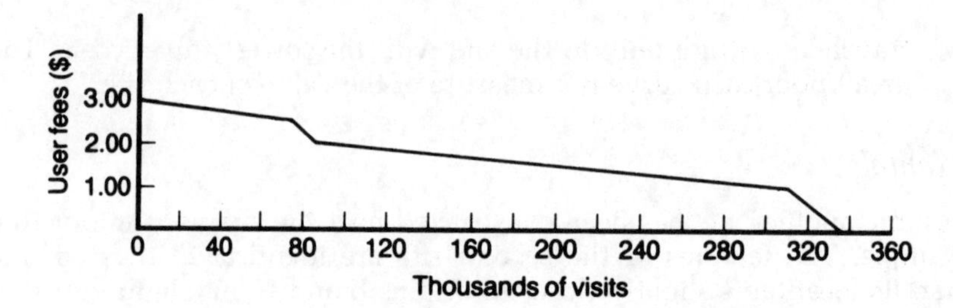

Figure 10.5 Aggregate site demand curve for competitive site II

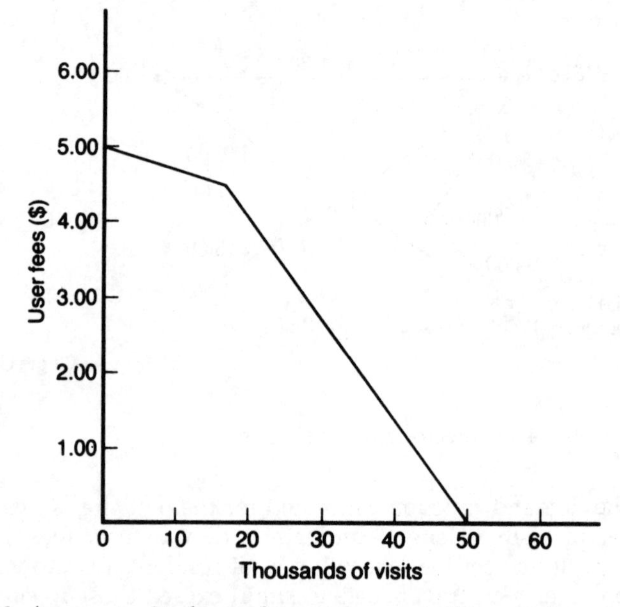

Figure 10.6 Aggregate site demand curve for competitive site I

Travel savings estimation

Description

The estimation of travel savings is a procedure for determining the value of a proposed site in terms of reduced travel costs for travellers. This estimate of value is thus independent of any measure of consumer surplus (CVM or TCM). As you will recall, consumer surplus is a measure of the hypothetical willingness of a traveller to pay for some non-market tourism commodity. This is, therefore, an abstract value. The value estimated with the travel savings approach is, in contrast, directly linked to real money.

This relatively simple procedure does not indicate anything about the inherent values of competing sites. Rather, it simply measures the net travel savings attributable to the creation of a new destination. This measure applies only to the saving accrued by current travellers. It should not be used to estimate the value of a site attributable to the stimulation of new travel.

Procedure

1. Determine the existing number of trips from all origins to some destinations and the round-trip costs for each origin.
2. Locate the alternative site and estimate new travel costs to this alternative site from each origin. If the new travel costs are less than the current costs of travel from any origin to the existing destination, subtract the costs from existing costs to estimate net savings per trip from the origin.
3. Estimate the total number of existing trips that are likely to be diverted to the new facility. This may be done either by assuming that travellers will always go to the site with the lowest overall costs or by conducting a survey to estimate the percentage of travellers that will be diverted by the new site.
4. Multiply the per trip savings by the total number of trips diverted from each origin. Sum these to determine the total savings.

Example

Using data from the previous example, we can compare travel costs associated with trips from each of the three origins to either site (Table 10.6). Assuming 100 per cent diversion to the nearest site, we observe that all travellers from origins B and C will switch to site II. We multiply the number of observed trips currently made by residents at each of these origins by the net savings (travel costs to site II minus travel costs to site I) to determine the total savings travellers will enjoy.

In this example, 75 000 travellers saved an average of $2.50 at origin C while another 200 000 travellers saved an average of $1.00 each at origin B, for a total savings of $387 500. The proximity of site II to origins B and C also stimulates an additional 3440 trips from origin B

Table 10.6 Calculation of travel savings attributable to site II

Site	A	B	C	
I ($)	11.50	5.00	15.00	
II ($)	16.00	4.00	12.50	
Savings attributable to site II ($) (per trip)	0	1.00	2.50	
Number of trips diverted	50 000	200 000	75 000	
Travel savings ($)	0	+200 000	+187 500	= 387 500

and 54 000 trips from origin C. No travel savings are attributed to these new trips.

Summary

Methods of valuation of a resource are based on expression of the willingness of travellers to pay for either continued access to that resource or for some degree of enhancement. These estimates are obtained by asking users directly or by inferring indirectly from travel costs. The direct method – contingent valuation – is most useful for making estimates for activities not associated with specific sites, when managers of sites are working to reduce congestion, or when trips to the study site have multiple purposes or destinations.

A problem with CVM is the potential for respondents to mislead the researcher by giving answers they perceive will offer some personal advantage. Indirect methods such as the TCM have been developed to, in part, avoid this potential bias.

TCM produces an aggregate site demand curve from actual travel costs and trips made. The area under the demand curve is a measure of the consumer surplus – the net economic value – of the resource. Consumer surplus reflects not only the intrinsic worth of a resource but also the number, quality and costs of alternatives.

The travel savings method defines value in terms of savings attributable to the creation of a new site. This method is useful when you wish to consider the potential benefits to be derived from opening a new site that will divert users from an existing and perhaps overcrowded site.

Further reading

Edwards S F and **Anderson G D** 1987 Overlooked biases in contingent valuation surveys: some considerations. *Land Economics* **63**: 168–78.

Martin L R G 1987 Economic impact analysis of a sport fishery on Lake Ontario: an appraisal of a method. *Transactions of the American Fisheries Society* **115**: 461–8.

Walsh R G 1986 *Recreation economic decisions*. Venture Publishing, State College, Pennsylvania.

CHAPTER 11

Impact and evaluation methods

Introduction

Community leaders expect tourism to increase incomes and create new jobs. Tourism managers as well as public officials expect marketing campaigns to generate increased visitation rates and tourism revenues. And most individuals – public officials, tourism industry leaders and local residents – expect that tourism will not harm the environment and may, in fact, help protect it.

The economic impact of tourism is a fundamental concern. Local business operators, of course, profit directly from tourists' spending. As they and their employees spend profits, salaries, and wages in the community, the money circulates through the local economy benefitting the community at large. Tourism commodities are often heavily taxed, not to discourage tourism spending (although that can be an unintended result), but to provide revenues for governments. Revenues gained from tourism are 'basic income' (Tiebout 1962) in that they result from export earnings. Money spent by tourists is new money in a local economy, not just a recycling of existing wealth. As a result, tourism revenues help pay for imported commodities and contribute to governmental income through taxes. At the same time, residents may be concerned about other changes tourism can precipitate. While economic growth is generally desirable, it can exact a price through social or environmental change and through reallocation of scarce resources.

Such observations are familiar to most tourism analysts, but they have not always been part of the conventional understanding of planners, policy-makers and developers. Tourism was long viewed as an unorganized collection of small businesses. The realization that this unorganized collection of predominantly small businesses comprises the largest industry in the world and is a leading earner of foreign exchange for many nations has prompted officials and planners to look more seriously at tourism.

Unfortunately, as local leaders looked for information about how tourism could be planned, developed, and managed in ways that would benefit the local community and protect the natural environment, they found little relevant research. They were offered, instead, anecdotal information, personal impressions, and the enthusiasms of promoters. Some extravagant claims were made about the economic impacts of tourism generally and about the effectiveness of advertising campaigns on generating tourists' revenues specifically. There is a need for objective, scientifically defensible data on economic and environmental aspects of tourism promotion and development, ranging from the micro or individual firm to the macro or national. This chapter examines methods that provide information on four issues: (1) the degree to which expenditure on tourism marketing translates into increased spending by tourists, (2) the magnitude of tourists' spending, (3) how the economic and other benefits from increased tourism compare with the costs imposed on the community by tourism development, and (4) whether a currently economically viable form of tourism development is sustainable in the long run.

Local expenditure model

Description

There are a number of different strategies for estimating the economic impact of tourism. Some economists collect supply side data (i.e. information related to receipts and expenditure of local businesses) to calculate the 'Keynesian multiplier', which is a measure of the level of regional economic growth stimulated by tourism. Another approach is to develop an impact model from an input–output matrix for a provincial, state or national economy. This allows the examination of the flows of tourism receipts and expenditure through an economy. Such information can be supplemented with data on tax structures to estimate the effect of tourism growth on government revenues. An example of this approach is the Conference Board's TEAM (Tourism Economic Assessment Model). Models such as TEAM can be used in a variety of contexts, such as estimating the overall impact of general tourists' spending, the economic impacts of ongoing business operations, or the impacts of major new capital projects. The outputs include the number of jobs supported by tourism, total wages and salaries, contributions to provincial or national GDP in various sectors (e.g. agriculture, construction, retail, business services), tax revenues, value of imports, and the international travel trade balance.

Such information comes at a cost. Models like TEAM require substantial investment in time and money for development, data collection, and analysis. As a result, they are usually developed by private consultants or government agencies who have access to the resources and the specialists required for impact assessment. As a further result, the models are prohibitively expensive for many local organizations. For example, in 1994, the Conference Board's TEAM carried a CDN $30 000 price tag, plus another $3000 annual maintenance fee. Clearly, there is a need for models that can be implemented at a lower cost in local communities and yet still provide basic impact information. One of these is the 'local expenditure model'.

The local impact model is based on techniques developed by Frechtling (1974) and by Wassenaar (1981). Versions of the model are now used by a number of different provinces and states. It is based on data derived from guest surveys and other data from the accommodation sector. There are two reasons for this (Wassenaar 1981). First, tourists who use commercial accommodation are responsible for the greatest portion of tourist expenditures. Second, the use of commercial accommodation to operationally define tourists is one of the least ambiguous and most practical techniques. However, there are a couple of problems with this approach. The expenditure of same-day visitors is not assessed even though it can be substantial. This omission can be overcome by conducting on-site surveys at major tourist attractions to estimate levels of expenditure and the percentage of visitors at each attraction that constitute same-day visitors.

Another problem with the use of commercial accommodation is the omission of travellers who stay with friends or relatives in private homes. This is a significant segment in many regions. Canadian statistics suggest that about one in three travellers stay in private homes and that these travellers are responsible for about half of all visitor-nights generated nationally. Although tourists who visit friends and relatives VFR typically spend less per day than tourists using commercial accommodation, you will miss a significant portion of tourists expenditure if you do not take VFR travellers into consideration. Again, you can correct for this omission by conducting an on-site survey to estimate the percentage of visitors staying in private homes as well as their overall spending levels and patterns.

It should be emphasized that this method estimates the magnitude of tourists' spending only in the destination region. Spending on airline or automobile travel and other *en route* services is not assessed. This is not a relevant weakness because the model is designed to apply only to destination regions. You will want to remember, though, that the total value of tourists' spending is usually greater than just the sum of expenditure made at the destination. Finally, the local expenditure model assesses direct gross revenues only. It tells us nothing about net economic impacts (the balance of benefits and costs) or the size of the tourism multiplier. If this item is of interest, it is necessary to employ more sophisticated and costly methods.

Procedure

1 Define the geographical limits of the region to be studied. This will usually be a municipality or county. Conduct an accommodation inventory of all commercial accommodation establishments in the region. This inventory should include the entire range of accommodation establishments: hotels, motels, motor inns,

bed-and-breakfasts, rental cottages, and so on. The total number of rooms or beds in each establishment should be recorded.

2 Conduct a survey of the accommodation establishments in the study region. You can survey all establishments if the number is small or a sample if the number is large. Collect the following information on either an annual or seasonal basis, as appropriate:

   (a) average room rate;
   (b) average number of guests per room;
   (c) average occupancy rate;
   (d) average length of stay.

   You may also wish to collect the following information to provide a more complete picture of the local industry. These statistics are not essential for the local expenditure model:

   (e) clientele mix (convention, other business, pleasure, personal or family matters, specific festivals or events);
   (f) tourists' origins by county, province and nation.

3 Prepare a regional summary of the survey results, either by totalling all responses if you conducted a comprehensive survey or by generalizing to the population of establishments if you used a representative sample.

4 Perform the following series of calculations:

   (a) Multiply the total room capacity in each category by the average occupancy rate, and that product by 365 nights per year to obtain the total number of rooms let annually. If you obtained seasonal or quarterly information, calculate totals separately for each season, and then sum as appropriate.
   (b) Multiply the average annual number of rooms let by the average room rate to obtain annual receipts from guests. If you used seasonal data, perform this step for each season and then add to obtain the annual total. Next, obtain the total receipts for all accommodation categories combined.
   (c) Multiply the total number of rooms let annually by the average number of persons per room to obtain an estimate of the total visitor-nights generated.
   (d) Divide the number of visitor-nights by the average length of stay to obtain an estimate of the number of visitors per year in commercial accommodation in each category. Sum these to obtain total visitors in all accommodation classes.

5 Divide the total number of visitors by the percentage of all visitors who stay in all types of commercial accommodation (obtained from national or regional surveys) to obtain an estimate of total visitors in all types of accommodation. At this point you have obtained the following information: (1) the total number of visitors who stay in various types of commercial accommodation; (2) the total number of visitors who stayed overnight regardless of the type of accommodation; and (3) total expenditure on accommodation. The

accommodation guest survey will also have provided you with information on occupancy rates, which are an indicator of the health of the local tourism industry as well as a barometer for potential expansion of the accommodation sector. The next series of calculations provides you with an estimate of tourism spending in other sectors of tourism.

6 With the support of the general managers of commercial accommodation establishments, conduct a survey of their guests to obtain estimates of their expenditure on:

   (a) accommodation;
   (b) food and beverages;
   (c) attractions, recreation and entertainment;
   (d) retail purchases, such as souvenirs;
   (e) gasoline and auto services;
   (f) other local transportation, such as taxis.

   Using the totals for each category of expenditure, determine the ratio between each category's total and the expenditure for accommodation. In other words, obtain an estimate of the amount spent by the average visitor in each category for every dollar or other currency unit spent on accommodation.

7 Using the expenditure ratios just calculated, multiply the ratio for each expenditure category by the total expenditure on accommodation obtained in step 4(b) above. This provides you with an estimate of total tourist spending in a variety of major sectors. Add the individual sums to obtain a grand total of local tourist spending.

8 Using national census information, obtain payroll/sales receipts ratios for each expenditure category. Obtain also the average annual wage-rate for employees in each expenditure category. Use local- or county-level information if available; otherwise, use provincial or national averages.

9 Multiply the total spending in each category by the payroll/sales receipts ratio to calculate approximate tourism-based payrolls.

10 Divide the payroll estimate by the average wage rate in each category to estimate the number of jobs supported by tourism.

Example

Data for the following example were drawn from a community economic impact study in a Canadian municipality. A summary of the accommodation inventory is shown in Table 11.1. The number of rooms in each category was multiplied by the annual occupancy rate and 365 days/year to determine the total number of rooms let per year. This figure was then used in step 4; the results of which are shown in Table 11.2. The total of all visitors in all accommodation categories was obtained and divided by the estimated percentage of travellers who stay in commercial accommodation (based on a provincial exit survey). This

Table 11.1 Local accommodation inventory

Accommodation category	Rooms	% of total rooms	Average occupancy	Average rate	Average persons/ room	Average length of stay
Hotels	512	62.7	57%	75.27	2.1	3.1
Motels	257	31.4	68%	51.88	2.9	3.2
Bed-and-breakfast	48	5.9	51%	42.09	2.0	2.9
Total	817	100.0				

Source: author's data

Table 11.2 Calculation of total visitors and visitor-nights

HOTELS
512 rooms × 57% × 365 nights = 106 522 room-nights let per year
106 522 room-nights × $75.27 = $8 017 881 in annual receipts
106 522 room-nights × 2.1 persons/room/night = 223 696 visitor-nights
223 696 visitor-nights ÷ 3.1 nights = 72 160 visitors

MOTELS
257 rooms × 68% × 365 nights = 60 973 room-nights let per year
60 973 room-nights × $51.88 = $3 163 292 in annual receipts
60 973 room-nights × 2.9 persons/room/night = 176 822 visitor-nights
176 822 visitor-nights ÷ 3.2 nights = 55 257 visitors

BED-AND-BREAKFAST
48 rooms × 51% × 365 nights = 8935 room-nights let per year
8935 room-nights × $42.09 = $376 083 in annual receipts
8935 room-nights × 2.0 persons/room/nights = 17 870 visitor-nights
17 870 visitor-nights ÷ 2.9 nights = 6162 visitors

72 160 + 55 257 + 6162 = 133 579 visitors in all commercial categories

133 579 ÷ 0.649 (ratio of visitors who stay in commercial accommodation) = 205 823 total visitors

produced an estimate of 205 823 visitors, 133 579 of whom stayed in commercial accommodation.

Results of a visitor survey on expenditure is shown in Table 11.3. This table also presents the accommodation expenditure ratios as described in step 6. The data from this table were then combined with total accommodation expenditure to estimate tourist spending in each of the six sectors of tourist spending (Table 11.4). Note that the expenditure in the first row of Table 11.4, 'Accommodation', is identical to the estimated receipts in the various accommodation categories in Table 11.2. Total expenditure was estimated to be $39 933 632 – or approximately $40 million.

Payroll/sales receipts were calculated from census sources for each

Table 11.3 Visitor-party expenditure survey data

	Hotels ($)	Motels ($)	Bed-and-Breakfast ($)
SPENDING ON:			
Accommodation	233	166	122
Food and beverages	421	261	171
Recreation, entertainment	25	38	21
Retail	91	68	70
Gasoline and auto service	14	15	15
Local transportation	21	3	2
ACCOMMODATION/SPENDING SECTOR RATIO			
Accommodation	1.00	1.00	1.00
Food and beverages	1.81	1.53	1.40
Recreation, entertainment	0.21	0.23	0.17
Retail	0.39	0.41	0.57
Gasoline and auto service	0.06	0.09	0.12
Local transportation	0.09	0.02	0.02

Table 11.4 Estimation of tourist-spending by sector

Sector	Hotels ($)	Motels ($)	Bed-and-breakfasts ($)	Total ($)
Accommodation	8 017 881	3 163 292	376 083	11 181 550
Food and beverages	14 512 365	4 966 368	526 516	20 005 249
Recreation, entertainment	1 683 755	727 557	53 934	2 475 246
Retail purchases	3 126 974	1 296 950	214 367	4 638 291
Gasoline and auto service	481 073	284 696	45 130	810 899
Local transportation	721 609	63 266	7 522	822 397
Total				39 933 632

Table 11.5 Tourism payroll estimation

Sector	Payroll/ sales ratio	Estimated annual payroll ($)	Average annual wages ($)	Number of jobs
Accommodation	0.35	3 913 542	15 700	249
Food and beverage	0.37	7 401 942	13 400	552
Recreation, entertainment	0.29	717 821	20 100	36
Retail purchases	0.25	1 159 573	18 100	64
Gas and auto services	0.22	178 398	18 100	10
Local transportation	0.30	246 719	16 200	15
Total		13 617 995		926

expenditure sector and are summarized in Table 11.5. When these ratios are multiplied by the expenditure in each sector from Table 11.4, the product is the estimated annual payroll in each sector attributable to tourism. Finally, dividing these annual payroll estimates by annual wages per worker in each sector yields estimated total jobs, by sector, supported by tourism (Table 11.5).

Conversion studies

Description

A question of enduring interest to destination marketing organizations (DMOs) is the effectiveness of advertising expenditure in terms of generating increased numbers of tourists. Research into techniques for assessing tourism advertising effectiveness dates back to at least Woodside and Reid's (1974) examination of print advertising for the state of South Carolina (USA). Since that study, numerous articles have appeared on the assumptions, techniques, and limitations used in studies of advertising effectiveness. Some of these are listed in the Further reading section at the end of this chapter.

The basic question addressed by advertising effectiveness studies is deceptively simple: how many tourists did a specific advertisement (or advertising campaign) bring to a destination? Answering this question, however, is far from simple. One of the main reasons for this is that the connection between a DMO's placing of an advertisement and the arrival of tourists involves several steps.

First, potential visitors must see the advertisement, comprehend it, and note the telephone number or address given for follow-up information. The advertisement may lead to new or increased awareness of a destination. The increased awareness may lead to a positive image of the destination either as a direct result of the advertisement or, more likely, as a result of a combination of the advertisement and other sources of information such as word-of-mouth. The recipient of the advertising message must also be in the market for a tourism experience; the advertisement must appear at a time when its message can influence the potential tourist's awareness of possible destinations.

If the content and timing of the advertisement is appropriate, the potential tourist may seek additional information by contacting the organization placing the advertisement. The organization must respond with the expected information in a timely way. The information must then be read and comprehended. If sufficiently positive and consistent with the potential tourist's interests, the individual may select the destination being advertised.

Research designs – advertising tracking studies – can be developed to assess the performance of an advertisement and follow-up information at each stage of this lengthy process. Siegal and Ziff-Levine (1990) describe some of the basic procedures and issues involved, based on their work as private consultants specializing in tracking studies. Such studies, however, are time-consuming, labour-intensive and expensive. Most DMOs lack the resources to conduct full-scale tracking research, and must consider alternatives. The most common alternative is a conversion study.

A conversion study is designed to estimate the percentage of individuals responding to an advertisement by requesting additional information (through a coupon, telephone call, or other verifiable mechanism) who actually visited the destination advertised. Conversion studies, in effect, look at only two stages in the process described above: the number of people responding to an advertisement and the number of those who actually visit the destination.

Conversion studies are increasingly popular with DMOs, especially those supported by public funds, as a tool to justify budgets. Because of the survival-motivation, many conversion studies fall short of full adherence to rigorous scientific procedures. Muha (1976) noted in his early review of conversion studies that returns on investments of between 23 and 100 to 1 were commonly claimed. Burke and Gitelson (1990) believe that more recent conversion studies have produced more realistic (i.e. lower) conversion rates. However, conversion studies sometimes still carry the stigma of shoddy, self-serving research techniques. This section examines some of the problems associated with conversion studies and describes some of the principles that should be followed for credible conversion research. The following comments are based, in part, on Burke and Gitelson's (1990) review of conversion studies for the US Department of Commerce's task force on accountability in tourism research.

Problems

One of the fundamental weaknesses associated with much conversion research is the inaccurate assumption that there is a direct, causal relationship between an information request and a destination visit. DMOs might wish to assert that every individual who visited their destination after requesting information came as a result of the information provided. Gitelson (1986) cast doubt on this assertion in a survey of visitors to North Carolina (USA). He found that only 68 per cent of people requesting information packages were actually considering a vacation in North Carolina and that 46 per cent of these had already decided to visit the state before receiving the information kit. The information request thus had no effect on the decision to visit.

One-quarter of the people requesting information kits were planning to visit friends or relatives – a trip purpose that is generally not influenced by advertising. Other requesters included people considering moving to the state, planning a business trip, conducting school projects, or engaging in vicarious travel (by reading travel brochures). Some respondents already lived in North Carolina, but were considering vacations in a part of the state different from where they lived.

Conversion studies typically assume that the potential visitors' only

...mpaign. However, most ...ty of sources, including previous ...club books, popular media sources, and ...oring the effects of these other sources of information ...e analyst to conclude that an advertising campaign is more effective than it really is.

Conversion studies are based on surveys. A common problem in tourism surveys is non-response bias. The usual sampling frame for a conversion study is all those who requested additional information. However, people who actually visited a destination tend to be more likely to respond to questionnaires from that destination than those who did not. This bias produces unrealistically high conversion ratios. Hunt and Dalton (1983) performed a study to assess the impact of non-response bias on conversion estimates. They conducted both a mail survey and a telephone survey of a sample of potential skiers in the western US. Their mail survey, with a relatively low response rate, yielded a conversion rate of 33.2 per cent. A telephone survey, which achieved a higher response rate, yielded a conversion rate of 23.1 per cent. The results of this study suggest that non-response bias exaggerated the conversion rate by over 40 per cent.

Another issue associated with survey sampling in conversion studies is the precision of the results. Precision is a function of sample size. Most DMOs, especially those at local levels, have little experience and less budget for survey research. A tendency is to put as little time and money as possible into conducting a survey, with the result that response rates are usually quite low, sometime below 10 per cent. Such low response rates not only permit non-response bias, they lead to a deterioration in the reliability of the results. For example, assuming a conversion ratio of 20 per cent, a sample of at least 1000 respondents is needed to achieve a sampling tolerance of ±2 per cent, 95 times out of 100. If you have a sample of 500 people, the tolerance is ±4 per cent; at 250 respondents, the tolerance increases to ±5 per cent, and at 100 respondents, the sampling tolerance is ±8 per cent.

Care must be taken in determining the actual destination visited by tourists during a conversion study. This is a special concern for conversion studies conducted by local or regional DMOs. Consider the following hypothetical example. An American living in Dayton, Ohio (USA) writes to the Chamber of Commerce, Niagara-on-the-Lake for information about the annual Shaw Festival. This individual also contacts Festive Stratford, the DMO for Stratford, Ontario, for information on the annual Shakespearean Festival. The tourist and his family decides to visit Stratford, but not Niagara-on-the-Lake. The Chamber of Commerce in Niagara-on-the-Lake conducts a conversion study and draws this individual's name for the sample. He notes in his response to the questionnaire that he did visit Ontario during the period in question. However, the individual's trip to Ontario should not be counted as a successful conversion by the Chamber because no trip was made to that community. Generally, the more inclusive the destination area described in a conversion questionnaire, the higher the conversion rate.

Conversion studies typically include information about the magnitude of visitors' spending as well as advertising costs in order to express conversion rates in terms of financial returns on advertising investment. There are two potential problems associated with this approach. The first is the difficulty of obtaining reliable estimates of visitor spending in the destination. Recall time for expenditure information is very short. In fact, it is likely that most travellers are unable to provide accurate estimates of spending at all, unless they maintain a diary of actual expenditures. Most researchers believe that tourists tend to *under-estimate* actual expenditures, which would lead to an artificially low estimate of the return on investment. The problem is complicated by the difficulty in separating out expenditures in the destination as opposed to total trip expenditures. For example, transportation costs to a destination can be substantial, but unless these costs were paid to a firm located in the destination, travel costs *to* the destination should not be counted as revenue for that destination.

Finally, there is sometimes a tendency for DMOs to ignore the total costs associated with an advertising campaign. Many DMO advertisements are co-op ads (co-operative advertisements) in which the DMO shares costs with local tourism firms. As a typical example, a local DMO in Canada recently conducted a mass media campaign involving television, radio and print advertisements. The DMO allocated $20 000 to the campaign, which generated an identified 234 visits. The average visitor-party to the area spent $130 per trip. This translates into $30 420 income from an outlay of $20 000 – or a return of 52 per cent. However, the campaign actually involved several local business partners, with a total outlay of nearly $85 000. Thus the return on investment was actually a loss of 64 per cent. Further, the $85 000 represented only production costs and direct media purchase (print space and air time); it did not include staff time spent handling enquiries, postage, and handling of information kits sent to potential tourists. If these costs had been included, the loss would have been even greater.

Recommendations

The following points should be kept in mind when conducting a conversion study. Some of these are general principles and some are specific, practical guidelines.

1 Remember that the decision to travel is usually influenced by many factors; only rarely will an advertising campaign be the sole, causal factor. The language of a conversion study should be descriptive and simply suggest correlations rather than state firm cause-and-effect relationships. Recall, in particular, that increasing repeat visits and promoting conventions and meetings have the effect of reducing the net conversion rate for discretionary travel to a destination.
2 Select as large a sample as the agency can afford. Ensure the sample is representative and unbiased. Work to minimize

non-response bias. Hunt and Dalton (1983) suggest at least an 80 per cent response rate is needed to avoid non-response bias. If your response rate is below that level, you may want to follow-up with non-respondents to assess the extent, if any, of non-response bias. Equation [11.1] defines the confidence limit for binomial distributions.

$$p = \hat{p} \pm [1.96\sqrt{(\hat{p}\,\hat{q}/n)} + 1/(2n)] \quad [11.1]$$

where: p = confidence limits;
$\hat{p}$ = r/n;
r = number of respondents who visited destination;
n = number of respondents; and
$\hat{q}$ = $1.00 - \hat{p}$.

3 If the conversion study is being done for an advertising campaign with multiple advertisements, the sample selection should reflect the distribution of enquiries among the advertisements. For example, if 40 per cent of enquiries came from a TV spot, another 40 per cent from a newspaper advertisement, and the balance of 20 per cent from a radio spot, the sample should include 40 per cent of its respondents from each of TV and newspaper advertisements, and 20 per cent from radio.
4 Include questions in the conversion study to determine why the respondents were seeking information, whether or not they had already decided to visit before receiving the requested information, and how the information affected the decision (e.g. did it influence their decision to visit; did it help them select accommodation).
5 Be explicit about defining the destination area. You might ask respondents to mention the name of a specific community or destination they visited to ensure they actually visited the destination you are interested in.
6 Remember that recall about trip expenditure is unreliable unless some form of diary was kept; respondents tend to under-report actual expenditure.
7 Incorporate all costs (production, placement, postage and handling, etc.) in estimating return on investment.

Benefit–cost analysis

Introduction

Benefit–cost analysis (BCA) is a hypothetical experiment. This resource management tool is intended to answer the question of whether or not society would be better off after the implementation of a proposed project. Although BCA superficially appears to be nothing more than intuitive common sense – a project is justifiable only when the potential benefits outweigh the potential costs – its application involves issues and conventions that are not necessarily intuitive. This section briefly describes the history of benefit–cost analysis as well as the major conventions and issues associated with the method.

Benefit–cost analysis is nearly a century old. In the US (benefit–cost analysis or cost–benefit analysis has a long history in Canada and Europe, as well), it was first authorized as a decision-making tool by the US Rivers and Harbors Act of 1902 and further refined under the US Flood Control Act of 1936. The 1936 Act introduced a phrase that is central in benefit–cost analysis: benefits and costs are assessed and compared regardless of 'to whomsoever they may accrue'. In other words, benefit–cost analysis (in principle) adopts a social or public perspective, rather than an agency or firm perspective. Thus benefit–cost analyses are rarely ever done by a private firm, because private firms are normally interested only in the costs and benefits *they* accrue and generally are not interested in any costs or benefits accruing to other entities.

As you might have noted from the two American acts cited above, BCA was originally developed to guide decisions about the development of water resources by public agencies. The technique, however, was quickly extended to other natural resource developments. The common elements in these projects include the following: (1) they involve capital investments in tangible facilities, such as the construction of a dam and reservoir, (2) costs and benefits are easily defined, and (3) the projects directly contribute to the gross domestic product through improvements in productivity or economic efficiency.

However, since the 1960s, BCA has been extended to a wider range of public projects, including criminal justice, foreign aid, and welfare (Walsh 1986: 556). These extensions have not been without controversy. Part of the debate concerns whether or not monetary equivalents can be meaningfully defined for costs and benefits of social programmes. There is also the perception that benefit–cost analysis is intended to provide an irrefutable, scientifically objective answer that removes decisions from the realm of political judgement. While some may wish for that, the fact is that benefit–cost analysis is only one tool to provide public officials and managers with information to assist them in exercising their own judgement.

The application of benefit–cost analysis in tourism involves seeking answers to three broad questions:

1 What are things like now?
2 What benefits will the project give us that we do not currently have?
3 How much do we have to pay to obtain those benefits?

The time frame for the answer to the first question is the present, but the answers to the second and third questions involve looking into the future, usually up to 20 or 30 years into the future. Benefit–cost analysis focuses only on new benefits and new costs; any existing benefits and sunk costs are not incorporated into the evaluation.

BCA is used in four different contexts. First, it indicates whether a specific project is feasible. For example, the technique could determine whether a proposed convention centre with specific characteristics would be economically justifiable from a community perspective. Second, BCA can assist in determining the scale at which a single project is to be developed. The method could be applied to examining the relative economic efficiency of different sizes of the proposed convention centre. Third, BCA can assist an agency in deciding which of several competing projects is most efficient. A resource management agency might be considering proposals for dams or reclamation projects at several different locations. BCA can help identify which of these should be pursued. Finally, BCA can also help the agency set priorities or a development schedule for the order of development of a series of projects.

Benefit–cost analysis is founded on several assumptions. We assume that we can identify and measure in commensurate quantitative terms (i.e. money). Benefits or costs that cannot be expressed in terms of money, whether through market valuation or non-market valuation techniques such as contingent valuation, are ignored. The need to ignore non-quantifiable benefits or costs is one reason why benefit–cost analysis can never replace informed, wise judgement of public officials.

Projects have value only to the extent that a need exists for their services. Generally, option or existence values are not included as benefits. Each project, if it is developed, should be developed at the scale that provides the greatest economic efficiency (the greatest return on benefits for the costs). Development priorities should follow the order of economic efficiency, i.e. the most efficient projects should be developed first.

Several other principles guide benefit–cost analysis. BCA examines only the marginal or new costs and benefits arising from a project. These are real or allocative effects rather than distributional effects. Allocative effects are those that represent real increases in total income, production or consumption and not simply a shift of wealth from one group to another. For example, a large-scale construction project might prompt new labour agreements that result in higher hourly wages for construction workers. While these higher wages are a benefit to the employee, they are a cost to the investors in the project. Unless the higher wages are justified on the basis of better productivity resulting from new training, skills, or more efficient technology, the benefits to employees of wage increases will be offset by the cost to employers of having to pay those increases – and thus they should not be included in a benefit–cost analysis.

Basic concepts

One of the basic concepts to be resolved in a benefit–cost analysis is the definition of benefits and costs (Figure 11.1). Benefits consist of (1) primary or direct and (2) secondary or indirect benefits. Direct benefits are the gains that accrue to individuals who make use of the new

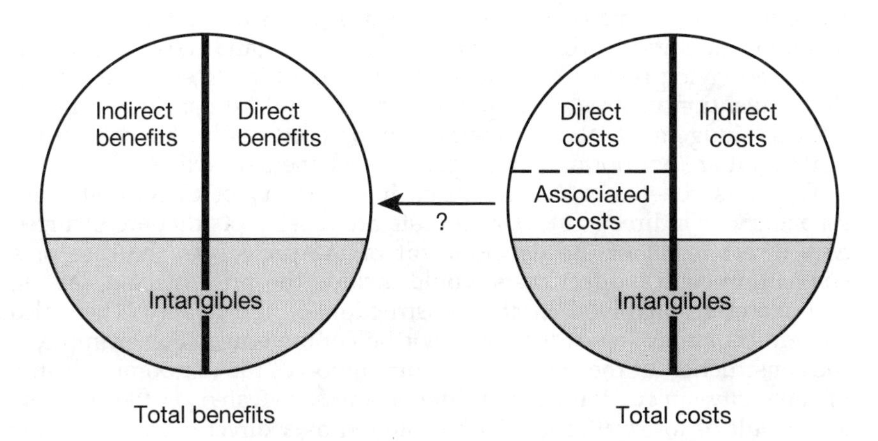

Figure 11.1 Classification of benefits and costs

project. For example, the construction of a convention centre may increase the number of visitors coming to a community for meetings and conventions. The net increase in consumer spending by these new visitors would be a direct benefit of the project. Indirect benefits are those induced by the project but are not the intended result of a project. In the case of a new convention centre, a new hotel might be built to serve the needs of business travellers visiting the area. The net income of that hotel would be calculated as an indirect benefit.

A word of caution is needed here. You need to ensure that the income gains you claim as benefits are indeed gains and not simply a transfer of income from one location or business to another. For example, if the scale of analysis for the convention centre in our example is national (perhaps because it is being assessed by a federal agency in connection with a possible economic development grant), the net incomes generated by the new facility may simply represent a transfer of meeting expenditure from one community to another and thus there would be no national net gain. On the other hand, if the convention centre is supported only by local funds, the business it draws is not simply drawn away from existing local hotels and meeting facilities, and the scale of the analysis is local, net income generated is a benefit.

Benefits, whether primary or secondary, may be tangible or intangible. Tangible benefits are those exchanged through the market-place, e.g. the revenue generated by room rentals in a convention centre. Intangible benefits are those that are not exchanged through the market. The development of co-operative education job placements at the convention centre for local university students might be counted as a direct intangible benefit. The increase in recreation opportunities provided by the development of a public park across the street from a new convention centre would be an intangible indirect benefit. Be sure to note that 'intangible' is not synonymous with 'non-quantitative'. Intangible benefits (or costs) are simply those whose value must be estimated through indirect means such as the contingent valuation method. 'Non-quantitative' benefits (and costs), such as enhanced civic pride or a reduction (or increase) in street crime around the convention centre, are important and should be considered in the political judgement about whether to proceed with the convention centre, but are not normally included in the benefit–cost analysis.

Costs, as benefits, may be divided into primary or direct costs, and secondary or indirect costs. Direct costs are those costs that are incurred as a direct result of the development of the project. In the case of a convention centre, direct costs would include the price of land, labour, and materials involved in the construction of the facility. They also include economic losses that may not be compensated. For example, if the construction of the convention centre involves the temporary closure of some city streets, the losses to the businesses located on those streets as a result of lost traffic should be counted as a direct cost, even if the project developers do not compensate the businesses. Intangible direct costs should also be included. As with intangible benefits, intangible costs are those that do not normally appear in the market-place. A potential example of an intangible cost for our hypothetical convention centre would be the loss of the recreation opportunities of a public park that was lost because of the construction of the facility.

There is a special category of direct costs that should also be noted: associated costs. These are costs incurred by the private, primary beneficiaries of a project in order to acquire the direct benefits of the project. The classic example of associated costs are the costs to farmers who must pay for pumps and pipes in order to use irrigation water provided by a public project. The purpose of the concept of associated costs is to allow the analyst to distinguish between the direct public costs absorbed by the agency constructing the irrigation project and the direct private costs absorbed by the beneficiaries of the irrigation water.

You have a choice of two methods for dealing with associated costs. You may subtract the associated (private) costs from direct benefits, before the net benefits are compared to costs. Alternatively, associated costs may be added to other direct costs. It is suggested that you adopt the second strategy: adding associated costs to other direct costs. One of the principles of benefit–cost analysis is that all benefits and costs should be assessed regardless of to whomsoever they accrue. The allocation of costs to public versus private entities can be examined later, if desired; but initially all costs should be grouped together to be compared to all benefits. The two different tactics can produce radically different benefit–cost ratios. Thus, regardless of which tactic you adopt, be sure that the same tactic is used in all benefit–cost analyses that you compare.

Indirect costs are those costs associated with the production of indirect benefits. To return to our convention centre example, if the income from a new hotel is claimed as an indirect benefit, the costs of constructing and operating that hotel should be counted as indirect costs.

As was noted previously, sunk costs and existing benefits are excluded from any benefit–cost analysis: bygones are bygones. This principle is straightforward, but a related issue – that of joint costs – is less straightforward. Consider a dam that is constructed for a reservoir that will provide both flood protection during the spring thaw and recreational boating during the summer. The cost of the dam might be apportioned between the two general benefit sets: flood protection and recreation. In contrast, a spillway to handle flood overflow would be clearly attributable to the flood prevention function, whereas the cost of boat ramps would be directly linked to the recreational function of the new dam. How are joint costs to be handled?

In the context of a benefit–cost analysis, the apportioning of costs is irrelevant. All costs of a project must be considered, along with all benefits, regardless of the number or nature of costs and benefits.

One final word of caution: there seems to be a tendency in tourism for planners to overestimate benefits and to underestimate costs. In particular, there is a tendency to assign economic value to a wide range of intangible benefits without attempting to estimate – or even identify – intangible costs. The objective tourism analyst will guard against this. One possible strategy is to incorporate an estimate of an indirect cost for every indirect benefit claimed for a project. Such an approach does not guarantee total objectivity, but it does promote greater sensitivity to acknowledging both sides – costs and benefits – of the project ledger.

Benefits and costs do not accrue immediately. Costs often are concentrated in the start-up phase. Benefits of a project, in contrast, usually reach a maximum several years into the life of the project. Whatever the actual pattern of benefit and cost flows, it is necessary to adjust their values to reflect the fact that future costs and future benefits carry less value than immediate costs or benefits. For example, at an annual percentage rate of 5 per cent, a payment (to you or by you) of \$105 next year is equivalent to a \$100 payment today because you could put the \$100 in a bank today and draw out \$105 a year later. Or if you were expecting a \$105 payment next year, you could borrow \$100 today, and then pay back the loan with the future payment of \$105. The value of future payments or future receipts expressed in their worth today is known as their 'present value'. The conversion of future value into present value is done through a discounting factor:

$$b' = \frac{b}{(1+j)^i} \qquad [11.2]$$

where: i = the year in which the benefit or cost occurs;
j = discount rate;
b = value of future benefit or cost; and
b' = present value of a benefit or cost.

The value of costs and benefits in each year is weighted by this factor and then summed over the years of the project's life. Note that the higher the value of *i*, reflecting years further in the future, the lower the present value. Similarly, the higher the value of *j*, the lower the present value.

Two bases are available to set the discount rate. The first is the social opportunity cost rate, which reflects the cost of money in the market-place. The social opportunity cost rate approximates the interest the money invested in a project would earn if invested in some private venture. Advocates of this rate argue that the goal of economic efficiency is not likely to be achieved if projects are evaluated using a discount rate that does not reflect the potential earnings from the project's funds if they were invested in the private market-place.

The alternative is the social time preference rate. This rate is lower than the social opportunity cost rate because, as advocates of social time preference argue, investors generally find that loaning money to government is less risky than loaning to most private enterprises. Further, citizens feel (it is presumed) a sense of goodwill towards government as well as trust, and thus are willing to loan money to government at a lower rate. Finally, government obtains its funds from all sectors of the economy, so it is inappropriate to use return rates in one or two high-growth sectors as a discounting guide. Rather, some lower figure representing the total growth rate of the economy as a whole is more appropriate.

Either base is defensible. It is recommended that if you conduct a benefit–cost analysis, you examine the impact of different rates on the decision to support or reject the project. As noted, the social opportunity discount rate is usually higher than the social time preference rate. Recall, too, that the higher the discount rate and the more distant the future year, the lower the present value of future costs and benefits. Because costs tend to be greater in the early years of a project, and benefits tend to be greater in later years, the social opportunity cost rate tends to emphasize costs over benefits and thus reduces the probability of approving a project. The social time preference rate discounts benefits less than the social opportunity cost rate, and thus gives projects a somewhat greater probability of being approved.

In order to apply discounting, some economic lifetime for a project must be determined. This is defined as the time span over which benefits may be derived. For some projects, however, such as national parks, the benefits may be derived in perpetuity so a more practical definition is needed. A common rule of thumb is to define the lifetime of a project as that period over which 80 or 90 per cent of the present value of benefits will be achieved. This approach, however, means that the economic life of a project is greatly influenced by the discount rate. A discount rate of 5 per cent yields an economic lifetime of about 46 years, while a discount rate of 10 per cent reduces the lifetime to 25 years.

Cost estimates in BCA should also include opportunity costs. Opportunity costs are the value of benefits foregone that could have been derived from the next best alternative use of resources. This requires that you identify what other potential uses a parcel of land might have been put to. In the case of the construction of a new airport, some assessment of the potential returns from that land in agriculture or housing might be appropriate. Although the idea of including opportunity costs is relatively straightforward, some thought is necessary to determine what might be included as an opportunity cost. To continue with the airport example, if the project under consideration is an expansion of the runways, the opportunity costs from expansion may not be associated with agriculture or housing because the existing airport development cannot be realistically applied to any use other than as an airport. In fact, there may not be any opportunity costs at all for an airport expansion because the alternative uses of runways are largely non-existent. There are always, of course, the opportunity costs associated with the money that government takes from residents and businesses through taxes. Some assessment should be given to what benefits that money might have produced generally if it had been left in the hands of the private sector. This line of argument is, of course, the reasoning behind arguments for the use of the social opportunity cost as the discounting factor in BCA.

Project evaluation

The criterion for selecting the optimum scale of a project is maximum net benefits. This is the scale at which a marginal increase in benefits is equal to the marginal increase in costs. Figure 11.2 illustrates the criterion. The upper half of Figure 11.2 represents typical variations in total costs and benefits associated with increasing scale of development. Note that total costs are shown as increasing in a linear fashion with increasing size. Benefits, however, show a different pattern. Here benefits rise at an increasing rate, reflecting economies of scale, to some point when the rate of increase begins to slow down.

There are four points of interest in the upper half of Figure 11.2. Points *A* and *D* define the extremes of feasible projects. Between the scales represented by *A* and *D* lies the range of scales for which benefits are greater than costs. Point *B* indicates the scale for which the ratio between benefits and costs is at a maximum. Although this point is close to the optimum, one can continue to increase the scale of the project and gain even greater net benefits until point *C*. *C* is the point at which the net difference between benefits and costs is at a maximum. This is considered to be the ideal scale for a project.

The relationships just described may be more obvious in the lower half of Figure 11.2. Here the *Y*-axis represents the benefit/cost ratio. Note that *A* and *D* have benefit/cost ratios of 1.0 – the limits of a feasible project. *C*, the scale offering maximum net benefits, may also be seen in the lower half of the diagram as the point at which the ratio of marginal increments of benefits to marginal increments of costs (**not** the ratio of total benefits to total costs) is unity.

With respect to choosing among competing projects, the criterion for choice is ambiguous. In fact, four different measures are available. These are:

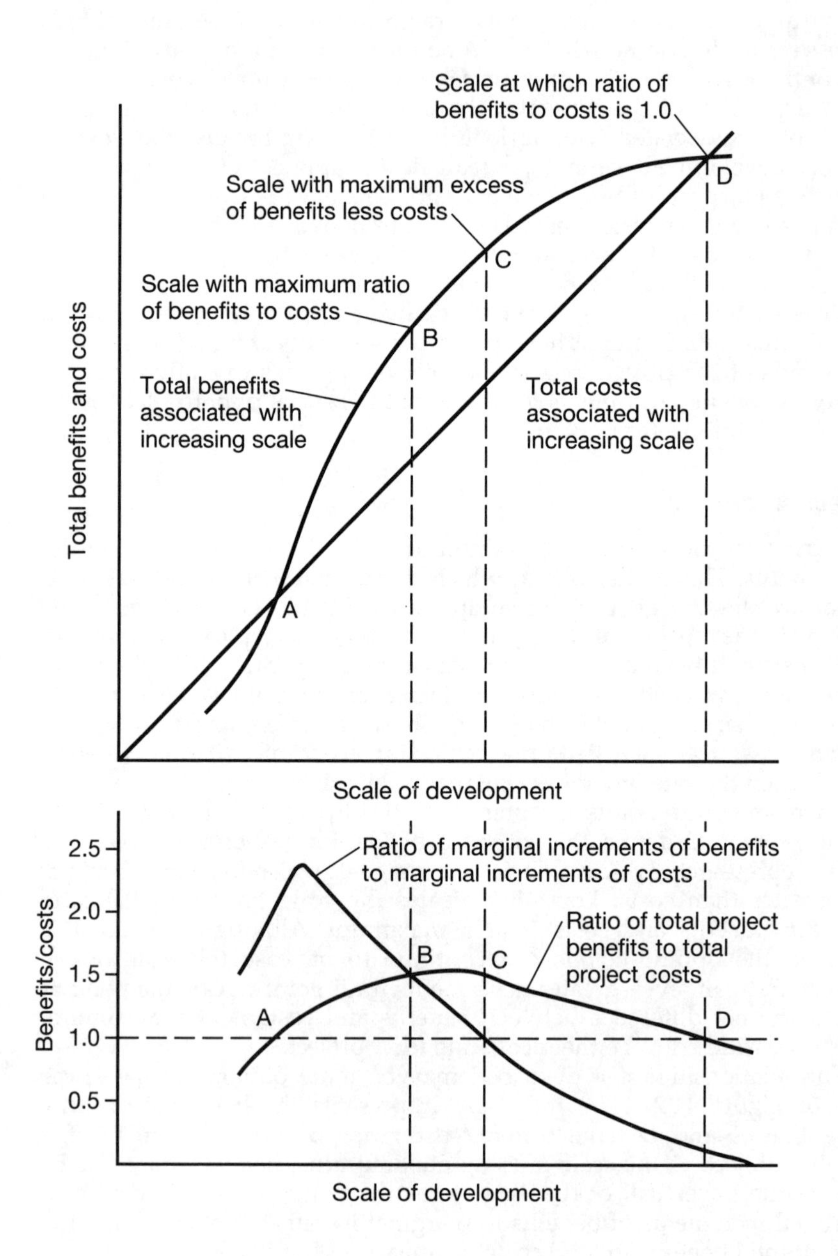

Figure 11.2 Selection of optimal project scale

Criterion	Equation	Range for feasible projects
Net benefits	$b - c$	>0
Benefit/cost ratio	$\frac{b}{c}$	>1
Net return on total costs	$\frac{b-c}{c}$	>0
Net return on operating costs	$\frac{b-oc}{ic}$	>1

Where: b = present value of benefits;
c = present value of total costs;
oc = present value of operating costs; and
ic = present value of investment costs.

Each has strengths and weaknesses. Net benefits provide an indication of the absolute magnitude of net benefits, but give no indication of the relative efficiency of a project. The benefit/cost ratio assesses relative efficiency, but is incapable of reflecting different absolute sizes in projects. Net return on total costs is roughly comparable to the second method. It provides a measure of the relative efficiency of a project in terms of the percentage rate of return on total costs. The final criterion, net return on investment, is viewed by some analysts as being the most realistic, but it is of limited usefulness when the costs of maintenance and operations are of secondary importance to investment costs. Any of these four measures is acceptable, although most analysts use either net benefits or the benefit/cost ratio. Whichever you use, be sure that the same measure is used for all projects to be compared.

Common errors associated with benefit–cost analysis

The Treasury Board of Canada reviewed benefit–cost analyses done for a range of public projects and identified a number of mistakes commonly made. The following represent those that have the greatest relevance for tourism.

1 *Adjusting for anticipated price changes when calculating future benefits and costs.* Price changes, including the general inflation rate, should not be incorporated into calculating the present value of benefits and costs. The analysis should focus on effective real prices in today's terms.
2 *Including increases in land values as a benefit.* Increased land value is usually a distributive effect, not an allocative effect. Increased land

values simply represent a potential transfer of wealth from one individual to another.

3 *Using average rather than marginal benefits and costs.* The assessment of allocative costs and benefits requires the use of marginal (net changes) benefits and costs. For example, highway construction will result in lowered operating costs on alternative routes due to lessened maintenance needs. However, operating costs tend to change less than proportionately with increased traffic, marginal benefits (from lowered operation costs) will be less than savings calculated on average costs. Unfortunately, average operating costs are more often collected and reported than marginal costs.

4 *Ignoring the costs of government grants.* Private projects are often assessed for financial feasibility net of government grants. However, benefit–cost analysis requires a consideration of full costs to society, so any government grants made to participating partners in a project must be counted as costs.

5 *Including interest payments on borrowed capital as a cost.* Although interest payments on loans represent a real financial obligation, they are irrelevant in determining the true costs of a project – which are the costs of resources. Resource costs are independent of the source of financing.

6 *Including depreciation as a cost.* Depreciation is an accounting device. It should not be included as a cost on top of the real costs paid for capital equipment.

7 *Estimating shadow prices of non-market goods at rates below market prices.* Shadow prices should be estimated at rates that match market forces as closely as possible, even if those resources are not being used for other purposes.

8 *Including the multiplier effect as a benefit.* There are a couple of problems with this practice. First, the practical problems of reliably estimating an economic multiplier and leakages associated with a project are substantial. Further, the value associated with the induced utilization of unemployed resources must be offset by the opportunity costs associated with the redirection of capital to the project and away from alternative uses.

9 *Claiming higher wages resulting from a project as a benefit.* Higher wages may be claimed as a benefit only if it can be empirically demonstrated that the employees were under-utilized or under-paid in previous positions or were under-employed. Alternatively, higher wages might be seen as a benefit if it can be empirically proven that the new jobs provided on-the-job training that, in turn, increased the skills and productivity of the workers.

10 *Including transfer payments in benefits or costs.* Transfer payments, as their name suggests, have no direct effect on production or consumption possibilities. They merely represent a transfer of wealth to one agency or government to another. They should not be included as either a benefit or a cost.

Limitations

Decisions about the implementation of public projects from a social perspective greatly benefit from an objective, logical, and easily understood evaluative tool. Benefit–cost analysis is such a tool. However, there are limitations. Critics argue that analysts are sometimes inconsistent in their adherence to the principles of BCA, particularly in overemphasizing benefits and under-estimating costs. Further, the distinction between allocative and distributional effects can blur in practice, with the result that some effects (particularly wealth-transfer effects claimed as benefits) might be included when they should not be. Some externalities, especially environmental costs, might be ignored or underestimated because of the difficulty of developing accurate shadow prices. Agencies are sometimes accused of using discounting rates that are too low in order to increase the chances that a proposed project will be approved.

In practice, BCA is often used merely to demonstrate economic feasibility and not to determine optimal scale or to decide among competing projects. In other words, BCA is sometimes used just to assure project planners that the project falls somewhere between points *A* and *D* on the upper graph in Figure 11.2.

Another limitation is that benefits and costs are disassociated from each other. BCA does not indicate anything about who pays the costs or who receives the benefits. The focus is on society as a whole, not on sub-groups within the society. Although BCA presumably examines all costs and benefits regardless of to whomsoever they shall accrue, this ideal is usually extended only to the political borders of the government doing the study. Consider a recent proposed project on the Red River, a stream that flows across the border between the US and Canada. The US Army Corps of Engineers, in their benefit–cost analysis of a proposed dam, focused only on US costs and benefits, and ignored the environmental and economic impacts that the project would have imposed on Canada.

As noted at the beginning of this section on benefit–cost analysis, the method is aimed at assisting public decision-makers arrive at the 'best' decision regarding a proposed project. BCA has limitations, but then so do all other decision-making tools, including the political process. In conclusion, BCA offer two advantages that support its widespread use. The conventions and procedures for conducting it are logical, well-understood and explicit. The findings of a BCA are not presented as an irrefutable dictate, but simply as one input into the resource allocation process.

Indicators for sustainable tourism

The tourism system, involving the interplay of tourists, businesses, agencies, and the environments in which tourism occurs, is so complex that no one can comprehend its totality. As a result, managers, planners, and analysts use selected pieces of information – indicators – to monitor the system. Indicators are empirical, quantitative measures that serve as windows allowing insights into a larger, complex reality. They are not comprehensive measures; rather they provide reliable, limited information to guide the decisions and actions of planners and managers. Quantitative measures of tourist arrivals, fluctuations in currency exchange rate, travel balance of payments, and job creation are familiar examples of indicators that reveal information about the broader progress and economic impacts of the industry. However, the industry needs not only economic indicators but also indicators of environmental and social change. Indicators do not simply measure current conditions but also serve as 'early warning' devices to alert managers of imminent problems. This section presents some possible indicators for assessing the long-term sustainability of tourism development.

Much has been written on the concept of sustainability in tourism (see, for example, McIntyre 1993; Inskeep 199; Nelson *et al.*, 1993). However, relatively little of this literature has provided specific, practical guidelines for guiding sustainable tourism development. In part, this is because the concept of sustainability is still an idealistic and vague goal. Some authors (e.g. Chipeniuk 1993) have proposed bio-indicators to assess 'sustainability' but often these indicators have little direct relevance to tourism. For example, Chipeniuk (1993: 277) suggests using the 'number and size of earthworms', 'presence or absence of bald eagles, ospreys, and loons during breeding season', and 'presence or absence of toads and tree frogs'.

Practical tourism indicators will include some physical environmental indicators but they must also include measures relevant to visitor satisfaction, economic goals, and social impacts. The measures must be sensitive to changes in the physical, economic, and social environment resulting from the passage of time, as well as to spatial variations in those environments. They must allow comparisons over time or between regions. Indicators should allow an analyst not only to understand current conditions but to anticipate the direction, probability, and the implications of change. Ideally, there should be some sort of bench-mark to determine critical levels of the indicator, i.e. what value indicates a 'safe' or 'dangerous' condition. Data for the indicators should be easy to obtain, easy to interpret, and capable of leading to a meaningful management response.

There is still need for developmental work on identifying and testing candidate indicators for the tourism industry. The following candidate indicators are based on work done by the International Working Group on Indicators of Sustainable Development (1993) for national and local conditions. The suggestions in the table follow the Working Group's recommendation although some wording has been changed for the sake of clarity. Further, some indicators have been excluded because they were not judged as useful for management. These included the existence of a guideline for environmental impact assessments. Other indicators, such as those related to economic benefits for local communities, have been added because they are important for planning and assessment purposes.

The indicators in Tables 11.6 and 11.7 include:

1. **Warning indicators** to alert decision-makers to emerging critical situations that require immediate action.
2. **Indicators of environmental stress** to help focus management actions.
3. **Base line indicators** to help decision-makers understand the degree and direction of change in the environment for which they are responsible.
4. **Indicators of impacts** to explicitly monitor problems arising from environmental deterioration or the success of new management actions.

Table 11.6 Suggested national indicators for sustainable tourism

	Indicator	Comments
1	Per cent of national area in protected categories – classification also by level or type of protection – classify also by types of tourism activities permitted	May need to include measures of enforcement and/or estimates of extent of significant areas
2	Endangered species or spaces inventory	A potentially useful measure of threats to tourism resources, especially those that are ecologically or aesthetically significant
3	Per cent of significant cultural sites protected by legislation	Identification of significant cultural sites may be difficult to achieve; this may be an especially difficult indicator to develop for multicultural nations
4	Number of visitors or visitor-nights – domestic – international	Provides a gross measure of demand on aggregate tourism resources; measures should conform to WTO definitions
5	Site stress levels: – per cent of tourist sites classified as degraded or threatened by over-use – relative concentration of total number of visitors at most popular sites or at UNESCO-designated sites	Identification of sites may be debatable, unless UNESCO-designated sites are the focus; provides a link between national indicators and local problems (and thus local indicators)
6	Resource consumption/depletion – energy – water – vehicle fuel – CO_2 emissions – other pollution emissions measured per tourist or tourist-night	Need to be able to separate consumption or pollution associated with tourists rather than residents

Table 11.6 *Continued*

	Indicator	Comments
7	Ratio of number of tourists to residents	Potential indicator of social stress; may need to be supplemented by some measure of the number of tourists who are obviously from cultural backgrounds different than local residents
8	Per cent of tourists victimized by/guilty of crime	May be a key indicator of problems with sustainability of local tourism industry or social stress
9	Per cent of tourists infected by communicable disease	Measure of hygiene standards; need to ensure disease was contracted as a result of visit to destination and not imported from other origin
10	Per cent of major tourist facilities (e.g. major hotels) owned by non-residents	May provide a measure of foreign influence in local tourism development; the figure may reflect either foreign interest in investing or governmental restrictions on foreign investment
11	Environmental quality – Per cent of tourist accommodation with potable water – Per cent of sewage discharged raw into watercourses, sea – Per cent of beaches safe for swimming	Also provides a direct measure of level of pollution in local communities
12	Infrastructure capacity utilization – Sewage and water systems – Transportation systems (roads, airports, ports, trains) – Energy sources	Indicates stress on infrastructure, potential need for maintenance, and constraints on growth. Useful as a regional or national indicator
13	Tourism employment – Per cent of labour force employed in tourism – Ratio of management to front-line or servile positions – Per cent of locals employed in tourism labour force – Wage rates in various tourism sectors, and comparison to averages in other economic sectors – Tourism job creation rate (per cent annual change)	Monitors one aspect of economic effects of tourism, as well as equity of distribution of job creation and wealth creation among various groups. Should be based on tourism satellite account data, if such a model is available
14	Foreign exchange leakage	Monitors another tourism economic impact. Provides a measure of the net benefit of inbound international tourism to host nation

Source: Adapted from International Working Group on Indicators of Sustainable Tourism (1993).

Table 11.7 Suggested local indicators

	Indicator	Comments
1	Destination attractiveness index	Could be based on attractiveness index methodology described in Chapter 9; a series of individual scores should be reported for landscape variety, cultural variety, uniqueness, level of maintenance, ease of access, and level of social stability and safety, as well as some overall assessment
2	Site stress index	A composite index including: – measures of spatial intensity of use (e.g. person per square metre of land, per cent of land in development) – temporal intensity of use (e.g. peaking index, described in Chapter 9) – potential degree of social impacts (e.g. no. of tourists/no. of locals; or Defert's *Tf*, Chapter 9); can be reported on a seasonal basis to reflect variations over a year – measures of environmental stress (e.g. per cent of site degraded; magnitude of expenditure needed to repair site damage) – measures of pollution (e.g. standard measures of water quality, volume of solid waste production) Scores for individual components should also be reported
3	Site protection – Level and type of protection – Per cent of site available for visitor use – Per cent of site hardened-off (paved)	Provides a measure of base condition of the site allowing for assessment of long-term changes
4	Endangered space – Is site considered an endangered space or under stress? – Does the site contain unique ecosystems or rare species?	Can provide a base for site categorization and indicate need for protection
5	Energy and water consumption per visitor-day	Particularly useful for destination resorts and for tourism development in water- or energy-poor regions

Table 11.7 *Continued*

	Indicator	Comments
6	Foreign ownership – Per cent of individual facilities totally owned by international investors – Per cent of investment in individual properties held by international investors	Provides a measure of degree of foreign control as well as potential economic leakage of revenues
7	Environmental quality measures – Per cent of days meeting standard – Availability of potable water – Per cent of sewage from tourism developments treated; types of water treatment available (primary, secondary, tertiary) – Days of beach closure due to contamination – Annual incidence of waterborne diseases (e.g. cholera, bilharzia) – Incidence of contamination in fish and seafood	Indicators provide information on site stress, health risks, aesthetics
8	Capacity utilization of local infrastructure: – Sewage – Water – Transportation – Energy	Highlights both potential stresses as well as limits to sustained economic contributions and growth. Most relevant to established sites but could be used to track impacts of new developments
9	Expenditures on environmental maintenance – Overall level of expenditure on maintenance – Amount spent to mitigate damage – Per cent of above relative to total amount needed to maintain site on a sustainable basis	Provides a measure of response of managers and government to environmental impacts caused by tourism; may be difficult to distinguish between tourism-related impacts and impacts
10	Critical habitats	A useful measure of environmental sensitivity of site
11	Economic impacts: – Total tourist-generated revenues – Jobs directly and indirectly supported by tourism – Tax revenues generated by tourism	A vital measure of economic sustainability of tourism; ideally, total revenues should be supplemented with some measure of leakage or income multiplier, if data are available

Summary

Tourism policy analysis, planning, and development require good information to guide decisions. We have examined in this chapter several methods that assist analysts, planners and decision-makers make more informed judgements. The methods address issues that range from the scale of an individual firm to broad national concerns. The most specific of the methods concerns the issue of advertising conversion rates – the degree to which spending on advertising translates into increased tourist arrivals and receipts. We have examined the assumptions, logic, limitations, and recommendations associated with conversion studies.

We then examined how a community might estimate the actual magnitude of tourists' spending. The approach presented here focuses on expenditures estimates only, and not on indirect or spin-off effects. Although methods exist for estimating tourism multipliers and other spin-off effects, they have significant data and technical requirements that place such methods beyond the scope of this book.

The next issue explored is how the benefits derived from tourism development can be compared to the costs associated with that development. The perspective adopted for this method was a social or policy perspective: benefit–cost analysis. The concept of BCA as well as its conventions, methods, and limitations were described.

Finally, we looked at an emerging area of work: the identification of indicators for sustainable development in tourism. The process of defining such indicators is a major task because of the breadth of issues associated with sustainable tourism and the fact that indicators are needed for levels ranging from site-specific to entire nations. Specific suggestions for both micro-scale and macro-scale indicators were presented.

The tourism industry is increasingly being held accountable for the results – intentional as well as unintentional – of its activities. Tourism is no different from other industries in needing to accept greater accountability. However, it is a highly visible industry as well as one that has impacts ranging from individual neighbourhoods to the international level. The methods described in this chapter are not the entire range of evaluation techniques relevant to tourism, but they do represent some of the more important ones with which a tourism analyst should be familiar.

Further reading

Ballman G, Burke J, Blank U and **Korte D** 1984 Towards higher quality conversion studies: refining the numbers game. *Journal of Travel Research* **26**(4): 28–32.

Burke J and **Lindblom L** 1989 Strategies for evaluating direct response tourism marketing. *Journal of Travel Research* **28**(2): 33–7.

Hammond R J 1966 Convention and limitation in benefit–cost analysis. *Natural Resources Journal* **6**: 195–219.

Harberger A C 1971 Three basic postulates for applied welfare economics: an interpretive essay. *Journal of Economic Literature* **IX**(September): 785–97.

Knapp M R J and **Vickerman R W** 1985 The conceptualization of output and cost–benefit evaluation for leisure facilities. *Environment and Planning A* **17**: 1217–29.

Mak J, Moncur J and **Yonamine D** 1977 How to and how not to measure visitor expenditures. *Journal of Travel Research* **16**(3): 2–5.

Mok H M 1990 A quasi-experimental measure of the effectiveness of destination advertising: some evidence from Hawaii. *Journal of Travel Research* **24**(1): 30–4.

Murphy P E 1983 *Tourism: a community approach*. Methuen, New York.

Ronkainen I A and **Woodside A G** 1987 Advertising conversion studies. In Ritchie J R B and Goeldner C R (eds) *Travel, Tourism, and Hospitality Research*. John Wiley, New York, pp. 481–8.

Smith V K 1988 Resource evaluation at the crossroads. *Resources* **90**: 2–6.

Weisbrod B A 1968 Income redistribution effects and benefit–cost analysis. In *Problems in public expenditure analysis*. Washington, DC: Brookings Institution, pp. 177–209.

Woodside A G and **Ronkainen I A** 1989 How serious is non-response bias in advertising conversion research? *Journal of Travel Research* **26**(4): 34–7.

Appendix

Significance graph for R_n values

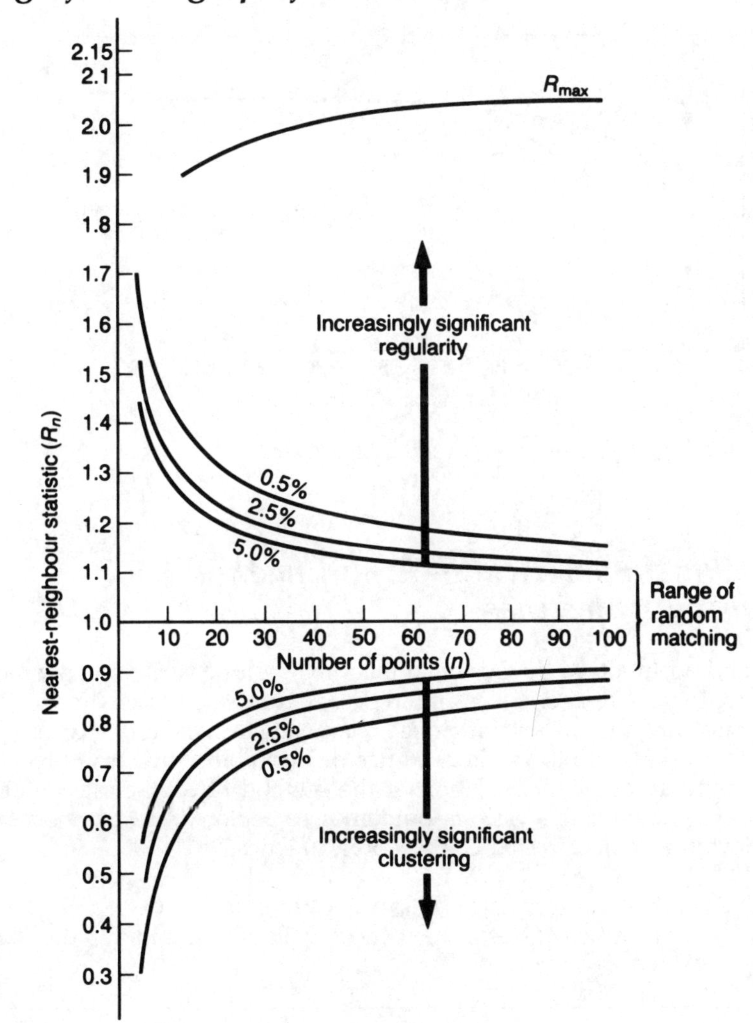

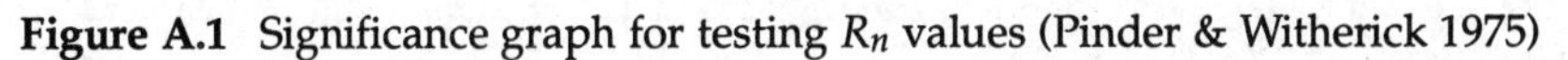
Figure A.1 Significance graph for testing R_n values (Pinder & Witherick 1975)

Significance graph for LR_n values

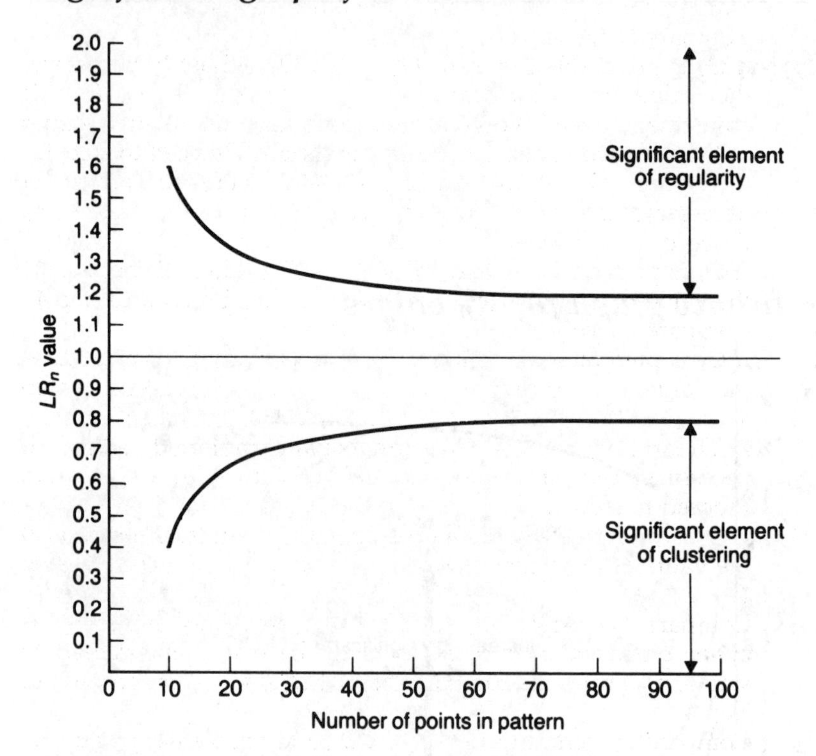

Figure A.2 Significance graph for testing *LRn* values

The Churchman–Ackoff–Arnoff method for weighting objectives[1]

This method involves the systematic comparison of various 'outcomes' (O_m) (in the context of assessing regional tourism attractiveness, the 'outcomes' are the various regional characteristics) in order to assign weights proportional to the importance or desirability of each outcome. The general symbolic formulation of the method of estimating weights (v_j) associated with the various outcomes or regional characteristics is quite formidable in appearance but not in practice.

1 Rank the outcomes in their order of value. Let O_1 represent the most valued, O_2, the next most valued,. . ., and O_m the least valued.

[1] The following material is reprinted from Churchman et al. (1957) with permission from John Wiley and Sons, Publishers

2 Assign the value of 1.00 to O_1 (i.e. $v_1 = 1.00$) and assign values that appear suitable to each of the other outcomes.

3 Compare O_1 versus $O_2 + \ldots + O_m$[2]

3.1 If O_1 is preferable to versus $O_2 + \ldots + O_m$, adjust (if necessary) the value of v_1 so that $v_1 > v_2 + v_3 + \ldots + v_m$. In this adjustment, as in all others, attempt to keep the relative values of the adjusted group (v_2, v_3, etc.) invariant. Proceed to step 4.

3.2 If O_1 and $O_2 + \ldots + O_m$ are equally preferred, adjust (if necessary) the value of v_1 so that $v_1 = v_2 + v_3 + \ldots + v_m$. Proceed to step 4.

3.3 If O_1 is preferred less than $O_2 + \ldots + O_m$, adjust (if necessary) the value of v_1 so that $v_1 < v_2 + v_3 + \ldots + v_m$. Proceed to step 4.

3.3.1 Compare O_1 versus $O_2 + \ldots + O_{m-1}$.

3.3.1.1 If O_1 is preferable to versus $O_2 + \ldots + O_m$, adjust (if necessary) the value of v_1 so that $v_1 > v_2 + v_3 + \ldots + v_{m-1}$. Proceed to step 4.

3.3.1.2 If O_1 and $O_2 + \ldots + O_{m-1}$ are equally preferred, adjust (if necessary) the value of v_1 so that $v_1 = v_2 + v_3 + \ldots + v_{m-1}$. Proceed to step 4.

3.3.1.3 If O_1 is preferred less than $O_2 + \ldots + O_{m-1}$, adjust (if necessary) the value of v_1 so that $v_1 < v_2 + v_3 + \ldots + v_{m-1}$. Proceed to step 4.

3.3.1.3.1 Compare O_1 versus $O_2 + \ldots + O_{m-2}$, etc., until either O_1 is either preferred over or is equal to the rest, then proceed to step 4, or until O_1 is compared to just $O_2 + O_3$. Then proceed to step 4.

4 Compare O_2 versus $O_3 + O_4 + \ldots O_m$ and proceed as in step 3.

5 Continue until the comparison of O_{m-2} versus $O_{m-1} + O_m$ is completed.

6 Convert each v_j into a normalized v'_j by dividing v_j by Σv_j. Then $\Sigma v'_j$ will equal 1.00.

It should be noted that the resulting values are relative; that is, deletion or addition of an outcome or regional characteristics will affect the values for all other outcomes.

*Significance test for R**

The test is based on the calculation of the standard error of the R^* estimate and on the calculation of a Z variate from the standard error:

$$Z = \frac{|d_o - d_r|}{SE_{R^*}}$$

[2] The + sign here designates the logical connective 'and'

where:

$$SE_R^{\,*} = \left\{ \left[\frac{m_1(n_1/a) + m_2(n_2/a)}{(n_1/a)\,(n_2/a)\,\pi} - \left(\frac{m_1\sqrt{(n_1/a)} + m_2\sqrt{(n_2/a)}}{2\sqrt{[(n_1/a)\,(n_2/a)]}} \right)^2 \right] \Big/ N \right\}$$

and all other variables are as defined in the text.

The probabilities for various values of Z are given in the following table A.1.

Table A.1 Probabilities for Z-tests

Probabilities are given for values as extreme as the values observed for Z in the normal distribution

Z		0.00	0.01	0.02	0.03	0.04	0.05	0.06	0.07	0.08	0.09
0.0	0.0	50 000	49 601	49 202	48 803	48 405	48 006	47 608	46 210	46 812	46 414
0.1		46 017	45 620	45 224	44 828	44 433	44 038	43 644	43 251	42 858	42 465
0.2		42 074	41 683	41 294	40 905	40 517	40 129	39 743	39 358	38 974	38 591
0.3		38 209	37 828	37 448	37 070	36 693	36 317	35 942	35 569	35 197	34 827
0.4		34 458	34 090	33 724	33 360	32 997	32 636	32 276	31 918	31 561	31 207
0.5		30 854	30 503	30 153	29 806	29 460	29 116	28 774	28 434	28 096	27 760
0.6		27 425	27 093	26 763	26 435	26 109	25 785	25 463	25 143	24 825	24 510
0.7		24 196	23 885	23 576	23 270	22 965	22 663	22 363	22 065	21 770	21 476
0.8		21 186	20 897	20 611	20 327	20 045	19 766	19 489	19 215	18 943	18 673
0.9		18 406	18 141	17 879	17 619	17 361	17 106	16 853	16 602	16 354	16 109
1.0		15 866	15 625	15 386	15 151	14 917	14 686	14 457	14 231	14 007	13 786
1.1		13 567	13 350	13 136	12 924	12 714	12 507	12 302	12 100	11 900	11 702
1.2		11 507	11 314	11 123	10 935	10 749	10 565	10 383	10 204	10 027	98 525
1.3	0.0	96 800	95 098	93 418	91 759	90 123	88 508	86 915	85 343	83 793	82 264
1.4		80 757	79 270	77 804	76 359	74 934	73 529	72 145	70 781	69 437	68 112
1.5		66 807	65 522	64 255	63 008	61 780	60 571	59 380	58 208	57 053	55 917
1.6		54 799	53 699	52 616	51 551	50 503	49 471	48 457	47 460	46 479	45 514
1.7		44 565	43 633	42 716	41 815	40 930	40 059	39 204	38 364	37 538	36 727
1.8		35 930	35 148	34 380	33 625	32 884	32 157	31 443	30 742	30 054	29 379
1.9		28 717	28 067	27 429	26 803	26 190	25 588	24 998	24 419	23 852	23 295
2.0		22 750	22 216	21 692	21 178	20 675	20 182	19 699	19 226	18 763	18 309
2.1		17 864	17 429	17 003	16 586	16 177	15 778	15 386	15 003	14 629	14 262
2.2		13 903	13 553	13 209	12 874	12 545	12 224	11 911	11 604	11 304	11 011
2.3		10 724	10 444	10 170	99 031	96 419	93 867	91 375	88 940	86 563	84 242
2.4	0.0^2	81 975	79 763	77 603	75 494	73 436	71 428	69 469	67 557	65 691	63 872
2.5		62 097	60 366	58 677	57 031	55 426	53 861	52 336	50 849	49 400	47 988
2.6		46 612	45 271	43 965	42 692	41 453	40 246	39 070	37 926	36 811	35 726
2.7		34 670	33 642	32 641	31 667	30 720	29 798	28 901	28 028	27 179	26 354
2.8		25 551	24 771	24 012	23 274	22 557	21 860	21 182	20 524	19 884	19 262
2.9		18 658	18 071	17 502	16 948	16 411	15 889	15 382	14 890	14 412	13 949

Table A.1 *Continued*

Z		0.00	0.01	0.02	0.03	0.04	0.05	0.06	0.07	0.08	0.09
3.0		13 499	13 062	12 639	12 228	11 829	11 442	11 067	10 703	10 350	10 008
3.1	0.0^3	96 760	93 544	90 426	87 403	84 474	81 635	78 885	76 219	73 638	71 136
3.2		68 714	66 367	64 095	61 895	59 765	57 703	55 706	53 774	51 904	50 094
3.3		48 342	46 648	45 009	43 423	41 889	40 406	38 971	37 584	36 243	34 946
3.4		33 693	32 481	31 311	30 179	29 086	28 029	27 009	26 023	25 071	24 151
3.5		23 263	22 405	21 577	20 778	20 006	19 262	18 543	17 849	17 180	16 534
3.6		15 911	15 310	14 730	14 171	13 632	13 112	12 611	12 128	11 662	11 213
3.7		10 780	10 363	99 611	95 740	92 010	88 417	84 957	81 624	78 414	75 324
3.8	0.0^4	72 348	69 483	66 726	64 072	61 517	59 059	56 694	54 418	52 228	50 122
3.9		48 096	46 148	44 274	42 473	40 741	39 076	37 475	35 936	34 458	33 037
4.0		31 671	30 359	29 099	27 888	26 726	25 609	24 536	23 507	22 518	21 569
4.1		20 658	19 783	18 944	18 138	17 365	16 624	15 912	15 230	14 575	13 948
4.2		13 346	12 769	12 215	11 685	11 176	10 689	10 221	97 736	93 447	89 337
4.3	0.0^5	85 399	81 627	78 015	74 555	71 241	68 069	65 031	62 123	59 340	56 675
4.4		54 125	51 685	49 350	47 117	44 979	42 935	40 980	39 110	37 322	35 612
4.5		33 977	32 414	30 920	29 492	28 127	26 823	25 577	24 386	23 249	22 162
4.6		21 125	20 133	19 187	18 283	17 420	16 597	15 810	15 060	14 344	13 660
4.7		13 008	12 386	11 792	11 226	10 686	10 171	96 796	92 113	87 648	83 391
4.8	0.0^6	79 333	75 465	71 779	68 267	64 920	61 731	58 693	55 799	53 043	50 418
4.9		47 918	45 538	43 272	41 115	39 061	37 107	35 247	33 476	31 792	30 190

Source: Fisher and Yates (1974).

References

Addelman S 1962 Orthogonal main-effect plans for asymmetrical factorial experiments. *Technometrics* **4**(i): 21–46.

Anon 1980 *Advertising Age* Ad beat, 27 Oct: 119.

American Express 1985 *World tourism overview, 1985*. American Express Publishing, New York.

Archer B H 1976 The uses and abuses of multipliers. In Gearing C E, Swart W W and Var T (eds) *Planning for tourism development*. Praeger, New York, pp. 115–32.

Arrow K J 1986 Comments. In Cummings R G, Brookshire D S and Schulze W D (eds) *Valuing environmental goods*. Rowman and Allanheld, Totawa, New Jersey.

Baligh H H and **Richartz L E** 1967 Variable-sum game models of marketing problems. *Journal of Marketing Research* **4** (May): 173–85.

Balmer, Crapo and Associates (no date) *The regional approach to tourism promotion and development in Saskatchewan*. Regina, Saskatchewan.

Barcun S D and **Jeming P** 1973 Airline seat share: a study in false optimization. *Management Science* **20** (Oct): 146–53.

Bass F M 1977 Analytical approaches in the study of purchase behaviour and brand choice. In Ferber R (ed.) *Selected aspects of consumer behavior*. National Science Foundation, Washington, DC, pp. 491–514.

Bekker P 1991 Tourism resources and protected areas along British Columbia's coastline. In *Tourism–environment–sustainable development: an agenda for research*, Proceedings of the Travel and Tourism Research Association – Canada Annual Conference. St Catherines, Ontario:

Brock University, Department of Recreation and Leisure Studies, pp. 83–8.

Berry B J L 1966 Interdependency of flows and spatial structure. In *Essays on commodity flows and the spatial structure of the Indian economy.* Department of Geography Research paper III, University of Chicago, Chicago, Illinois.

Bose R C and **Bush K A** 1952 Orthogonal arrays of strengths two and three. *The Annals of Mathematical Statistics* **23**: 508–24.

Brockhoff K 1975 The performance of forecasting groups in computer dialogue and face-to-face discussion. In Limestone H A and Turhoff M (eds) *The Delphi method: techniques and applications.* Addison-Wesley, Reading, Massachusetts, pp. 291–321.

Bull A 1991 *The economics of travel and tourism.* Longman Cheshire, Melbourne.

Burke J F and **Gitelson R** 1990 Conversion studies: assumptions, applications, accuracy, and abuse. *Journal of Travel Research* **28**(3): 46–50.

Calantone R J and **Sawyer A** 1978 The stability of benefit segments. *Journal of Marketing Research* **15** (Aug): 395–404.

Chipenluk R 1993 Vernacular bio-indicators and citizen monitoring of environmental change. In Nelson J G, Butler R and Wall G (eds) *Tourism and sustainable development: monitoring, planning, managing.* Heritage Resources Centre, University of Waterloo, Waterloo, Ontario, pp. 269–78.

Christaller W 1933 *Die zentralen orte in Suddentdeutschland.* Gustav Fischer, Jena.

Christaller W 1964 Some considerations of tourism locations in Europe: the peripheral regions – underdeveloped countries – recreation areas. *Papers of the Regional Science Association* **12**: 95–105.

Churchill G A 1979 A paradigm for developing better measures of marketing constructs. *Journal of Marketing Research* **16**: 64–73.

Churchman C W, **Ackoff R L** and **Arnoff E L** 1957 Weighing objectives. In *Introduction to operations research.* John Wiley, New York, Chap. 6.

Cicchetti C J and **Smith V K** 1973 Congestion, quality deterioration and optimal use: Wilderness and recreation in the Spanish Peaks Primitive area. *Social Sciences Research* **2**: 15–30.

Ciriacy-Wantrup S V 1947 Capital returns from soil-conservation practices. *Journal of Farm Economics* **29**: 1181–96.

Clark P J and **Evans F C** 1955 On some aspects of spatial patterns in biological populations. *Science* **121**: 397–8.

Clawson M C and **Knetsch J L** 1966 *Economics of outdoor recreation.* Johns Hopkins Press, Baltimore, Maryland.

Cochran W C and **Cox G M** 1957 *Experimental designs,* 2nd edn. John Wiley, New York.

Coffey W J 1981 *Geography: towards a general spatial systems approach.* Methuen and Company, London.

Cohen K 1968 Multiple regression as a general data-analytic system. *Psychological Bulletin* **70**(6): 426–43.

Cole J P and **King C A M K** 1968 *Quantitative geography.* Oxford University Press, London.

Coombs D B and **Thie J** 1979 The Canadian land inventory system. In *Planning the uses and management of land.* ASA/CSSA/SSSA, Madison, Wisconsin, pp. 909–33.

Crampon L J 1966 A new technique to analyze tourist markets. *Journal of Marketing* **30** (April): 27–31.

Cronbach L J 1951 Coefficient alpha and internal structure of tests. *Psychometrika* **16**: 297–334.

Dalkey N and **Helmer O** 1963 An experimental application of the Delphi method of the use of experts. *Management Sciences* **9**(3): 459–67.

Darling A 1972 Measuring benefits generated by urban water parks. *Land Economics* **49**: 22–34.

Davis R K 1963 Recreation planning as an economic problem. *Natural Resources Journal* **3**: 239–49.

Defert P 1967 Le taux de fonction touristique: mise au point et critique. *Les cahiers du tourisme.* Centre des Hautes Etudes Touristiques, Aix-en-Provence, C–13.

den Hoedt E 1994 *Concepts, definitions and classifications for tourism statistics: a technical manual.* World Tourism Organization, Madrid.

Donnelly D M and Nelson L J 1986 *Net economic value of deer hunting in*

Idaho, Resource Bulletin RM–13. Rocky Mountain Forest and Range Experiment Station, Fort Collins, CO.

Dorney R S 1976 Biophysical and cultural–historic land classification and mapping for Canadian urban and urbanizing land. In *Proceedings of the workshop on ecological land classification*. Commission on Ecological Land Classification, Toronto, pp. 57–71.

Draper N and **Smith H** 1966 *Applied regression analysis*. John Wiley, New York.

Dunn-Rankin P 1983 *Scaling methods*. Erlbaum, Hillsdale, New Jersey.

Ebdon D 1977 *Statistics in geography*. Basil Blackwell, Oxford.

Edwards A 1957 *Techniques of attitude scale construction*. Appleton-Century-Crofts, New York.

Edwards S L and **Dennis S J** 1976 Long distance day-tripping in Great Britain. *Journal of Transport Economics and Policy* **10**: 237–56.

Embacher J and **Buttle F** 1989 A repertory grid analysis of Austria's image as a summer vacation destination. *Journal of Travel Research* **27**(3): 3–7.

Engel J F, Blackwell R D and **Kollat D T** 1978 *Consumer behavior*, 3rd edn. Dryden Press, Hinsdale, Illinois.

Ezekial R S 1970 Authoritarianism, acquiescence, and field behaviour. *Journal of Personality* **38**(1): 31–42.

Fechner G T 1889 *Elemente der psychophysik*. Breitkopf und Hertel, Leipzig.

Fesenmaier D R and **Roehl W S** 1986 Location analysis in campground development. *Journal of Travel Research* **24**(3): 18–22.

Fishbein M 1963 An investigation of the relationship between the beliefs about an object and the attitude towards that object. *Human Relationships* **16**(2): 232–40.

Fishbein M 1966 The relationship between beliefs, attitudes and behaviour. In Feldman S (ed.) *Cognitive consistency*. Academic Press, New York, pp. 477–92.

Fishbein M 1967 Attitude and the prediction of behaviour. In Fishbein M (ed.) *Readings in attitude theory and measurement*. John Wiley, New York, pp. 477–92.

Fleiss J L 1973 Statistical methods for rates and proportions. John Wiley, New York.

Frechtling D C 1974 A model for estimating travel expenditures. *Journal of Travel Research* **12**(4): 9–12.

Fridgen J D 1987 Use of cognitive maps to determine perceived tourism regions. *Leisure Sciences* **9**: 101–17.

Fridgen J D, Udd E and **Deale C** 1983 Cognitive maps of tourism regions in Michigan. In *Proceedings of the Applied Geography Conference*, 1983. Ryerson Polytechnic Institute, Toronto, pp. 262–72.

Gartner W C and **Hunt J D** 1987 An analysis of state image change over a twelve-year period. *Journal of Travel Research* **26**(2): 15–19.

Gearing C E, Swart W W and **Var T** 1974 Establishing a measure of touristic attractiveness. *Journal of Travel Research* **12**(3): 1–8.

Gee C Y, Choy D J L and **Maken J C** 1984 *The travel industry*. AVI Publishing, Westport, Connecticut.

Gitelson R 1986 *1985 North Carolina Coupon Conversion Study: an analysis of those individuals who requested the North Carolina Information Packet*. Raleigh, North Carolina: North Carolina Division of Tourism.

Goddard J E 1970 Functional regions within the city centre. *Transactions of the Institute of British Geographers* **49**: 161–81.

Green P E 1974 On the design of choice experiments involving multifactor alternatives. *Journal of Consumer Research* **1**: 61–8.

Green P E and **Devita M T** 1973 A complementarity model of consumer utility for item collections. *Journal of Consumer Research* **1** (Dec): 56–67.

Green P E and **Rao V R** 1971 Conjoint measurement for quantifying judgemental data. *Journal of Marketing Research* **8** (Aug): 355–63.

Green P E and **Srinivasan V** 1978 *Research for marketing decisions*, 3rd edn. Prentice-Hall, Englewood Cliffs, NJ.

Green P E and **Tull D S** 1975 *Research for marketing decisions*, 3rd edn. Prentice-Hall, Englewood Cliffs, New Jersey.

Green P E and **Wind Y** 1975 New way to measure consumers' judgements. *Harvard Business Review* **53** (July–Aug): 107–17.

Greenhut M L 1956 *Plant location in theory and practice*. University of North Carolina Press, Chapel Hill, North Carolina.

Grether E T 1983 Regional-spatial analysis in marketing. *Journal of Marketing* **47** (Fall): 36–43.

Grieg-Smith P 1952 The use of random and contiguous quadrats in the study of the structure of plant communities. *Annals of Biology* NS **16**: 292–316.

Grigg D 1965 The logic of regional systems. *Annals of the Association of American Geographers* **55**: 465–91.

Guilford J P 1954 *Psychometric methods*. McGraw-Hill, New York.

Gunn C A 1972 *Vacationscape: designing tourist regions*. University of Texas Bureau of Business Research, Austin, Texas.

Gunn C A 1979 *Tourism planning*. Crane–Russah, New York.

Gunn C A 1982 Destination zone fallacies and half-truths. *Tourism Management* **3**(4): 263–9.

Gunn C A 1994 *Tourism planning*, 3rd edn. Taylor and Frances, Washington, DC.

Gunn C A and **Worms A J** 1973 *Evaluating and developing tourism*. Texas Agricultural Experiment Station, Texas A&M University, College Station, Texas.

Hammack J and **Brown G M** 1974 *Waterfowl and wetland: toward bioeconomic analysis*. Johns Hopkins Press, Baltimore, Maryland.

Hannermann W M 1978 *A methodical and empirical study of the recreation benefits from water quality improvement*. Unpublished PhD dissertation, Harvard University.

Holder J 1991 Tourism, the world and the Caribbean. *Tourism Management* **12**: 291–300.

Hoover E M 1948 *The location of economic activity*. McGraw-Hill, New York.

Hu Y and **Ritchie J R B** 1993 Measuring destination attractiveness: a contextual approach. *Journal of Travel Research* **32**(2): 25–34.

Hunt J D and **Dalton M J** 1983 Comparing mail and telephone for conducting coupon conversion studies. *Journal of Travel Research* **21**(3): 16–20.

Inskeep E 1991 *Tourism planning: an integrated and sustainable development approach*. Van Nostrand Reinhold, New York.

International Working Group on Indicators of Sustainable Tourism 1993 Indicators for the sustainable management of tourism. Unpublished report, Environment Committee, World Tourism Organization, Winnipeg, Manitoba.

Isard W 1956 *Location and space economy*. MIT Press, Cambridge, Massachusetts.

Jafari J 1992 The scientification of tourism. In El-Wahab S A and El-Roby N (eds) *Scientific tourism*. Egyptian Society of Scientific Experts on Tourism, Cairo, pp. 43–75.

Jefferson A and **Lickorish L** 1988 *Marketing tourism: a practical guide*. Longman Group, Harlow, Essex.

Johnson R 1973 Pairwise nonmetric multidimensional scaling. *Psychometrika* **38**: 313–22.

Johnson R 1974 Trade-off analysis of consumer values. *Journal of Marketing Research* **11**: 261–3.

Johnson R 1976 *Trade-off analysis: a method for quantifying consumer values*. Market Facts, Toronto.

June L P and **Smith S L J** 1987 Service attributes and situational effects on customer preferences for restaurant dining. *Journal of Travel Research* **26**(2): 20–7.

Kakkar P and **Lutz R J** 1975 Toward a taxonomy of consumption situations. In Maze E M (ed.) *Combined Proceedings*, Series 37. American Marketing Association, Chicago, pp. 206–10.

Kellerman A 1981 *Centrographic measures in geography*. Concepts and techniques in modern geography, No. 32. GeoAbstracts, Norwich, East Anglia.

Keogh B 1982 On measuring spatial variations in tourist activity. Unpublished manuscript.

King L J 1969 *Statistical analysis in geography*. Prentice-Hall, Englewood Cliffs, New Jersey.

Klaric Z 1992 Establishing tourist regions: the situation in Croatia. *Tourism Management* **13**: 305–11.

Kruskal J B 1964 Multidimensional scaling by optimizing goodness of fit to a nonmetric hypothesis. *Psychometrika* **29**(10): 1–27.

Kruskal J B 1965 Analysis of factorial experiments by estimating monotone transformations of the data. *Journal of the Royal Statistical Society*, Series B **27**: 251–63.

Lankford S V 1994 Attitudes and perceptions toward tourism and rural regional development. *Journal of Travel Research* **32**(3): 35–43.

Lee Y 1979 A nearest neighbour spatial association measure for the analysis of firm interdependence. *Environment and Planning A* **11**: 169–76.

Lefever D W 1926 Measuring geographical concentration by means of the standard deviational ellipse. *American Journal of Sociology* **32**: 88–94.

Leiper N 1993 Industrial entropy in tourism systems. *Annals of Tourism Research* **20**: 221–6.

Likert R 1932 *A technique for the measure of attitudes*. Archives of psychology, No. 140. Columbia University Press, New York.

Likert R 1967 The method of constructing an attitude scale. In Fishbein J (ed.) *Readings in attitude theory and measurement*. John Wiley, New York, pp. 90–5.

Linley B 1993 Cross-border study of inns and their financial performance. In *Regional marketing partnerships and adventure tourism*, Proceedings of the 1993 Joint TTRA-Canada and TTRA-New England Conference. Georgian College School of Hospitality and Tourism, Barrie, Ontario, pp. 24–7.

Liu J 1986 Relative economic contributions of visitor groups in Hawaii. *Journal of Travel Research* **25**(1): 2–9.

Lopez E M 1980 The effect of leadership style on satisfactions of tour quality. *Journal of Travel Research* **18**(4): 20–3.

Lösch A 1944 *Die raumliche ordnung der wirtschaft*. Gustav Fischer, Jena.

Luce R D 1959 *Individual choice behavior*. John Wiley, New York.

Luce R D and **Tukey J W** 1964 Simultaneous conjoint measurement: a new type of fundamental measurement. *Journal of Mathematical Psychology* **1**: 1–27.

Maddox R N 1985 Measuring satisfaction with tourism. *Journal of Travel Research* **23**(3): 2–5.

Market Facts of Canada 1989 *Pleasure travel markets to North America: United Kingdom*. Industry, Science and Technology, Canada, Ottawa.

Martin B, **Memmott F** and **Bone A** 1961 *Principles and techniques of predicting future demand for urban area travel.* MIT Press, Cambridge, Massachusetts.

McCool S and **Menning N L** 1993 The environment as tourism product: the case of fall travellers to Montana. In *Expanding responsibilities: A blueprint for the travel industry*, Proceedings of the 24th Annual TTRA-International Conference. TTRA-International, Wheat Ridge, Colorado, pp. 220–9.

McIntire G 1993 *Sustainable tourism development: guide for local planners.* World Tourism Organization, Madrid.

Middleton V T C 1988 *Marketing in travel and tourism.* Heinemann, Oxford.

Milman A, **Reichel A** and **Pizam A** 1990 The impact of tourism on ethnic attitudes: the Israeli–Egyptian case. *Journal of Travel Research* **29**(2): 45–9.

Miossec J M 1977 Un modèle de l'espace touristique. *L'Espace Géographique* **6**: 41–8.

Mitchell R C and **Carson R J** 1989 *Using surveys to value public goods.* Resources for the Future, Washington, DC.

Moser C and **Kalton G** 1974 *Survey methods in social investigation*, 2nd edn. Basic Books, New York.

Moutinho L and **Currey B** 1994 Modelling site location decisions in tourism. *Journal of Travel and Tourism Marketing* **3**(2): 35–58.

Muha S L 1976 Evaluating travel advertising: a survey of existing studies. Paper presented at the Fifth Annual Educational Seminar for State Travel Officials. Lincoln, Nebraska.

National Task Force on Tourism Data 1985 *Final joint report of the working groups on user needs and current data bases.* Statistics Canada, Ottawa.

National Task Force on Tourism Data 1985b Progress report, unpublished mimeograph. Statistics Canada, Ottawa.

Neft D S 1966 *Statistical analysis for spatial distributions*, mimeograph no. 2. Regional Science Institute, Philadelphia, Pennsylvania.

Nelson J G, **Butler R** and **Wall G** 1993 *Tourism and sustainable development: monitoring, planning, managing.* Heritage Resources Centre, University of Waterloo, Waterloo, Ontario.

Niedercorn J H and **Bechdoldt B V** 1966 An economic derivation of the 'gravity law' of spatial interaction. *Journal of Regional Science* **9**: 273–82.

Nunnally J C 1978 *Psychometric theory*, 2nd edn. McGraw-Hill, New York.

OECD Tourism Committee 1973 *Tourism policy and international tourism in OECD member countries*. Organization for Economic Cooperation and Development, Paris.

Ohlin B 1935 *Interregional and international trade*. Harvard University Press, Cambridge, Massachusetts.

Opperman M 1993 Regional market segmentation analysis in Australia. *Journal of Travel and Tourism Marketing* **2**(4): 59–74.

Osgood C E, **Suci G J** and **Tannenbaum P H** 1957 *The measurement of meaning*. University of Illinois Press, Urbana, Illinois.

Philbrick A K 1957 Principles of areal functional organization in regional human geography. *Economic Geography* **33**: 299–336.

Pinder D A 1978 Correcting underestimation in nearest-neighbour analysis. *Area* **10**: 379–85.

Pinder D A and **Witherick M E** 1975 A modification of nearest-neighbour analysis for use in linear situations. *Geography* **60**: 16–23.

Pisarski A 1992 *Standard international classification of tourism activities, Rev. 2*. World Tourism Organization, Madrid.

Plackett R L and **Burman J P** 1946 The design of optimum multifunctional experiments. *Biometrika* **33**: 305–25.

Rao R C 1981 *Advertising decisions in oligopoly: an industry equilibrium analysis*. Krannert Graduate School of Management, Series 752. Purdue University, West Lafayette, Indiana.

Robelek L 1994 *Destination image as a factor in meeting site selection*. Unpublished MA thesis, Department of Recreation and Leisure Studies, University of Waterloo, Waterloo, Ontario.

Rosenberg M 1956 Cognitive structure and attitudinal effect. *Journal of Abnormal and Social Psychology* **53**: 367–72.

Scott A 1965 The valuation of game resources: some theoretical aspects. *Canadian Fisheries Report* **4**: 27–47.

Seymour D L 1968 The polygon of forces and the Weber problem. *Journal of Regional Science* **8**: 243–6.

Shafer E L, Moeller G H and **Getty R E** 1974 *Future leisure environments*. Forest Research Paper NE–301. USDA Forest Experiment Station, Upper Darby, Pennsylvania.

Shepard R N 1957 Stimulus and response generalization: a stochastic model of consumer spatial behaviour. *Regional Science Perspectives* **7**(2): 122–34.

Shepard R N 1962 The analysis of proximities: multidimensional scaling with an unknown distance function. *Psychometrika* **27**(2): 125–40.

Sheth J N 1972 Reply to comments on the nature and uses of expectancy-value models in consumer attitude research. *Journal of Marketing Research* **9**: 462–5.

Siegal W and **Ziff-Levine W** 1990 Evaluating tourism advertising campaigns: conversion vs advertising tracking studies. *Journal of Travel Research* **28**(3): 51–5.

Smith R V 1992 Tourism's role in the economy and landscape of Luxembourg. *Tourism Management* **13**: 423–8.

Smith S L J 1983 *Recreation geography*. Longman, Harlow, Essex.

Smith S L J 1987 Regional analysis of tourism resources. *Annals of Tourism Research* **14**: 253–73.

Smith S L J and **Thomas D C** 1983 Assessment of regional potentials of rural recreation businesses. In Lieber S P and Fesenmaier D R (eds) *Recreation planning and management*. Venture Publishers, State College, Pennsylvania, and E and F N Spon, London, pp. 66–86.

Smith S L J, Frechtling D, Rach L and **Stevens B** 1993 *The 1993 TTRA research agenda*. Travel and Tourism Research Association International, Wheat Ridge, Colorado.

Smith W R 1956 Product differentiation and market segmentation as alternative marketing strategies. *Journal of Marketing* **21** (July): 3–8.

Sonquist J A and **Dunkelberg W C** 1977 *Survey and opinion research*. Prentice-Hall, Englewood Cliffs, New Jersey.

Statistics Canada 1984 *Urban family expenditures, 1984*. Department of Supply and Services, Ottawa.

Stewart J Q 1948 Demographic gravitation: evidence and applications. *Sociometry* **11**: 30–58.

Stutz F P 1973a Distance and network effects on urban social travel fields. *Economic Geography* **49**: 134–44.

Stutz F P 1973b Interactance communities: Transportation's role in urban social geography. *Proceedings of the Association of American Geographers* **5**: 257–61.

Stutz F P 1974 Interactance communities versus named communities. *Professional Geographer* **26**: 407–11.

Stynes D J 1978 The peaking problem in outdoor recreation: measurement and analysis. Unpublished paper presented at the Annual Meeting of the National Recreation and Parks Association, Miami, Florida.

Stynes D J 1983 An introduction to recreation forecasting. In Lieber S and Fesenmaier D (eds) *Recreation planning and management*. Venture Publishing, State College, Pennsylvania, pp. 87–95.

Taylor P J 1977 *Quantitative methods in geography*. Houghton Mifflin Company, Boston, Massachusetts.

Thompson J A 1982 *Site selection*. Lebhar-Freidman, New York.

Thurstone L L 1927 A law of comparative judgement. *Psychological Review* **34**: 273–86.

Tiebout C M 1962 *The community economc base study*. Committee for Economic Development, New York.

Tull D S and **Hawkins D I** 1980 *Marketing research*, 2nd edn. Macmillan, New York.

US Army Corps of Engineers 1974 *Plan formulation and evaluation studies – recreation, Vol. II, Appendix A, Estimating initial reservoir recreation use-project data*. IWR Report 74–R1. USAE Institute for Water Resources, Fort Belvoir, Virginia.

US Water Resources Council 1983 *Economic and environmental principles for water and related land resources implementation studies*. US Government Printing Office, Washington, DC.

Uysal M and **McDonald C D** 1989 Visitor segmentation by trip index. *Journal of Tourism Research* **27**(3): 38–42.

Vanhove N 1989 Tourist market segmentation. In Witt S F and Moutinho L (eds) *Tourism marketing and management handbook.* Prentice-Hall International, Hemel Hempstead, UK, pp. 563–8.

Vickerman R W 1975 The leisure sector in urban areas. In *The economics of leisure and recreation.* Macmillan Press, London, Chap. 8.

Wahab S 1972 An introduction to tourism theory. *IUOTO Research Journal* (1): 34–43.

Walsh R G 1977 Effects of improved research methods on the value of recreational benefits. In *Outdoor recreation: advances in the application of economics,* USDA Forest Service General Technical Report WO-2. Washington, DC, pp. 145–53.

Walsh R G 1986 *Recreation economic decisions.* Venture Publishing, State College, PA.

Wassenaar D J 1981 *California visitor impact model.* Office of Tourism, Department of Economic and Business Development, Sacramento, California.

Weber A 1928 *Alfred Weber's theory of the location of industries* (Friedrich C J, trans.). University of Chicago Press, Chicago, Illinois.

Weber A 1931 Die standorslehre und die landes politik. *Archive für sozialwissenschaft und Socialpolitik* **32**: 677–88.

WEFA Group 1993 *Measuring the size of the global travel and tourism industry.* The World Travel and Tourism Council, Brussels.

Wennergren E B and **Nielsen D B** 1968 *A probabilistic approach to estimating demand for outdoor recreation,* Bulletin 470. Utah Agricultural Experiment Station, Logan, Utah.

Whitehead J I 1965 Road traffic growth and capacity in a holiday district (Dorset). *Proceedings of the Institute of Civil Engineers* **30**: 589–608.

Wilson T P 1971 A critique of ordinal variables. *Social Forces* **49**: 432–44.

Winer B J 1971 *Statistical principles in experimental design.* McGraw-Hill, New York.

Wolfe R I 1966 *Parameters of recreational travel in Ontario,* DHO Report RB111. Ontario Department of Highways, Downsview, Ontario.

Wolfe R I 1972 The inertia model. *Journal of Leisure Research* **4**: 73–6.

Woodside A G and **Reid D M** 1974 Tourism profiles vs audience profiles: are upscale magazines really upscale? *Journal of Travel Research* **12**(4): 17–23.

World Tourism Organization 1991 *International conference on travel and tourism statistics: resolutions*. World Tourism Organization, Madrid.

World Tourism Organization 1994 *Recommendations on tourism statistics*. World Tourism Organization, Madrid.

WTTC 1993 *Travel and tourism: a new economic perspective*. WTTC, Brussels.

Yokeno N 1974 The general equilibrium system of space-economic for tourism. *Reports for the Japan Academic Society of Tourism* **8**: 38–44.

Young C W and **Smith R W** 1979 Aggregated and disaggregated outdoor recreation participation models. *Leisure Sciences* **2**: 143–54.

Zeller R E, **Achabal D D** and **Brown L A** 1980 Market penetration and locational conflict in franchise systems. *Decision Sciences* **11**: 58–80.

Zipf G K 1946 The P_1P_2/D hypothesis: an inter-city movement of persons. *American Sociological Review* **11**: 677–86.

Name index

Subject index